LIBRARY & MEDIA SERVICES
WESTERN NEVADA COLLEGE
2201 WEST COLLEGE PARKWAY
CARSON CITY, NV 89703

STRUCTURAL WOOD DESIGN

STRUCTURAL WOOD DESIGN

A PRACTICE-ORIENTED APPROACH USING THE ASD METHOD

Abi Aghayere
Jason Vigil

JOHN WILEY & SONS, INC.

This book is printed on acid-free paper. ∞

Copyright © 2007 by John Wiley & Sons, Inc. All rights reserved

Published by John Wiley & Sons, Inc., Hoboken, New Jersey
Published simultaneously in Canada

Wiley Bicentennial Logo: Richard J. Pacifico

No part of this publication may be reproduced, stored in a retrieval system, or transmitted in any form or by any means, electronic, mechanical, photocopying, recording, scanning, or otherwise, except as permitted under Section 107 or 108 of the 1976 United States Copyright Act, without either the prior written permission of the Publisher, or authorization through payment of the appropriate per-copy fee to the Copyright Clearance Center, 222 Rosewood Drive, Danvers, MA 01923, (978) 750-8400, fax (978) 646-8600, or on the web at www.copyright.com. Requests to the Publisher for permission should be addressed to the Permissions Department, John Wiley & Sons, Inc., 111 River Street, Hoboken, NJ 07030, (201) 748-6011, fax (201) 748-6008, or online at www.wiley.com/go/permissions.

Limit of Liability/Disclaimer of Warranty: While the publisher and the author have used their best efforts in preparing this book, they make no representations or warranties with respect to the accuracy or completeness of the contents of this book and specifically disclaim any implied warranties of merchantability or fitness for a particular purpose. No warranty may be created or extended by sales representatives or written sales materials. The advice and strategies contained herein may not be suitable for your situation. You should consult with a professional where appropriate. Neither the publisher nor the author shall be liable for any loss of profit or any other commercial damages, including but not limited to special, incidental, consequential, or other damages.

For general information about our other products and services, please contact our Customer Care Department within the United States at (800) 762-2974, outside the United States at (317) 572-3993 or fax (317) 572-4002.

Wiley also publishes its books in a variety of electronic formats. Some content that appears in print may not be available in electronic books. For more information about Wiley products, visit our web site at www.wiley.com.

Library of Congress Cataloging-in-Publication Data:
  Aghayere, Abi O.
    Structural wood design: a practice-oriented approach using the ASD method/
      by Abi Aghayere, Jason Vigil.
        p. cm.
    ISBN: 978-0-470-05678-3
     1. Wood. 2. Building, Wooden. I. Vigil, Jason, 1974– II. Title.
TA419.A44 2007
624.1′84—dc22

                                               2006033934

Printed in the United States of America

10 9 8 7 6 5 4 3 2 1

Contents

Preface xi

chapter one **INTRODUCTION: WOOD PROPERTIES, SPECIES, AND GRADES**   **1**

    1.1 Introduction   1
        The Project-based Approach   1
    1.2 Typical Structural Components of Wood Buildings   2
    1.3 Typical Structural Systems in Wood Buildings   8
        Roof Framing   8
        Floor Framing   9
        Wall Framing   9
    1.4 Wood Structural Properties   11
        Tree Cross Section   11
        Advantages and Disadvantages of Wood as a Structural Material   11
    1.5 Factors Affecting Wood Strength   12
        Species and Species Group   12
        Moisture Content   13
        Duration of Loading   14
        Size Classifications of Sawn Lumber   14
        Wood Defects   15
        Orientation of the Wood Grain   16
        Ambient Temperature   16
    1.6 Lumber Grading   16
        Types of Grading   17
        Stress Grades   18
        Grade Stamps   18
    1.7 Shrinkage of Wood   19
    1.8 Density of Wood   19
    1.9 Units of Measurement   19
    1.10 Building Codes   20
        NDS Code and NDS Supplement   22

vi | CONTENTS

    References 23
    Problems 23

chapter two **INTRODUCTION TO STRUCTURAL DESIGN LOADS** 25

  2.1 Design Loads 25
    Load Combinations 25
  2.2 Dead Loads 26
    Combined Dead and Live Loads on Sloped Roofs 27
    Combined Dead and Live Loads on Stair Stringers 28
  2.3 Tributary Widths and Areas 28
  2.4 Live Loads 30
    Roof Live Load 30
    Snow Load 32
    Floor Live Load 35
  2.5 Deflection Criteria 39
  2.6 Lateral Loads 42
    Wind Load 43
    Seismic Load 45
    References 54
    Problems 54

chapter three **ALLOWABLE STRESS DESIGN METHOD FOR SAWN LUMBER AND GLUED LAMINATED TIMBER** 57

  3.1 Allowable Stress Design Method 57
    NDS Tabulated Design Stresses 58
    Stress Adjustment Factors 59
    Procedure for Calculating Allowable Stress 66
    Moduli of Elasticity for Sawn Lumber 66
  3.2 Glued Laminated Timber 66
    End Joints in Glulam 67
    Grades of Glulam 67
    Wood Species Used in Glulam 68
    Stress Class System 68
  3.3 Allowable Stress Calculation Examples 69
  3.4 Load Combinations and the Governing Load Duration Factor 69
    Normalized Load Method 69
    References 78
    Problems 78

chapter four **DESIGN AND ANALYSIS OF BEAMS AND GIRDERS** 80

  4.1 Design of Joists, Beams, and Girders 80
    Definition of Beam Span 80

Layout of Joists, Beams, and Girders   80
Design Procedure   80
4.2 Analysis of Joists, Beams, and Girders   86
Design Examples   100
Continuous Beams and Girders   117
Beams and Girders with Overhangs or Cantilevers   117
4.3 Sawn-Lumber Decking   118
4.4 Miscellaneous Stresses in Wood Members   121
Shear Stress in Notched Beams   121
Bearing Stress Parallel to the Grain   122
Bearing Stress at an Angle to the Grain   122
Sloped Rafter Connection   123
4.5 Preengineered Lumber Headers   126
4.6 Flitch Beams   128
4.7 Floor Vibrations   131
Floor Vibration Design Criteria   131
Remedial Measures for Controlling Floor Vibrations in Wood Framed Floors   136
References   142
Problems   143

chapter five WOOD MEMBERS UNDER AXIAL AND BENDING LOADS   145

5.1 Introduction   145
5.2 Pure Axial Tension: Case 1   146
Design of Tension Members   146
5.3 Axial Tension plus Bending: Case 2   151
Euler Critical Buckling Stress   153
5.4 Pure Axial Compression: Case 3   153
Built-up Columns   159
P–Delta Effects in Members Under Combined Axial Compression and Bending Loads   162
5.5 Axial Compression plus Bending: Case 4   162
Eccentrically Loaded Columns   178
5.6 Practical Considerations for Roof Truss Design   178
Types of Roof Trusses   179
Bracing and Bridging of Roof Trusses   179
References   180
Problems   181

chapter six ROOF AND FLOOR SHEATHING UNDER VERTICAL AND LATERAL LOADS (HORIZONTAL DIAPHRAGMS)   183

6.1 Introduction   183
Plywood Grain Orientation   183
Plywood Species and Grades   183

Span Rating   185
6.2 Roof Sheathing: Analysis and Design   186
6.3 Floor Sheathing: Analysis and Design   186
  Extended Use of the IBC Tables for Gravity Loads on Sheathing   188
6.4 Panel Attachment   189
6.5 Horizontal Diaphragms   190
  Horizontal Diaphragm Strength   192
  Openings in Horizontal Diaphragms   197
  Chords and Drag Struts   200
  Nonrectangular Diaphragms   213
  References   214
  Problems   214

chapter seven **VERTICAL DIAPHRAGMS UNDER LATERAL LOADS (SHEAR WALLS) 216**
7.1 Introduction   216
  Wall Sheathing Types   216
  Plywood as a Shear Wall   217
7.2 Shear Wall Analysis   219
  Shear Wall Aspect Ratios   219
  Shear Wall Overturning Analysis   220
  Shear Wall Chord Forces: Tension Case   224
  Shear Wall Chord Forces: Compression Case   226
7.3 Shear Wall Design Procedure   227
7.4 Combined Shear and Uplift in Wall Sheathing   242
  References   245
  Problems   245

chapter eight **CONNECTIONS   248**
8.1 Introduction   248
8.2 Design Strength   249
8.3 Adjustment Factors for Connectors   249
8.4 Base Design Values: Laterally Loaded Connectors   257
8.5 Base Design Values: Connectors Loaded in Withdrawal   268
8.6 Combined Lateral and Withdrawal Loads   270
8.7 Preengineered Connectors   273
8.8 Practical Considerations   273
  References   276
  Problems   276

chapter nine **BUILDING DESIGN CASE STUDY   278**
9.1 Introduction   278

9.2 Gravity Loads   279
9.3 Seismic Lateral Loads   283
9.4 Wind Loads   284
9.5 Components and Cladding Wind Pressures   287
9.6 Roof Framing Design   291
   Analysis of a Roof Truss   292
   Design of Truss Web Tension Members   293
   Design of Truss Web Compression Members   293
   Design of Truss Bottom Chord Members   295
   Design of Truss Top Chord Members   297
   Net Uplift Load on a Roof Truss   299
9.7 Second Floor Framing Design   299
   Design of a Typical Floor Joist   300
   Design of a Glulam Floor Girder   301
   Design of Header Beams   307
9.8 Design of a Typical Ground Floor Column   311
9.9 Design of a Typical Exterior Wall Stud   312
9.10 Design of Roof and Floor Sheathing   317
   Gravity Loads   317
   Lateral Loads   317
9.11 Design of Wall Sheathing for Lateral Loads   319
9.12 Overturning Analysis of Shear Walls: Shear Wall Chord Forces   322
   Maximum Force in Tension Chord   325
   Maximum Force in Compression Chord   327
9.13 Forces in Horizontal Diaphragm Chords, Drag Struts, and Lap Splices   331
   Design of Chords, Struts, and Splices   331
   Hold-Down Anchors   337
   Sill Anchors   338
9.14 Design of Shear Wall Chords   339
9.15 Construction Documents   344
   References   345

appendix A **Weights of Building Materials   347**

appendix B **Design Aids   350**

Index   391

Preface

The primary audience for this book are students of civil and architectural engineering, civil and construction engineering technology, and architecture in a typical undergraduate course in wood or timber design. The book can be used for a one-semester course in structural wood or timber design and should prepare students to apply the fundamentals of structural wood design to typical projects that might occur in practice. The practice-oriented and easy-to-follow but thorough approach to design that is adopted, and the many practical examples applicable to typical everyday projects that are presented, should also make the book a good resource for practicing engineers, architects, and builders and those preparing for professional licensure exams.

The book conforms to the 2005 *National Design Specification for Wood Construction,* and is intended to provide the essentials of structural design in wood from a practical perspective and to bridge the gap between the design of individual wood structural members and the complete design of a wood structure, thus providing a holistic approach to structural wood design. Other unique features of this book include a discussion and description of common wood structural elements and systems that introduce the reader to wood building structures, a complete wood building design case study, the design of wood floors for vibrations, the general analysis of shear walls for overturning, including all applicable loads, the many three- and two-dimensional drawings and illustrations to assist readers' understanding of the concepts, and the easy-to-use design aids for the quick design of common structural members, such as floor joists, columns, and wall studs.

Chapter 1 The reader is introduced to wood design through a discussion and description of the various wood structural elements and systems that occur in wood structures as well as the properties of wood that affect its structural strength.

Chapter 2 The various structural loads—dead, live, snow, wind, and seismic—are discussed and several examples are presented. This succinct treatment of structural loads gives the reader adequate information to calculate the loads acting on typical wood building structures.

Chapter 3 Calculation of the allowable stresses for both sawn lumber and glulam in accordance with the 2005 *National Design Specification* as well as a discussion of the various stress adjustment factors are presented in this chapter. Glued laminated timber (glulam), the various grades of glulam, and determination of the controlling load combination in a wood building using the normalized load method are also discussed.

Chapter 4 The design and analysis of joists, beams, and girders are discussed and several examples are presented. The design of wood floors for vibrations, miscellaneous stresses in wood members, the selection of preengineered wood flexural members, and the design of sawn-lumber decking are also discussed.

Chapter 5 The design of wood members subjected to axial and bending loads, such as truss web and chord members, solid and built-up columns, and wall studs, is discussed.

Chapter 6 The design of roof and floor sheathing for gravity loads and the design of roof and floor diaphragms for lateral loads are discussed. Calculation of the forces in diaphragm chords and drag struts is also discussed, as well as the design of these axially loaded elements.

Chapter 7 The design of exterior wall sheathing for wind load perpendicular to the face of a wall and the design of wood shear walls or vertical diaphragms parallel to the lateral loads are discussed. A general analysis of shear walls for overturning that takes into account all applicable lateral and gravity loads is presented. The topic of combined shear and uplift in wall sheathing is also discussed, and an example presented.

Chapter 8 The design of connections is covered in this chapter in a simplified manner. Design examples are presented to show how the connection capacity tables in the NDS code are used. Several practical connections and practical connection considerations are discussed.

Chapter 9 A complete building design case study is presented to help readers tie together the pieces of wood structural element design presented in earlier chapters to create a total building system design, and a realistic set of structural plans and details are also presented. This holistic and practice-oriented approach to structural wood design is the hallmark of the book. The design aids presented in Appendix B for the quick design of floor joists, columns, and wall studs subjected to axial and lateral loads are utilized in this chapter.

In conclusion, we would like to offer the following personal dedications and thanksgiving:

To my wife, Josie, the love of my life and the apple of my eye, and to my precious children, Osa, Itohan, Odosa, and Eghosa, for their support and encouragement. To my mother for instilling in me the discipline of hard work and excellence, and to my Lord and Savior, Jesus Christ, for His grace, wisdom, and strength.

Abi Aghayere
Rochester, New York

For Adele and Ivy; and for Michele, who first showed me that "I can do all things through Christ which strengtheneth me" (Phil. 4:13)

Jason Vigil
Rochester, New York

STRUCTURAL WOOD DESIGN

chapter *one*

INTRODUCTION: WOOD PROPERTIES, SPECIES, AND GRADES

1.1 INTRODUCTION

The purpose of this book is to present the design process for wood structures in a quick and simple way, yet thoroughly enough to cover the analysis and design of the major structural elements. In general, building plans and details are defined by an architect and are usually given to a structural engineer for design of structural elements and to present the design in the form of structural drawings. In this book we take a project-based approach covering the design process that a structural engineer would go through for a typical wood-framed structure.

The intended audience for this book is students taking a course in timber or structural wood design and structural engineers and similarly qualified designers of wood or timber structures looking for a simple and practical guide for design. The reader should have a working knowledge of statics, strength of materials, structural analysis (including truss analysis), and load calculations in accordance with building codes (dead, live, snow, wind, and seismic loads). Design loads are reviewed in Chapter 2. The reader must also have available:

1. *National Design Specification for Wood Construction,* 2005 edition, ANSI/AF&PA (hereafter referred to as the NDS code) [1]
2. *National Design Specification Supplement: Design Values for Wood Construction,* 2005 edition, ANSI/AF&PA (hereafter referred to as NDS-S) [2]
3. *International Building Code,* 2006 edition, International Code Council (ICC) (hereafter referred to as the IBC) [3]
4. *Minimum Design Loads for Buildings and Other Structures,* 2005 edition, American Society of Civil Engineers (ASCE) (hereafter referred to as ASCE 7) [4]

The Project-based Approach

Wood is nature's most abundant renewable building material and a widely used structural material in the United States, where more than 80% of all buildings are of wood construction. The number of building configurations and design examples that could be presented is unlimited. Some applications of wood in construction include residential buildings, strip malls, offices, hotels, schools and colleges, healthcare and recreation facilities, senior living and retirement homes, and religious buildings. The most common wood structures are residential and multifamily dwellings as well as hotels. Residential structures are usually one to three stories in height, while multifamily and hotel structures can be up to four stories in height. Commercial, industrial, and other structures that have higher occupancy loads and factors of safety are not typically constructed with wood, although wood may be used as a secondary structure, such as a storage mezzanine. The structures that support amusement park rides are mostly built out of wood because of the relatively low maintenance cost of exposed wood structures and its unique ability to resist the repeated cycles of dynamic loading (fatigue) imposed on the structure by the amusement park rides. The approach taken here is a simplified version of the design process required for each major structural element in a timber structure. In Figures 1.1 and 1.2 we identify the typical structural elements in a wood building. The elements are described in greater detail in the next section.

1.2 TYPICAL STRUCTURAL COMPONENTS OF WOOD BUILDINGS

The majority of wood buildings in the United States are typically *platform construction*, in which the vertical wall studs are built one story at a time and the floor below provides the platform to build the next level of wall that will in turn support the floor above. The walls usually span vertically between the sole or sill plates at a floor level and the top plates at the floor or roof level above. This is in contrast to the infrequently used *balloon-type construction*, where the vertical

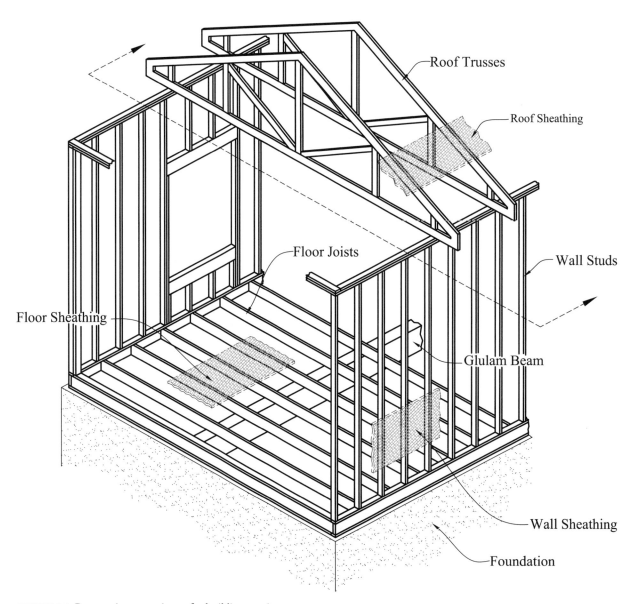

FIGURE 1.1 Perspective overview of a building section.

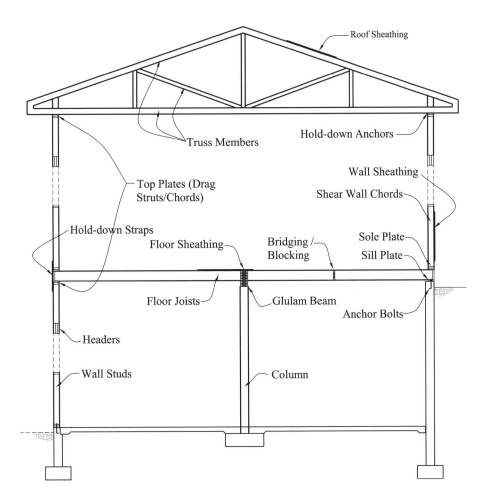

FIGURE 1.2 Overview of major structural elements.

studs are continuous for the entire height of the building and the floor framing is supported on brackets off the face of the wall studs. The typical structural elements in a wood-framed building system are described below.

Rafters (Figure 1.3) These are usually sloped sawn-dimension lumber roof beams spaced at fairly close intervals (e.g., 12, 16, or 24 in.) and carry lighter loads than those carried by the roof trusses, beams, or girders. They are usually supported by roof trusses, ridge beams, hip beams, or walls. The span of rafters is limited in practice to a maximum of 14 to 18 ft. Rafters of varying spans that are supported by hip beams are called *jack rafters* (see Figure 1.6). Sloped roof rafters with a nonstructural ridge, such as a 1× ridge board, require ceiling tie joists or collar ties to resist the horizontal outward thrust at the exterior walls that is due to gravity loads on the sloped rafters. A rafter-framed roof with ceiling tie joists acts like a three-member truss.

Joists (Figure 1.4) These are sawn-lumber floor beams spaced at fairly close intervals of 12, 16, or 24 in. that support the roof or floor deck. They support lighter loads than do floor beams or girders. Joists are typically supported by floor beams, walls, or girders. The spans are usually limited in practice to about 14 to 18 ft. Spans greater than 20 ft usually require the use of preengineered products, such as I-joists or open-web joists, which can vary from 12 to 24 in. in depth. Floor joists can be supported on top of the beams, either in-line or lapped with other joists framing into the beam, or the joist can be supported off the side of the beams using joist hangers. In the former case, the top of the joist does not line up with the top of the beam as it does in the latter case. Lapped joists are used more commonly than in-line joists because of the ease of framing and the fact that lapped joists are not affected by the width (i.e., the smaller dimension) of the supporting beam.

FIGURE 1.3 Rafter framing options.

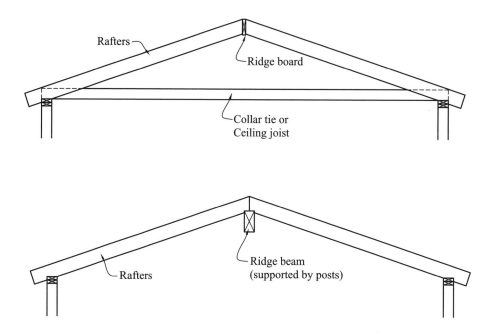

Double or Triple Joists These are two or more sawn-lumber joists that are nailed together to act as one composite beam. They are used to support heavy concentrated loads or the load from a partition wall or a load-bearing wall running parallel to the span of the floor joists, in addition to the tributary floor loads. They are also used to frame around stair openings (see header and trimmer joists).

Header and Trimmer Joists These are multiple-dimension lumber joists that are nailed together (e.g., double joists) and used to frame around stair openings. The trimmer joists are parallel to the long side of the floor opening and support the floor joists and the wall at the edge of the stair. The header joists support the stair stringer and floor loads and are parallel to the short side of the floor opening.

Beams and Girders (Figure 1.5) These are horizontal elements that support heavier gravity loads than rafters and joists and are used to span longer distances. Wood beams can also be built from several joists nailed together. These members are usually made from beam and stringer

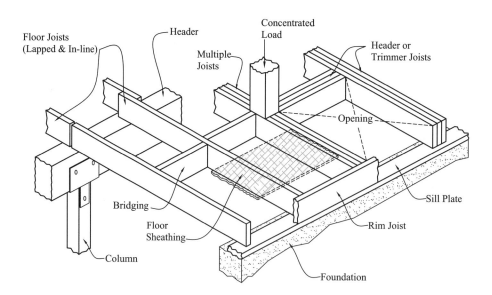

FIGURE 1.4 Floor framing elements.

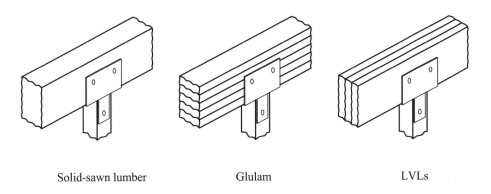

FIGURE 1.5 Types of beams and girders.

(B&S) sawn lumber, glued laminated timber (glulam) or parallel strand lumber (PSL), or laminated veneer lumber (LVL).

Ridge Beams These are roof beams at the ridge of a roof that support the sloped roof rafters. They are usually supported at their ends on columns or posts (see Figure 1.3)

Hip and Valley Rafters These are sloped diagonal roof beams that support sloped jack rafters in roofs with hips or valleys, and support a triangular roof load due to the varying spans of the jack rafters (see Figure 1.6). The hip rafters are simply supported at the exterior wall and on the sloped main rafter at the end of the ridge. The jack or varying span rafters are supported on the hip rafters and the exterior wall. The top of a hip rafter is usually shaped in the form of an inverted V, while the top of a valley rafter is usually V-shaped. Hip and valley rafters are designed like ridge beams.

Columns or Posts These are vertical members that resist axial compression loads and may occasionally resist additional bending loads due to lateral wind loads or the eccentricity of the gravity loads on the column. Columns or posts are usually made from post and timber (P&T) sawn lumber or glulam. Sometimes, columns or posts are built up using dimension-sawn lumber. Wood posts may also be used as the chords of shear walls, where they are subjected to axial

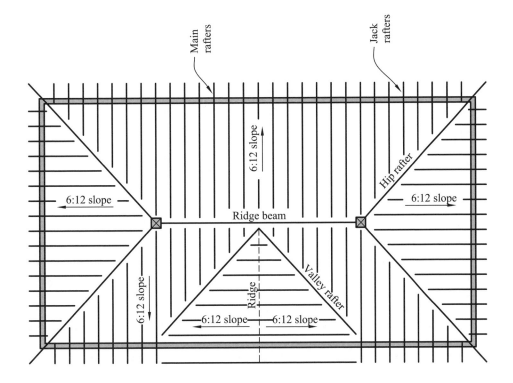

FIGURE 1.6 Hip and Valley rafters.

tension or compression forces from the overturning effect of the lateral and seismic loads on the building.

Roof Trusses (Figure 1.7) These are made up typically of dimension-sawn lumber top and bottom chords and web members that are subject to axial tension or compression plus bending loads. Trusses are usually spaced at not more than 48 in. on centers and are used to span long distances up to 120 ft. The trusses usually span from outside wall to outside wall. Several truss configurations are possible, including the Pratt truss, the Warren truss, the scissor truss, the Fink truss, and the bowstring truss. In building design practice, prefabricated trusses are usually specified, for economic reasons, and these are manufactured and designed by truss manufacturers rather than by the building designer. Prefabricated trusses can also be used for floor framing. These are typically used for spans where sawn lumber is not adequate. The recommended span-to-depth ratios for wood trusses are 8 to 10 for flat or parallel chord trusses, 6 or less for pitched or triangular roof trusses, and 6 to 8 for bowstring trusses [16].

Wall Studs (Figure 1.8) These are axially loaded in compression and made of dimension lumber spaced at fairly close intervals (typically, 12, 16, or 24 in.). They are usually subjected to concentric axial compression loads, but exterior stud walls may also be subjected to a combined concentric axial compression load plus bending load due to wind load acting perpendicular to the wall. Wall studs may be subjected to eccentric axial load: for example, in a mezzanine floor with single-story stud and floor joists supported off the narrow face of the stud by joist hangers. Interior wall studs should, in addition to the axial load, be designed for the minimum 5 psf of interior wind pressure specified in the IBC.

Wall studs are usually tied together with plywood sheathing that is nailed to the narrow face of studs. Thus, wall studs are laterally braced by the wall sheathing for buckling about their weak axis (i.e., buckling in the plane of the wall). Stud walls also act together with plywood sheathing as part of the vertical diaphragm or shear wall to resist lateral loads acting parallel to the plane of the wall. *Jack studs* (also called *jamb* or *trimmer studs*) are the studs that support the ends of window or door headers; *king studs* are full-height studs adjacent to the jack studs and *cripple studs* are the stubs or less-than-full-height stud members above or below a window or door opening and are usually supported by header beams. The wall frame consisting of the studs, wall sheathing, top and bottom plates are usually built together as a unit on a flat horizontal surface and then lifted into position in the building.

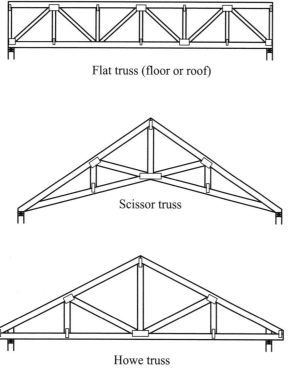
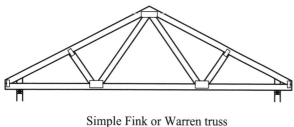
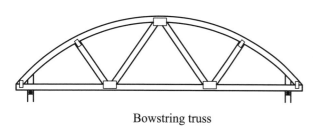

FIGURE 1.7 Truss profiles.

Header Beams (Figure 1.7) These are the beams that frame over door and window openings, supporting the dead load of the wall framing above the door or window opening as well as the dead and live loads from the roof or floor framing above. They are usually supported with beam hangers off the end chords of the shear walls or on top of jack studs adjacent to the shear wall end chords. In addition to supporting gravity loads, these header beams may also act as the chords and drag struts of the horizontal diaphragms in resisting lateral wind or seismic loads. Header beams can be made from sawn lumber, parallel strand lumber, linear veneer lumber, or glued laminated timber, or from built-up dimension

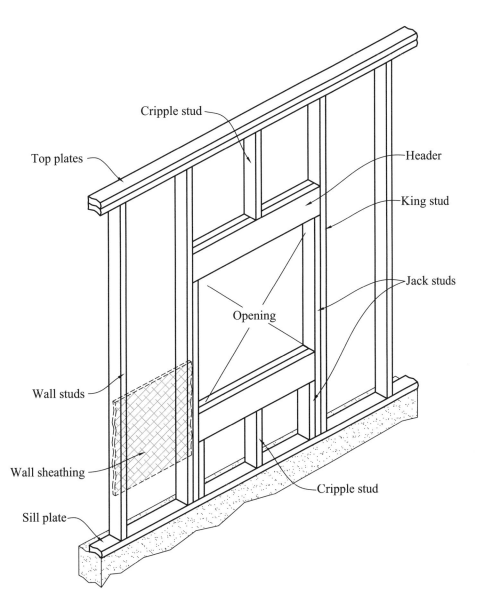

FIGURE 1.8 Wall framing elements.

lumber members nailed together. For example, a 2 × 10 double header beam implies a beam with two 2 × 10's nailed together.

Overhanging or Cantilever Beams (Figure 1.9) These beams consist of a back span between two supports and an overhanging or cantilever span beyond the exterior wall support below. They are sometimes used for roof framing to provide a sunshade for the windows and to protect the exterior walls from rain, or in floor framing to provide a balcony. For these types of beams it is more efficient to have the length of the back span be at least three times the length of the overhang or cantilever span. The deflection of the tip of the cantilever or overhang and the uplift force at the back-span end support could be critical for these beams. They have to be designed for unbalanced or skip or pattern live loading to obtain the worst possible load scenario. It should be noted that roof overhangs are particularly susceptible to

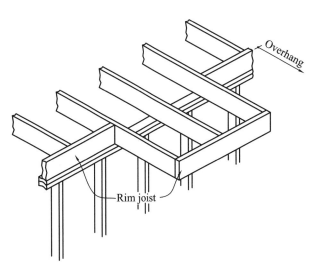

FIGURE 1.9 Cantilever framing.

large wind uplift forces, especially in hurricane-prone regions.

Blocking or Bridging These are usually 2× solid wood members or x-braced wood members spanning between roof or floor beams, joists, or wall studs, providing lateral stability to the beams or joists. They also enable adjacent flexural members to work together as a unit in resisting gravity loads, and help to distribute concentrated loads applied to the floor. They are typically spaced at no more than 8 ft on centers. The bridging (i.e., cross-bracing) in roof trusses is used to prevent lateral-torsional buckling of the truss top and bottom chords.

Top Plates These are continuous 2× horizontal flat members located on top of the wall studs at each level. They serve as the chords and drag struts or collectors to resist in-plane bending and direct axial forces due to the lateral loads on the roof and floor diaphragms, and where the spacing of roof trusses rafters or floor joists do not match the stud spacing, they act as flexural members spanning between studs and bending about their weak axis to transfer the truss, rafter or joist reactions to the wall studs. They also help to tie the structure together in the horizontal plane at the roof and floor levels.

Bottom Plates These continuous 2× horizontal members or sole plates are located immediately below the wall studs and serve as bearing plates to help distribute the gravity loads from the wall studs. They also help to transfer lateral the loads between the various levels of a shear wall. The bottom plates located on top of the concrete or masonry foundation wall are called sill plates and these are usually pressure treated because of the presence of moisture since they are in direct contact with concrete or masonry. They also serve as bearing plates and help to transfer the lateral base shear from the shear wall into the foundation wall below by means of the sill anchor bolts.

1.3 TYPICAL STRUCTURAL SYSTEMS IN WOOD BUILDINGS

The above-grade structure in a typical wood-framed building consists of the following structural systems: roof framing, floor framing, and wall framing.

Roof Framing

Several schemes exist for the roof framing layout:

1. Roof trusses spanning in the transverse direction of the building from outside wall to outside wall (Figure 1.10a).

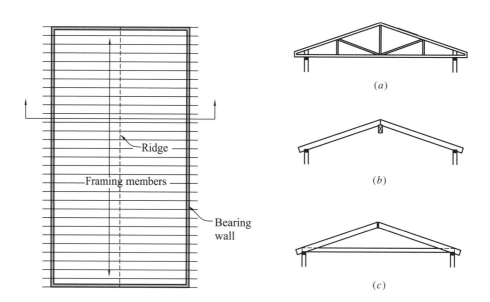

FIGURE 1.10 Roof truss layout: (a) roof truss; (b) vaulted ceiling; (c) rafter and collar tie.

2. Sloped rafters supported by ridge beams and hip or valley beams or exterior walls, used to form cathedral or vaulted ceilings (Figure 1.10b).

3. Sloped rafters with a 1× ridge board at the roof ridge line, supported on the exterior walls by the outward thrust resisted by collar or ceiling ties (Figure 1.10c). The intersecting rafters at the roof ridge level support each other by providing a self-equilibrating horizontal reaction at that level. This horizontal reaction results in an outward thrust at the opposite end of the rafter at the exterior walls, which has to be resisted by the collar or ceiling ties.

4. Wood framing, which involves using purlins, joists, beams, girders, and interior columns to support the roof loads such as in panelized flat roof systems as shown in Figure 1.11. Purlins are small sawn lumber members such as 2 × 4s and 2 × 6s that span between joists, rafter, or roof trusses in panelized roof systems with spans typically in the 8 to 10 ft range, and a spacing of 24 inches.

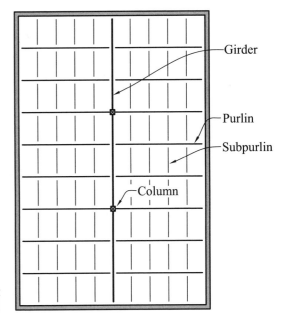

FIGURE 1.11 Typical roof framing layout.

Floor Framing

The options for floor framing basically involve using wood framing members, such as floor joists, beams, girders, interior columns, and interior and exterior stud walls, to support the floor loads. The floor joists are either supported on top of the beams or supported off the side faces of the beams with joist hangers. The floor framing supports the floor sheathing, usually plywood or oriented strand board (OSB), which in turn provides lateral support to the floor framing members and acts as the floor surface, distributing the floor dead and live loads. In addition, the floor sheathing acts as the horizontal diaphragm that transfers the lateral wind and seismic loads to the vertical diaphragms or shear walls. Examples of floor framing layouts are shown in Figure 1.12.

Wall Framing

Wall framing in wood-framed buildings consists of repetitive vertical 2 × 4 or 2 × 6 wall studs spaced at 16 or 24 in. on centers, with plywood or OSB attached to the outside face of the

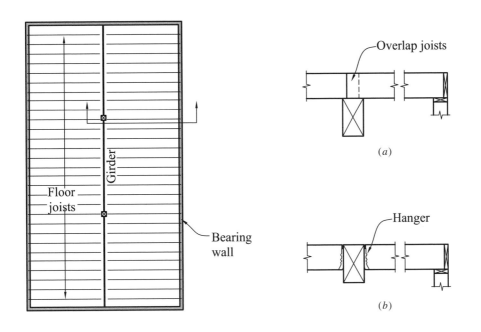

FIGURE 1.12 Typical floor framing layout: (a) framing over girder; (b) face-mounted joists.

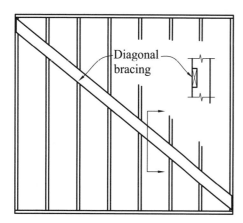

FIGURE 1.13 Diagonal let-in bracing.

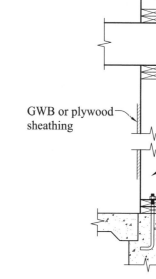

FIGURE 1.14 Typical wall section.

wall. It also consists of a top plate at the top of the wall, a sole or sill plate at the bottom of the wall, and header beams supporting loads over door and window openings. These walls support gravity loads from the roof and floor framing and resist lateral wind loads perpendicular to the face of the wall as well as acting as a shear wall to resist lateral wind or seismic loads in the plane of the wall. It may be necessary to attach sheathing to both the interior and exterior faces of the wall studs to achieve greater shear capacity in the shearwall. Occasionally, diagonal let-in bracing is used to resist lateral loads in lieu of structural sheathing, but this is not common (see Figure 1.13). A typical wall section is shown in Figure 1.14 (see also Figure 1.8)

Shear Walls in Wood Buildings

The lateral wind and seismic forces acting on wood buildings result in sliding, overturning, and racking of a building, as illustrated in Figure 1.15. Sliding of a building is resisted by the friction between the building and the foundation walls, but in practice this friction is neglected

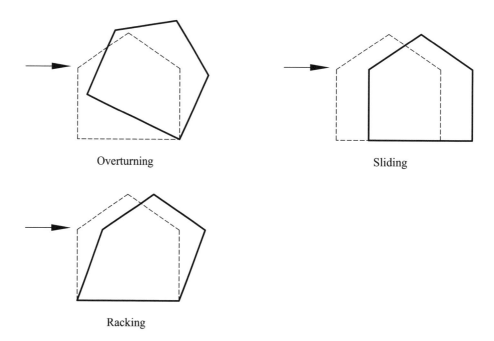

FIGURE 1.15 Overturning, sliding, and racking in wood buildings.

and sill plate anchors are usually provided to resist the sliding force. The overturning moment, which can be resolved into a downward and upward couple of forces, is resisted by the dead weight of the structure and by hold-down anchors at the end chords of the shear walls. Racking of a building is resisted by let-in diagonal braces or by plywood or OSB sheathing nailed to the wall studs acting as a shear wall.

The uplift forces due to upward vertical wind loads (or suction) on the roofs of wood buildings are resisted by the dead weight of the roof and by using toenailing or hurricane or hold-down anchors. These anchors are used to tie the roof rafters or trusses to the wall studs. The uplift forces must be traced all the way down to the foundation. If a net uplift force exists in the wall studs at the ground-floor level, the sill plate anchors must be embedded deep enough into the foundation wall or grade beam to resist this uplift force, and the foundation must also be checked to ensure that it has enough dead weight, from its self weight and the weight of soil engaged, to resist the uplift force.

1.4 WOOD STRUCTURAL PROPERTIES

Wood is a biological material and is one of the oldest structural materials in existence. It is nonhomogeneous and orthotropic, and thus its strength is affected by the direction of load relative to the direction of the grain of the wood, and it is naturally occurring and can be renewed by planting or growing new trees. Since wood is naturally occurring and nonhomogeneous, its structural properties can vary widely, and because wood is a biological material, its strength is highly dependent on environmental conditions. Wood buildings have been known to be very durable, lasting hundreds of years, as evidenced by the many historic wood buildings in the United States. In this chapter we discuss the properties of wood that are of importance to architects and engineers in assessing the strength of wood members and elements.

Wood fibers are composed of small, elongated, round or rectangular tubelike cells (see Figure 1.16) with the cell walls made of cellulose, which gives the wood its load-carrying ability. The cells or fibers are oriented in the longitudinal direction of the tree log and are bound together by a material called *lignin,* which acts like glue. The chemical composition of wood consists of approximately 60% cellulose, 30% lignin, and 12% sugar end extractives. The water in the cell walls is known as *bound water,* and the water in the cell cavities is known as *free water.* When wood is subjected to drying or seasoning, it loses all its free water before it begins to lose bound water from the cell walls. It is the bound water, not the free water, that affects the shrinking or swelling of a wood member. The cells or fibers are usually oriented in the vertical direction of the tree. The strength of wood depends on the direction of the wood grain. The direction parallel to the tree trunk or longitudinal direction is referred to as the *parallel-to-grain direction;* the radial and tangential directions are both referred to as the *perpendicular-to-grain direction.*

Tree Cross Section

There are two main classes of trees: hardwood and softwood. This terminology is not indicative of how strong a tree is because some softwoods are actually stronger than hardwoods. *Hardwoods* are broad-leaved, whereas *softwoods* have needlelike leaves and are mostly evergreen. Hardwood

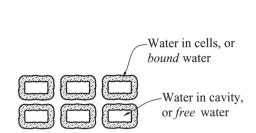

FIGURE 1.16 Cellular structure of wood.

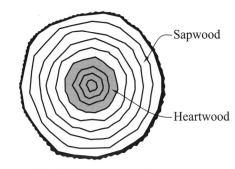

FIGURE 1.17 Typical tree cross-section.

trees take longer to mature and grow than softwoods, are mostly tropical, and are generally more dense than softwoods. Consequently, they are more expensive and used less frequently than softwood lumber or timber in wood building construction in the United States. Softwoods constitute more than 75% of all lumber used in construction in the United States [6], and more than two-thirds of softwood lumber are western woods such as douglas fir-larch and spruce. The rest are eastern woods such as southern pine. Examples of hardwood trees include balsa, oak, birch, and basswood.

A typical tree cross section is shown Figure 1.17. The growth of timber trees is indicated by an annual growth ring added each year to the outer surface of the tree trunk just beneath the bark. The age of a tree can be determined from the number of annual rings in a cross section of the tree log at its base. The tree cross section shows the two main sections of the tree, the sapwood and the heartwood. *Sapwood* is light in color and may be as strong as heartwood, but it is less resistant to decay. *Heartwood* is darker and older and more resistant to decay. However, sapwood is lighter and more amenable than heartwood to pressure treatment. Heartwood is darker and functions as a mechanical support for a tree, while sapwood contains living cells for nourishment of the tree.

Advantages and Disadvantages of Wood as a Structural Material

Some advantages of wood as a structural material are as follows:

- Wood is renewable.
- Wood is machinable.
- Wood has a good strength-to-weight ratio.
- Wood will not rust.
- Wood is aesthetically pleasing.

The disadvantages of wood include the following:

- Wood can burn.
- Wood can decay or rot and can be attacked by insects such as termites and marine borers. Moisture and air promote decay and rot in wood.
- Wood holds moisture.
- Wood is susceptible to volumetric instability (i.e., wood shrinks).
- Wood's properties are highly variable and vary widely between species and even between trees of the same species. There is also variation in strength within the cross section of a tree log.

1.5 FACTORS AFFECTING THE STRENGTH OF WOOD

Several factors that affect the strength of a wood member are discussed in this section: (1) species group, (2) moisture content, (3) duration of loading, (4) size and shape of the wood member, (5) defects, (6) direction of the primary stress with respect to the orientation of the wood grain, and (7) ambient temperature.

Species and Species Group

Structural lumber is produced from several species of trees. Some of the species are grouped together to form a *species group,* whose members are "grown, harvested and manufactured together." The NDS code's tabulated stresses for a species group were derived statistically from the results of a large number of tests to ensure that all the stresses tabulated for all species within a species group are conservative and safe. A species group is a combination of two or more species. For example, Douglas fir-larch is a species group that is obtained from a combination of Douglas fir and western larch species. Hem-fir is a species group that can be obtained from a combination of western hemlock and white fir.

Structural wood members are derived from different stocks of trees, and the choice of wood species for use in design is typically a matter of economics and regional availability. For a given location, only a few species groups might be readily available. The species groups that have the highest available strengths are Douglas fir-larch and southern pine, also called southern yellow pine. Examples of widely used species groups (i.e., combinations of different wood species) of structural lumber in wood buildings include Douglas fir-larch (DF-L), hem-fir, spruce-pine-fir (SPF), and southern yellow pine (SYP). Each species group has a different set of tabulated design stresses in the NDS-S, and wood species within a particular species group possess similar properties and have the same grading rules.

Moisture Content

The strength of a wood member is greatly influenced by its *moisture content*, which is defined as the percentage amount of moisture in a piece of wood. The *fiber saturation point* (FSP) is the moisture content at which the free water (i.e., the water in cell cavities) has been fully dissipated. Below the FSP, which is typically between 25 and 35% moisture content for most wood species, wood starts to shrink by losing water from the cell walls (i.e., the bound water). The *equilibrium moisture content* (EMC), the moisture content at which the moisture in a wood member has come to a balance with that in the surrounding atmosphere, occurs typically at between 10 and 15% moisture content for most wood species in a protected environment. The moisture content in wood can be measured using a hand held moisture meter. As the moisture content increases up to the FSP (the point where all the free water has been dissipated), the wood strength decreases, and as the moisture content decreases below the FSP, the wood strength increases, although this increase may be offset by some strength reduction from the shrinkage of the wood fibers. The moisture content (MC) of a wood member can be calculated as

$$\text{MC}(\%) = \frac{\text{weight of moist wood} - \text{weight of oven-dried wood}}{\text{weight of oven-dried wood}} \times 100\%$$

There are two classifications of wood members based on moisture content: green and dry. *Green lumber* is freshly cut wood and the moisture content can vary from as low as 30% to as high as 200% [6]. *Dry* or *seasoned lumber* is wood with a moisture content no higher than 19% for sawn lumber and less than 16% for glulam (see Table 1.1) Wood can be seasoned by air drying or by kiln drying. Most wood members are used in dry or seasoned conditions where the wood member is protected from excessive moisture. An example of a building where wood will be in a moist or green condition is an exposed bus garage or shed. The effect of the moisture content is taken into account in design by use of the moisture adjustment factor, C_M, which is discussed in Chapter 3.

Seasoning of Lumber

The *seasoning* of lumber, the process of removing moisture from wood to bring the moisture content to an acceptable level, can be achieved through air drying or kiln drying. *Air drying* involves stacking lumber in a covered shed and allowing moisture loss or drying to take place naturally over time due to the presence of air. Fans can be used to accelerate the seasoning process. *Kiln drying* involves placing lumber pieces in an enclosure or kiln at significantly higher temperatures. The kiln temperature has to be strictly controlled to prevent damage to the wood members from seasoning defects such as warp, bow, sweep, twists, or crooks. Seasoned wood is recommended for building construction because of its dimensional stability. The shrinkage that occurs when unseasoned wood is used can lead to problems in the structure as the shape changes

TABLE 1.1 Moisture Content Classifications for Sawn Lumber and Glulam

Lumber Classification	Moisture Content (%)	
	Sawn Lumber	Glulam
Dry	≤19	<16
Green	>19	≥16

TABLE 1.2 Size Classifications for Sawn Lumber

Lumber Classification	Size
Dimension lumber	Nominal *thickness:* from 2 to 4 in. Nominal *width:* ≥2 in. but ≤16 in. *Examples:* 2 × 4, 2 × 6, 2 × 8, 4 × 14, 4 × 16
Beam and stringer (B&S)	Rectangular cross section Nominal *thickness:* ≥5 in. Nominal *width:* >2 in. + nominal thickness *Examples:* 5 × 8, 5 × 10, 6 × 10
Post and timber (P&T)	Approximately square cross section Nominal *thickness:* ≥5 in. Nominal *width:* ≤2 in. + nominal thickness *Examples:* 5 × 5, 5 × 6, 6 × 6, 6 × 8
Decking	Nominal thickness: 2 to 4 in. Wide face applied directly in contact with framing Usually, tongue-and-grooved Used as roof or floor sheathing *Example:* 2 × 12 lumber used in a flatwise direction

upon drying out. The amount of shrinkage in a wood member varies considerably depending on the direction of the wood grain.

Duration of Loading

The longer a load acts on a wood member, the lower the strength of the wood member, and conversely, the shorter the duration, the stronger the wood member. This is because wood is susceptible to creep or the tendency for continuously increasing deflections under constant load because of the continuous loss of water from the wood cells due to drying shrinkage. The effect of load duration is taken into account in design by use of the load duration adjustment factor, C_D, discussed in Chapter 3.

Size Classifications of Sawn Lumber

As the size of a wood member increases, the difference between the actual behavior of the member and the ideal elastic behavior assumed in deriving the design equations becomes more pronounced. For example, as the depth of a flexural member increases, the deviation from the assumed elastic properties increases and the strength of the member decreases. The various size classifications for structural sawn lumber are shown in Table 1.2, and it should be noted that for sawn lumber, the *thickness* refers to the smaller dimension of the cross section and the *width* refers to the larger dimension of the cross section. Different design stresses are given in the NDS-S for the various size classifications listed in Table 1.2.

Dimension lumber is typically used for floor joists or roof rafters, and 2 × 8, 2 × 10, and 2 × 12 are the most frequently used floor joist sizes. For light-frame residential construction, dimension lumber is generally used. Beam and stringer lumber is used as floor beams or girders and as door or window headers, and post and timber lumber is used for columns or posts.

Nominal Dimension versus Actual Size of Sawn Lumber

Wood members can come in dressed or undressed sizes, but most wood structural members come in dressed form. When rough wood is dressed on two sides, it is denoted as S2S; rough wood that is dressed on all four sides is denoted as S4S. Undressed 2 × 6 S4S lumber has an actual or nominal size of 2 × 6 in., whereas the dressed size is $1\frac{1}{2} \times 5\frac{1}{2}$ in. (Figure 1.18). The lumber size is usually called out on structural and architectural drawings using the nominal dimensions of the lumber. The reader is reminded that for a sawn lumber cross section, the thickness is the smaller dimension and the width is the larger dimension of the cross section. In this book we assume that all wood is dressed on four sides (i.e., S4S). Section properties for wood members are given in Tables 1A and 1B of NDS-S [2].

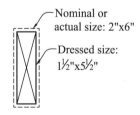

FIGURE 1.18 Nominal versus dressed size of a 2 × 6 sawn lumber.

Wood Defects

The various categories of defects in wood are natural, conversion, and seasoning defects. The nature, size, and location of defects affect the strength of a wood member because of the stress concentrations that they induce in the member. They also affect the finished appearance of the member. Some examples of natural defects are knots, shakes, splits, and fungal decay. Conversion defects occur due to unsound milling practices, one example being wanes. Seasoning defects result from the effect of uneven or unequal drying shrinkage, examples being various types of warps, such as cups, bows, sweep, crooks, or twists [6–8, 12, 14]. The most common types of defects in wood members are illustrated in Figure 1.19 and include the following:

- *Knots*. These are formed where limbs grow out from a tree stem.
- *Split or check*. This occurs due to separation of the wood fibers at an angle to annual rings and is caused by drying of the wood.
- *Shake*. This occurs due to separation of the wood fibers parallel to the annual rings.
- *Decay*. This is the rotting of wood due to the presence of wood-destroying fungi.
- *Wane*. In this defect the corners or edges of a wood cross section lack wood material or have some of the bark of the tree as part of the cross section. This leads to a reduction in the cross-sectional area of the member which affects the structural capacity of the member.

Defects lead to a reduction in the net cross section, and their presence introduces stress concentrations in the wood member. The amount of strength reduction depends on the size and location of the defect. For example, for an axially loaded tension member, a knot anywhere in the cross section would reduce the tension capacity of the member. On the other hand, a knot at the neutral axis of the beam would not affect the bending strength but may affect the shear strength if it is located near the supports. For visually graded lumber, the grade stamp, which indicates the design stress grade assigned by the grading inspector, takes into account the number and location of defects in that member.

It is recommended that lumber not be cut indiscriminately on site, as this could affect the strength of a member adversely [8]. Let us illustrate with an example. A 20-ft-long piece of 2 × 14 sawn lumber with a knot at the neutral axis at midspan has been delivered to a site to be used as a simply supported beam. The contractor would like to cut this member to use as a joist on a 12-ft span. To avoid reducing the shear strength of the member, it would need to be cut equally at both ends to maintain the relative location of the knot with respect to both ends of the member. Failure to do this would result in lower strength than that assigned by the grading inspector.

Other types of defects include warping and compression or reaction wood. *Warping* results from uneven drying shrinkage of wood, leading the wood member to deviate from the horizontal

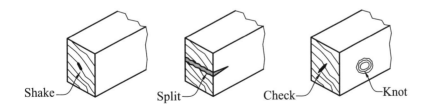

Cup (section)

Bow (elevation)

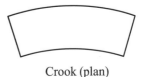
Crook (plan)

FIGURE 1.19 Common defects in wood.

or vertical plane. Examples of warping include members with a bow, cup, sweep, or crook. This defect does not affect the strength of the wood member but affects the constructability of the member. For example, if a bowed member is used as a joist or beam, there will be an initial sag or deflection in the member, depending on how it is oriented. This could affect the construction of the floor or roof in which it is used.

Compression or *reaction wood* is caused by a tree that grows abnormally in bent shape due either to natural effects or bending due to the effect of wind and snow loads. In a leaning tree trunk, one side of the tree cross section is subject to combined compression stresses from bending due to the crookedness of the tree trunk and axial load on the cross section from the self-weight of the tree. The wood fibers in the compression zone of the tree trunk will be more brittle and hard and will possess very little tensile strength, due to the existing internal compressive stresses. Compression wood should not be used for structural members.

Orientation of the Wood Grain

Wood is an orthotropic material with strengths that vary depending on the direction of the stress applied relative to the grain of the wood. As a result of the tubular nature of wood, three independent directions are present in a wood member: longitudinal, radial, and tangential. The variation in strength in a wood member with the direction of loading can be illustrated by a group of drinking straws glued tightly together. The group of straws will be strongest when the load is applied parallel to the length of the straws (i.e., longitudinal direction); loads applied in any other direction (i.e., radial or tangential) will crush the walls of the straws or pull apart the glue. The longitudinal direction is referred to as the *parallel-to-grain direction,* and the tangential and radial directions are both referred to as the *perpendicular-to-grain direction.* Thus, wood is strongest when the load or stress is applied in a direction parallel to the direction of the wood grain, is weakest when the stress is perpendicular to the direction of the wood grain, and has the least amount of shrinkage in the longitudinal or parallel-to-grain direction. The various axes in a wood member with respect to the grain direction are shown in Figure 1.20.

Axial or Bending Stress Parallel to the Grain This is the strongest direction for a wood member, and examples of stresses and loads acting in this direction are illustrated in Figure 1.21*a*.

Axial or Bending Stress Perpendicular to the Grain The strength of wood in compression parallel to the grain is usually stronger than wood in compression perpendicular to the grain (see Figure 1.21*b*). Wood has zero strength in tension perpendicular to the grain since only the lignin or glue is available to resist this tension force. Consequently, the NDS code does not permit the loading of wood in tension perpendicular to the grain.

Stress at an Angle to the Grain This case lies between the parallel-to-grain and perpendicular-to-grain directions and is illustrated in Figure 1.21*c*.

Ambient Temperature

Wood is affected adversely by temperature beyond 100°F. As the ambient temperature rises beyond 100°F, the strength of the wood member decreases. The structural members in most insulated wood buildings have ambient temperatures of less than 100°F.

1.6 LUMBER GRADING

Lumber is usually cut from a tree log in the longitudinal direction, and because it is naturally occurring, it has quite variable mechanical and structural properties, even for members cut from the same tree log. Lumber of similar mechanical and structural properties is grouped into a single category known as a *stress grade*. This simplifies the lumber selection process and increases economy. The higher the stress grade, the stronger and more expensive the wood member is. The classification of lumber with regard to strength, usage, and defects according to the grading rules of an approved grading agency is termed *lumber grading*.

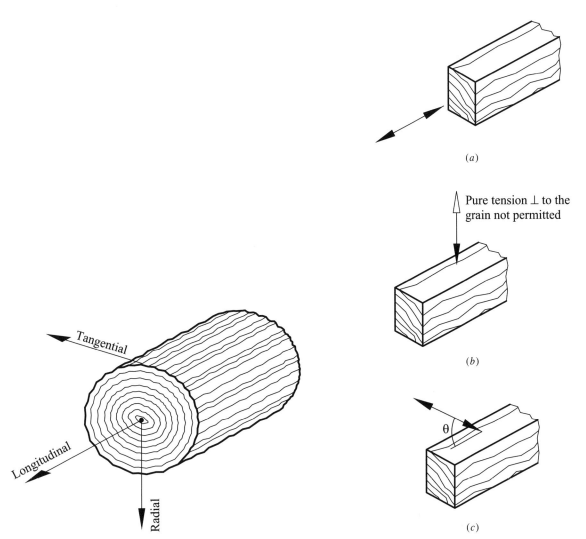

FIGURE 1.20 Longitudinal, radial, and tangential axes in a wood member.

FIGURE 1.21 Stress applied (*a*) parallel to the grain, (*b*) perpendicular to the grain, and (*c*) at an angle to the grain in a wood member.

Types of Grading

The two types of grading systems for structural lumber are visual grading and mechanical grading. The intent is to classify the wood members into various stress grades such as Select Structural, No. 1 and Better, No. 1, No. 2, Utility, and so on. A grade stamp indicating the stress grade and the species or species group is placed on the wood member, in addition to the moisture content, the mill number where the wood was produced, and the responsible grading agency. The grade stamp helps the engineer, architect, and contractor be certain of the quality of the lumber delivered to the site and that it conforms to the contract specifications for the project. Grading rules may vary among grading agencies, but minimum grading requirements are set forth in the American Lumber Product Standard US DOC PS-20 developed by the National Institute for Standards and Technology (NIST). Examples of grading agencies in the United States [2] include the Western Wood Products Association (WWPA), the West Coast Lumber Inspection Bureau (WCLIB), the Northern Softwood Lumber Bureau (NSLB), the Northeastern Lumber Manufacturers Association (NELMA), the Southern Pine Inspection Bureau (SPIB), and the National Lumber Grading Authority (NLGA).

Visual Grading

Visual grading, the oldest and most common grading system, involves visual inspection of wood members by an experienced and certified grader in accordance with established grading rules of a grading agency and application of a grade stamp. In visual grading, the lumber quality is reduced by the presence of defects, and the effectiveness of the grading system is very dependent on the experience of the professional grader. Grading agencies usually have certification exams that lumber graders have to take and pass annually to maintain their certification and to ensure accurate and consistent grading of sawn lumber. The stress grade of a wood member decreases as the number of defects increases and as their locations become more critical.

Machine Stress Rating

Mechanical grading is a nondestructive grading system that is based on the relationship between the stiffness and deflection of wood members. Each piece of wood is subjected to a nondestructive test in addition to a visual check. The grade stamp on machine-stress-rated (MSR) lumber includes the value of the tabulated bending stress and the pure bending modulus of elasticity. Because of the lower variability of material properties for MSR lumber, it is used in the fabrication of engineered wood products such as parallel strand lumber and laminated veneer lumber. *Machine-evaluated lumber* (MEL) relies on a relatively new grading process that uses a nondestructive x-ray inspection technique to measure density in addition to a visual check. The variability of MEL lumber is even lower than that of MSR lumber.

Stress Grades

The various lumber stress grades are listed below in order of decreasing strength. As discussed previously, the higher the stress grade, the higher the cost of the wood member.

- Dense Select Structural
- Select Structural
- No. 1 & Better
- No. 1
- No. 2
- No. 3
- Stud
- Construction
- Standard
- Utility

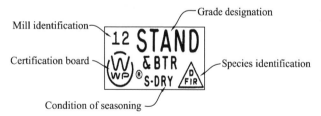

FIGURE 1.22 Typical grade stamp. (Courtesy of the Western Wood Products Association, Portland, OR.)

Grade Stamps

The use of a grade stamp on lumber assures the contractor and the engineer of record that the lumber supplied conforms to that specified in the contract documents. Lumber without a grade stamp should not be allowed on site or used in a project. A typical grading stamp on lumber might include the items shown in Figure 1.22.

1.7 SHRINKAGE OF WOOD

Shrinkage in a wood member takes place as moisture is dissipated from the member beyond the fiber saturation point. Wood shrinks as the moisture content decreases from its value at the installation of the member to the equilibrium moisture content, which can be as low as 8–10% in some protected environments. Shrinkage parallel to the grain of a wood member is negligible and much less than shrinkage perpendicular to the grain. Differential shrinkage is usually more critical than uniform shrinkage. Shrinkage effects in lumber can be minimized by using seasoned lumber or lumber with an equilibrium moisture content of 15% or less. To reduce the effects of shrinkage, minimize the use of details that transfer loads perpendicular to the grain. For wood members with two or more rows of bolts perpendicular to the direction of the wood grain, shrinkage across the width of the member causes tension stresses perpendicular to the grain in the wood member between the bolt holes, which could lead to the splitting of the member parallel to the grain [17]. Shrinkage can also adversely affect the functioning of hold-down anchors in shear walls by causing a gap between the anchor nut and the top of sill plate. As a result, the shear wall has to undergo excessive lateral displacement before the hold-down anchors can be engaged.

The effect of shrinkage on tie-down anchor systems can be minimized by pretensioning the anchors or by using proprietary shrinkage compensating anchor devices. [18] One method that has been used successfully to control the moisture content in wood during construction in order to achieve the required moisture threshold is by using portable heaters to dry the wood continuously during construction [19]. The effect of shrinkage can also be minimized by delaying the installation of architectural finishes to allow time for much of the wood shrinkage to occur. It is important to control shrinkage effects in wood structures by proper detailing and by limiting the change in moisture content of the member to avoid adverse effects on architectural finishes and to prevent the excessive lateral deflection of shear walls, and loosening of connections or splitting of wood members at connections.

The amount of shrinkage across the width or thickness of a wood member or element (i.e., perpendicular to the grain or to the longitudinal direction) is highly variable, but can be estimated using the following equation (adapted from ASTM D1990 [15]):

$$d_2 = d_1 \left[\frac{1 - (a - bM_2)/100}{1 - (a - bM_1)/100} \right] \quad (1.1)$$

where d_1 = initial member thickness or width at the initial moisture content M_1, in.
d_2 = final member thickness or width at the final moisture content M_2, in.
M_1 = moisture content at dimension d_1, %
M_2 = moisture content at dimension d_2, %

The variables a and b are obtained from Table 1.3. The total shrinkage of a wood building detail or section is the sum of the shrinkage *perpendicular to the grain* of each wood member or element in that detail or section; longitudinal shrinkage or the shrinkage *parallel to the grain* is negligible.

1.8 DENSITY OF WOOD

The density of wood is a function of the moisture content of the wood and the weight of the wood substance or cellulose present in a unit volume of wood. Even though the cellulose–lignin combination in wood has a specific gravity of approximately 1.50 and is heavier than water, most wood used in construction floats because of the presence of cavities in the hollow cells of a wood member. The density of wood can vary widely between species, from as low as 20 lb/in^3 to as high as 65 lb/in^3 [2, 6], and the higher the density, the higher the strength of the wood member. An average wood density of 31.2 lb/in^3 is used throughout this book.

1.9 UNITS OF MEASUREMENT

The U.S. system of units is used in this book, and accuracy to at most three significant figures is maintained in all the example problems. The standard unit of measurement for lumber in the United States is the *board foot* (bf), which is defined as the volume of 144 cubic inches of lumber

TABLE 1.3 Shrinkage Parameters

Wood Species	Width		Thickness	
	a	b	a	b
Redwood, western red cedar, northern white cedar	3.454	0.157	2.816	0.128
Other	6.031	0.215	5.062	0.181

Source: Ref. 15.

> ### EXAMPLE 1.1
>
> *Shrinkage in Wood Members*
>
> Determine the total shrinkage across (a) the width and (b) the thickness of two green 2 × 6 Douglas fir-larch top plates loaded perpendicular to the grain as the moisture content decreases from an initial value of 30% to a final value of 15%.
>
> *Solution:* For 2 × 6 sawn lumber, the actual width d_1 = 5.5 in. and the actual thickness = 1.5 in. The initial moisture content and the final equilibrium moisture content are M_1 = 30 and M_2 = 15, respectively.
>
> (a) *Shrinkage across the width of the two 2 × 6 top plates.* For shrinkage across the width of the top plate, the shrinkage parameters from Table 1.3 are obtained as follows:
>
> $$a = 6.031$$
> $$b = 0.215$$
>
> From Equation 1-1, the final width d_2 is given as
>
> $$d_2 = 5.5 \left[\frac{1 - [6.031 - (0.215)(15)]/100}{1 - [6.031 - (0.215)(30)]/100} \right] = 5.32 \text{ in.}$$
>
> Thus, the total shrinkage across the width of the two top plates = $d_1 - d_2$ = 5.5 in. − 5.32 in. = **0.18 in.**
>
> (b) *Shrinkage across the thickness of the two 2 × 6 top plates.* For shrinkage across the thickness of the top plate, the shrinkage parameters from Table 1.3 are:
>
> $$a = 5.062$$
> $$b = 0.181$$
>
> The final thickness d_2 of each top plate from Equation 1-1 is given as
>
> $$d_2 = 1.5 \left[\frac{1 - [5.062 - (0.181)(15)]/100}{1 - [5.062 - (0.181)(30)]/100} \right] = 1.46 \text{ in.}$$
>
> The total shrinkage across the *thickness* of the two top plates will be the sum of the shrinkage in each of the individual wood members:
>
> $$2 \text{ top plates} \times (d_1 - d_2) = (2)(1.5 \text{ in.} - 1.46 \text{ in.}) = \textbf{0.08 in.}$$

using nominal dimensions. The *Engineering News-Record,* the construction industry leading magazine, publishes the prevailing cost of lumber in the United States and Canada in units of 1000 board feet (Mbf). For example, 2 × 6 lumber that is 18 ft long is equivalent to 18 board feet or 0.018 Mbf. That is,

$$\frac{(2 \text{ in.})(6 \text{ in.})(18 \text{ ft} \times 12)}{144 \text{ in}^3} = 18 \text{ bf or } 0.018 \text{ Mbf}$$

1.10 BUILDING CODES

A *building code* is a minimum set of regulations adopted by a city or state that governs the design of building structures in that jurisdiction. The primary purpose of a building code is safety, and the intent is that in the worst-case scenario, even though a building is damaged beyond repair, it should stand long enough to enable its occupants to escape to safety. The most widely used building code in the United States is the *International Building Code* (IBC), first released in 2000

EXAMPLE 1.2

Shrinkage at Framed Floors

Determine the total shrinkage at each floor level for the typical wall section shown in Figure 1.23 assuming Hem Fir wood species, and the moisture content decreases from an initial value of 19% to a final value of 10%. How much gap should be provided in the plywood wall sheathing to allow for shrinkage?

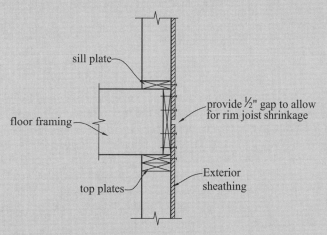

FIGURE 1.23 Wood shrinkage at a framed floor.

Solution: For a 2 × 6 sawn lumber, the actual thickness = 1.5 in.
For a 2 × 12 sawn lumber, the actual width, d_1 = 11.25 in.

The initial and final moisture contents are $M_1 = 19$ and $M_2 = 10$

(a) *Shrinkage across the width of the 2 × 12 continuous blocking.* The shrinkage parameters from Table 1.3 for shrinkage across the width of the 2 × 12 are

$$a = 6.031, \text{ and } b = 0.215$$

From Equation 1-1, the final width d_2 is given as,

$$d_2 = 11.25 \left[\frac{1 - [6.031 - (0.215)(10)]/100}{1 - [6.031 - (0.215)(19)]/100} \right] = 11.03 \text{ in.}$$

Thus, the total shrinkage across the width of the 2 × 12 is

$$d_1 - d_2 = 11.25 \text{ in.} - 11.03 \text{ in.} = 0.22 \text{ in.}$$

(b) *Shrinkage across the thickness of the two 2 × 6 top plates and one 2 × 6 sole plate.* The shrinkage parameters from Table 1.3 for shrinkage across the thickness of the 2 × 6 plates are

$$a = 5.062, \text{ and } b = 0.181$$

The final thickness d_2 of each plate from Equation 1-1 is given as,

$$d_2 = 1.5 \left[\frac{1 - [5.062 - (0.181)(10)]/100}{1 - [5.062 - (0.181)(19)]/100} \right] = 1.475 \text{ in.}$$

The total shrinkage across the *thickness* of the two top plates and one sill plate will be the sum of the shrinkage in each of the individual wood member calculated as

$$3 \text{ plates} \times (d_1 - d_2) = 3(1.5 \text{ in.} - 1.475 \text{ in.}) = 0.075 \text{ in.}$$

> The longitudinal shrinkage or shrinkage parallel to grain in the 2 × 6 studs is negligible. Therefore, the total shrinkage per floor, which is the sum of the shrinkage of all the wood members at the floor level, is
>
> $$0.075 \text{ in.} + 0.22 \text{ in.} = \mathbf{0.3 \text{ in.}} \text{ Therefore, use } \tfrac{1}{2} \text{ in. shrinkage gap.}$$
>
> An adequate shrinkage gap, typically about $\tfrac{1}{2}$ in. deep, is provided in the plywood sheathing at each floor level to prevent buckling of the sheathing panels due to shrinkage. It should also be noted that for multi-story wood buildings, the effects of shrinkage are even more pronounced and critical. For example, a five-story building with a typical detail as shown in Figure 1.23 will have a total accumulated vertical shrinkage of approximately five times the value calculated above!

[3]. The IBC contains, among such other things as plumbing and fire safety, up-to-date provisions on the design procedures for wind and seismic loads as well as for other structural loads. The IBC 2006 now references the ASCE 7 load standards [4] for the calculation procedures for all types of structural loads. The load calculations in this book are based on the ASCE 7 standards. In addition, the IBC references the provisions of the various material codes, such as the ACI 318 for concrete, the NDS code for wood, and the AISC code for structural steel. Readers should note that the building code establishes minimum standards that are required to obtain a building permit. Owners of buildings are allowed to exceed these standards if they desire, but this may increase the cost of the building.

NDS Code and NDS Supplement

The primary design code for the design of wood structures in United States is the *National Design Specification* (NDS) *for Wood Construction* [1] published by the American Forest & Paper Association (AF&PA), in addition to the *NDS Supplement* (or NDS-S) [2], which consist of the tables listed in Table 1.4. These NDS-S tables provide design stresses for the various stress grades of a wood member obtained from full-scale tests on thousands of wood specimens. It should be noted that the tabulated design stresses are not necessarily the allowable stresses; to obtain allowable stresses, the NDS-S stresses have to be multiplied by the product of applicable stress adjustment factors. This is discussed further in Chapter 3.

TABLE 1.4 Use of NDS-S Design Stress Tables

NDS-S Table	Applicability
4A	Visually graded dimension lumber (all species except southern pine)
4B	Visually graded southern pine dimension lumber
4C	Machine or mechanically graded dimension lumber
4D	Visually graded timbers (5 in. × 5 in. and larger)
4E	Visually graded decking
4F	Non–North American visually graded dimension lumber
5A	Structural glued laminated *softwood* timber (members stressed primarily in *bending*)
5A–Expanded	Structural glued laminated *softwood* timber combinations (members stressed primarily in *bending*)
5B	Structural glued laminated *softwood* timber (members stressed primarily in *axial tension* or *compression*)
5C	Structural glued laminated *hardwood* timber (members stressed primarily in *bending*)
5D	Structural glued laminated *hardwood* timber (members stressed primarily in *axial tension* or *compression*)

REFERENCES

1. ANSI/AF&PA (2005), *National Design Specification for Wood Construction,* American Forest & Paper Association, Washington, DC.
2. ANSI/AF&PA (2005), *National Design Specification Supplement: Design Values for Wood Construction,* American Forest & Paper Association, Washington, DC
3. ICC (2006), *International Building Code,* International Code Council, Washington, DC.
4. ASCE (2005), *Minimum Design Loads for Buildings and Other Structures,* American Society of Civil Engineers, Reston, VA.
5. Willenbrock, Jack H., Manbeck, Harvey B., and Suchar, Michael G. (1998), *Residential Building Design and Construction,* Prentice Hall, Upper Saddle River, NJ.
6. Faherty, Keith F., and Williamson, Thomas G. (1995), *Wood Engineering and Construction,* McGraw-Hill, New York.
7. Halperin, Don A., and Bible, G. Thomas (1994), *Principles of Timber Design for Architects and Builders,* Wiley, New York.
8. Stalnaker, Judith J., and Harris, Earnest C. (1997), *Structural Design in Wood,* Chapman & Hall, London.
9. NAHB (2000), *Residential Structural Design Guide—2000,* National Association of Home Builders Research Center, Upper Marlboro, MD.
10. Cohen, Albert H. (2002), *Introduction to Structural Design: A Beginner's Guide to Gravity Loads and Residential Wood Structural Design,* AHC, Edmonds, WA.
11. Kang, Kaffee (1998), *Graphic Guide to Frame Construction—Student Edition,* Prentice Hall, Upper Saddle River, NJ.
12. Kim, Robert H., and Kim, Jai B. (1997), *Timber Design for the Civil and Structural Professional Engineering Exams,* Professional Publications, Belmont, CA.
13. Hoyle, Robert J., Jr. (1978), *Wood Technology in the Design of Structures,* 4th ed., Mountain Press, Missoula, MT.
14. Kermany, Abdy (1999), *Structural Timber Design,* Blackwell Science, London.
15. ASTM (1990), *Standard Practice for Establishing Allowable Properties for Visually-Graded Dimension Lumber from In-Grade Tests of Full-Size Specimens,* ASTM D 1990, ASTM International, West Conshohocken, PA.
16. AITC (1994), *Timber Construction Manual,* 4th ed., Wiley, Hoboken, NJ.
17. Powell, Robert M. (2004), *Wood Design for Shrinkage,* STRUCTURE, pp. 24–25, November.
18. Nelson, Ronald F., Patel, Sharad T., and Avevalo Ricardo (2002), *Continuous Tie-Die System for Wood Panel Shear Walls in Multi-Story Structures,* Structural Engineers Association of California Convention, October 28.
19. Knight, Brian (2006), *High Rise Wood Frame Construction,* STRUCTURE, pp. 68–70, June.

PROBLEMS

1.1 List the typical structural components of a wood building.

1.2 What is moisture content, and how does it affect the strength of a wood member?

1.3 Define the terms *equilibrium moisture content* and *fiber saturation point*.

1.4 Describe the various size classifications for structural lumber, and give two examples of each size classification.

1.5 List and describe factors that affect the strength of a wood member.

1.6 How and why does the duration of loading affect the strength of a wood member?

1.7 What are common defects in a wood member?

1.8 Why does the NDS code not permit the loading of wood in tension perpendicular to the grain?

1.9 Describe the two types of grading systems used for structural lumber. Which is more commonly used?

1.10 Determine the total shrinkage across the width and thickness of a green triple 2×4 Douglas fir-larch top plate loaded perpendicular to grain as the moisture content decreases from an initial value of 30% to a final value of 12%.

FIGURE 1.24 Two-story exterior wall section.

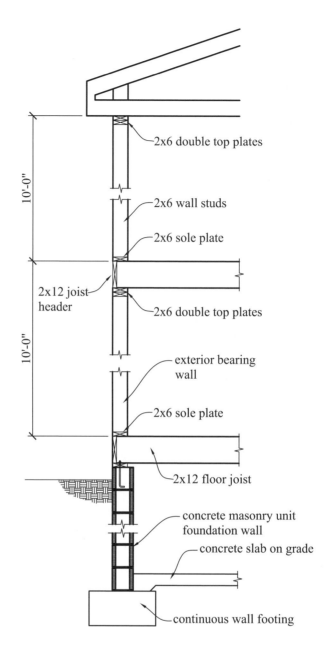

1.11 Determine the total shrinkage over the height of a two-story building that has the exterior wall cross section shown in Figure 1.24 as the moisture content decreases from an initial value of 25% to a final value of 12%.

1.12 How many board feet are there in a 4 × 16 × 36 ft-long wood member? How many Mbf are in this member? Determine how many pieces of this member would amount to 4.84 Mbf (4840 bf).

chapter *two*

INTRODUCTION TO STRUCTURAL DESIGN LOADS

2.1 DESIGN LOADS

Several types of loads can act on wood buildings: dead loads, live loads, snow loads, wind loads, and seismic loads. The combinations of these loads that act on any building structure is prescribed by the relevant building code, such as the *International Building Code* (IBC) [1] or the ASCE 7 load specifications [2].

Load Combinations

The various loads that act on a building do not act in isolation and may act on the structure simultaneously. However, these loads usually will not act on the structure simultaneously at their maximum values. The IBC and ASCE 7 load standards prescribe the critical combination of loads to be used for design; and for allowable stress design, two sets of load combinations, the basic and the alternate load combinations, are given. The basic load combinations shown in Section 1605.3.1 of the IBC are used in this book and are listed below for reference:

1. $D + F$ (IBC Equation 16-8)
2. $D + H + F + L + T$ (IBC Equation 16-9)
3. $D + H + F + (L_r$ or S or $R)$ (IBC Equation 16-10)
4. $D + H + F + 0.75(L + T) + 0.75(L_r$ or S or $R)$ (IBC Equation 16-11)
5. $D + H + F + (W$ or $0.7E)$ (IBC Equation 16-12)
6. $D + H + F + 0.75(W$ or $0.7E) + 0.75L + 0.75(L_r$ or S or $R)$ (IBC Equation 16-13)
7. $0.6D + W + H$ (IBC Equation 16-14)
8. $0.6D + 0.7E + H$ (IBC Equation 16-15)

where D = dead load
L = live load
L_r = roof live load
S = snow load
R = rain load
H = earth pressure, hydrostatic pressure, and pressure due to bulk materials
T = temperature change, shrinkage, or settlement
W = wind load
F = fluid load
E = seismic load
   = $E_h + E_v$ in load combinations 5 and 6
   = $E_h - E_v$ in load combinations 7 and 8
$E_h = \rho Q_E$ = horizontal seismic load effect (i.e., due to seismic lateral forces)
$E_v = 0.2S_{DS}D$ = vertical seismic load effect

ρ = redundancy coefficient (see Section 12.3.4 of Ref. 2) = 1.0 or 1.3

S_{DS} = design spectral response acceleration parameter at short period (see Section 11.4.4 of Ref. 2)

All structural elements must be designed for the most critical of these combinations. The use of these load combinations is described in greater detail later in the book.

Notes [1, 2]

- Where the flat roof snow load $P_f \leq 30$ psf, the snow load need not be combined with seismic loads E. Where $P_f > 30$ psf, only 20% of the snow load is combined with the seismic load.
- Where the load H counteracts the load W or E in combinations 7 and 8, set the load factor on H to zero (i.e., neglect H in load combinations 7 and 8).
- In load combinations 5 and 6, when E is included, the load factor on the floor live load L can be set to 0.5 for all occupancies where the basic floor live load L_0 is 100 psf or less, *except* for parking garages or areas of public assembly (see Section 12.4.2.3 of Ref. 2).

2.2 DEAD LOADS

Dead loads are the weights of all materials that are permanently and rigidly attached to a structure, including the self-weight of the structure, such that it will vibrate with the structure during a seismic or earthquake event (Figures 2.1 and 2.2). The dead load can be determined with more accuracy than other types of load and are not as variable as live loads. Typical checklists for the roof and floor dead loads in wood buildings follow.

Typical Roof Dead Load Checklist

- Weight of roofing material
- Weight of roof sheathing or plywood
- Weight of framing
- Weight of insulation
- Weight of ceiling
- Weight of mechanical and electrical fixtures (M&E)

Typical Floor Dead Load Checklist

- Weight of flooring (i.e., the topping: hardwood, lightweight concrete, etc.)
- Weight of floor sheathing or plywood
- Weight of floor framing
- Weight of partitions (15 psf minimum; not required when the floor live load is greater than 80 psf; see Section 4.2.2 of Ref. 2)
- Weight of ceiling

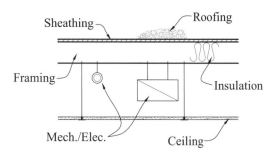

FIGURE 2.1 Typical roof dead load.

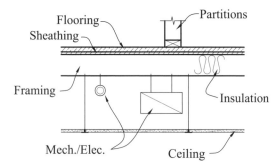

FIGURE 2.2 Typical floor dead load.

- Weight of mechanical and electrical fixtures (M&E)

To aid in the calculation of dead loads, weights of various building materials, such as those given in Appendix A, are typically used. The weights of framing provided in Appendix A are based on Douglas Fir Larch with a specific gravity of 0.5 and density of 31.2 pcf, which is conservative for most wood buildings. The following are sample roof and floor dead-load calculations for typical wood buildings that can serve as a guide to the reader. In the calculations below, the density of wood is assumed to be 31.2 pcf, as stated previously.

Sample Roof Dead-Load Calculation

Roofing (five-ply with gravel)	= 6.5 psf
Reroofing (i.e., future added roof)	= 2.5 psf (assumed)
$\frac{1}{2}$-in. plywood sheathing (= 0.4 psf/$\frac{1}{8}$ in. × 4)	= 1.6 psf
Framing (e.g., assuming 2 × 12 at 16 in. o.c.)	= 2.8 psf
Insulation (2 in. loose insulation: 0.5 psf/in. × 2 in)	= 1.0 psf
Channel-suspended system (steel)	= 2.0 psf
Mechanical and Electrical (M&E)	= 5.0 psf (typical for wood buildings)
Total roof dead load D_{roof}	= 21.4 psf ≈ **22 psf**

Sample Floor Dead-Load Calculation

Floor covering (e.g., assuming $1\frac{1}{2}$-in. lightweight concrete at 100 pcf)	= 12.5 psf
$1\frac{1}{8}$-in. plywood sheathing (0.4 psf/$\frac{1}{8}$ in. × 9)	= 3.6 psf
Framing (assuming 4 × 12 at 4ft 0 in. o.c.)	= 2.2 psf
$\frac{1}{2}$-in. drywall ceiling (= 5 psf/in. × $\frac{1}{2}$ in.	= 2.5 psf
Ceiling supports (say, 2 × 4 at 24 in. o.c.)	= 0.6 psf
Mechanical and electrical (M&E)	= 5.0 psf
Partition loads	= 20.0 psf (assumed)
Total floor dead load D_{floor}	= 46.4 psf ≈ **47 psf**

It should be noted that for buildings with floor live loads less than or equal to 80 psf, the partition dead load must be at least 15 psf (see Section 4.2.2 of Ref. 2), while for buildings with floor live loads greater than 80 psf, no partition loads have to be considered since for such assembly occupancies, there is less likelihood that partition walls will be present.

Combined Dead and Live Loads on Sloped Roofs

Since most wood buildings have sloped roofs, we discuss next how to combine the dead loads acting on the sloped roof surface with the live loads (i.e. snow, rain, or roof live load) acting on a horizontal projected plan area of the roof surface (Figure 2.3). Most building codes give live loads in units of pounds per square foot of the horizontal projected plan area, while the

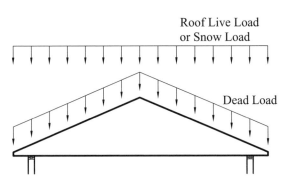

FIGURE 2.3 Loads on a sloped roof.

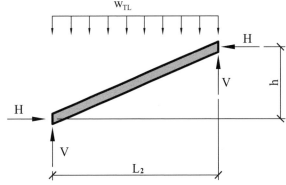

FIGURE 2.4 Free-body diagram of a sloped rafter.

dead load of a sloped roof is in units of pounds per square foot of the sloped roof area. Therefore, to combine the dead and live loads in the same units, two approaches are possible:

1. Convert the dead load from units of pounds per square foot of sloped roof area to units of psf of horizontal projected plan area and then add to the live load, which is in units of psf of horizontal projected plan area. When using this approach, the reader should not forget the horizontal thrust or force acting at the exterior wall from the component of the dead and live loads acting parallel to the roof surface. These lateral thrusts must be considered in the design of the walls, and collar or ceiling ties should be provided to resist this lateral force.

2. Convert the live load from pounds per square foot of horizontal projected plan area to psf of sloped roof area and then add to the dead load, which is in units of psf of sloped roof area.

Option 1 is most commonly used in design practice and is adopted in this book. Using the load combination equations presented earlier in this chapter, the total dead plus live load w_{TL} in psf of horizontal plan area will be

$$w_{TL} = D\frac{L_1}{L_2} + (L_r \text{ or } S \text{ or } R) \qquad \text{psf of horizontal plan area} \qquad (2.1)$$

where D = roof dead load in psf of sloped roof area
S = roof snow load in psf of horizontal plan area
L_r = roof live load in psf of horizontal plan area
R = rain load in psf of horizontal plan area (usually not critical for sloped roofs)
L_1 = sloped length of rafter
L_2 = horizontal projected length of rafter

Calculation of the Horizontal Thrust at an Exterior Wall

Summing moments about the exterior wall support of the rafter yields (Figure 2.4)

$$-H(h) + w_{TL}L_2\frac{L_2}{2} = 0$$

Thus, the horizontal thrust H becomes

$$H = \frac{w_{TL}L_2\,(L_2/2)}{h} \qquad (2.2)$$

Combined Dead and Live Loads on Stair Stringers

The same equation, (2.1), used for calculating the total load on sloped roofs can be applied to stair stringers. Using the load combinations in Section 2.1, the total load on the stair stringer is given as

$$w_{TL} = D\frac{L_1}{L_2} + L \qquad \text{psf of horizontal plan area} \qquad (2.3)$$

where D = stair dead load in psf of sloped roof area
L = stair live load in psf of horizontal plan area (**100 psf** according to IBC Table 1607.1 or ASCE 7 Table 4-1)
L_1 = sloped length of rafter
L_2 = horizontal projected length of rafter

2.3 TRIBUTARY WIDTHS AND AREAS

In this section we introduce the concept of tributary widths and tributary areas (Figure 2.8). These concepts are used to determine the distribution of floor and roof loads to the various structural elements. The *tributary width* (TW) of a beam or girder is defined as the width of floor

EXAMPLE 2.1

Design Loads for a Sloped Roof

Given the following design parameters for a sloped roof (Figure 2.5), (a) calculate the uniform total load and the maximum shear and moment on the rafter. (b) Calculate the horizontal thrust on the exterior wall if rafters are used.

Roof dead load $D = 10$ psf (of sloped roof area)

Roof snow load $S = 66$ psf (of horizontal plan area)

Horizontal projected length of rafter $L_2 = 18$ ft

Sloped length of rafter $L_1 = 20.12$ ft

Rafter or truss spacing = 4 ft 0 in.

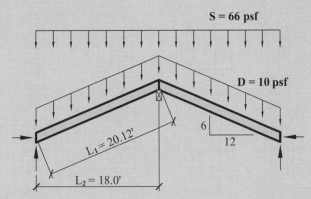

FIGURE 2.5 Cross section of a building with a sloped roof.

Solution: (a) Using the load combinations in Section 2.1, the total load in psf of horizontal plan area will be (see Figure 2.6)

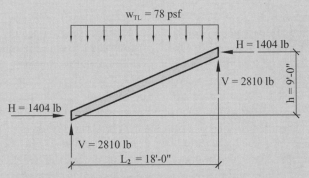

FIGURE 2.6 Free-body diagram of a sloped rafter.

$$w_{TL} = D\left(\frac{L_1}{L_2}\right) + (L_T \text{ or } S \text{ or } R), \text{ psf of horizontal plan area}$$

$$= (10 \text{ psf})\left(\frac{20.12 \text{ ft}}{18 \text{ ft}}\right) + 66 \text{ psf}$$

$$= \mathbf{78 \text{ psf}} \text{ of horizontal plan area}$$

The total load in pounds per horizontal linear foot (lb/ft) is given as

$$w_{TL}(\text{lb/ft}) = w_{TL}(\text{psf}) \times \text{tributary width or spacing of rafters}$$

$$= (78 \text{ psf})(4 \text{ ft}) = \mathbf{312 \text{ lb/ft}}$$

The horizontal thrust H is

$$H = \frac{w_{TL} L_2 (L_2/2)}{h} = \frac{(312 \text{ lb/ft})(18 \text{ ft})(18 \text{ ft}/2)}{9 \text{ ft}} = 5616 \text{ lb}$$

The collar or ceiling ties must be designed to resist this horizontal thrust.

(b) $L_2 = 18$ ft. The maximum shear force in the rafter is

$$V_{\max} = w_{TL} \frac{L_2}{2} = (312)\left(\frac{18 \text{ ft}}{2}\right) = 2810 \text{ lb}$$

The maximum moment in the rafter is

$$M_{\max} = \frac{w_{TL}(L_2)^2}{8} = \frac{(312)(18 \text{ ft})^2}{8} = 12{,}700 \text{ ft-lb} = 12.70 \text{ ft-kips}$$

or roof supported by the beam or girder, and is equal to the sum of one-half the distance to the adjacent beam on the right of the beam in question plus one-half the distance to the adjacent beam on the left of the beam in question. Thus, the tributary width is given as

$$\text{TW} = \tfrac{1}{2}(\text{distance to adjacent beam on the right}) + \tfrac{1}{2}(\text{distance to adjacent beam on the left})$$

The *tributary area* of a beam, girder, or column is the floor or roof area supported by the structural member. For beams and girders, the tributary area is the product of the span and the tributary width of the beam or girder. For columns, it is the plan area bounded by lines located at one-half the distance to the adjacent columns surrounding the column in question. It should be noted that a column does not have a tributary width, only a tributary area.

2.4 LIVE LOADS

Any item not rigidly attached to a structure is usually classified as a *live load*. Live loads are short-term nonstationary gravity forces that can vary in magnitude and location. There are three main types of live loads: roof live load (L_r), snow load (S), and floor live load (L). Rain load (R) is also a roof live load but is rarely critical compared to roof live loads and snow load because of the pitched roofs in most wood buildings. These live loads are usually specified in the IBC [1] and the ASCE 7 load standards [2]. The magnitude of live loads depends on the occupancy and use of the structure and on the tributary area of the structural element under consideration. Live loads as specified in the IBC and ASCE 7 are given in units of pounds per square foot of the horizontal projected plan area, but it should be noted that the codes also specify alternative concentrated live loads (in pounds), as we discuss later. The live loads specified in the codes are minimum values, and the owner of a building could decide to use higher live load values than those given in the code, although this will lead to a more expensive structure.

Roof Live Load

Roof live loads occur due to the weight of equipment and maintenance personnel during the servicing of a roof. The magnitude is a function of the roof slope and the tributary area of the structural element under consideration. The larger the tributary area, the smaller the probability that the entire tributary area (TA) will be loaded with the maximum roof live load, resulting in a smaller design roof live load. Similarly, the steeper the roof slope, the smaller the design roof live load.

EXAMPLE 2.2

Design Loads for Stair Stringers

Determine the total dead plus live load on a stair stringer assuming a dead load of 50 psf and a live load of 100 psf. The plan and elevation of the stair are shown in Figure 2.7.

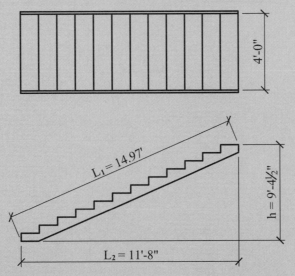

FIGURE 2.7 Stair plan and elevation.

Solution: The stair dead load $D = 50$ psf of the sloped area and the stair live load $L = 100$ psf of the horizontal plan area. The total uniform load on the stair in psf of horizontal plan area is given as

$$w_{TL} = D\frac{L_1}{L_2} + L \quad \text{psf of horizontal plan area}$$

$$= (50 \text{ psf})\left(\frac{14.97 \text{ ft}}{11.67 \text{ ft}}\right) + 100 \text{ psf}$$

$$= \mathbf{164 \text{ psf}} \text{ of horizontal plan area}$$

The tributary width for each stair stringer = 4 ft/2 = 2 ft. Thus, the total load in pounds per horizontal linear foot (lb/ft) on each stair stringer is given as

$$w_{TL} \text{ (lb/ft)} = w_{TL}(\text{psf}) \times \text{tributary width or spacing of rafters}$$
$$= (164 \text{ psf})(2 \text{ ft}) = \mathbf{328 \text{ lb/ft}}$$

The maximum shear force in the stair stringer is

$$V_{max} = w_{TL}\frac{L_2}{2} = (328)\left(\frac{11.67 \text{ ft}}{2}\right) = 1914 \text{ lb}$$

The maximum moment in the stair stringer is

$$M_{max} = \frac{w_{TL}(L_2)^2}{8} = \frac{(328)(11.67 \text{ ft})^2}{8} = 5584 \text{ ft-lb} = 5.6 \text{ ft-kips}$$

FIGURE 2.8 Tributary width and area.

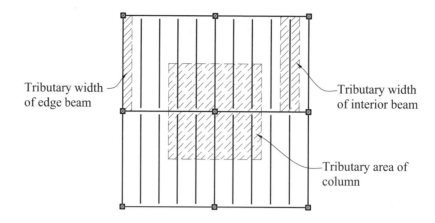

For ordinary flat, pitched or curved roofs, the ASCE 7 load standards [2] specify the design roof live load in pounds per square foot as

$$L_r = 20 R_1 R_2 \qquad (2.4)$$

$$12 \leq L_r \leq 20$$

where $R_1 = \begin{cases} 1.0 \text{ for TA} \leq 200 \text{ ft}^2 \\ 1.2 - 0.001 \text{TA for } 200 \text{ ft}^2 < \text{TA} < 600 \text{ ft}^2 \\ 0.6 \text{ for TA} \geq 600 \text{ ft}^2 \end{cases}$

$R_2 = \begin{cases} 1.0 \text{ for } F \leq 4 \\ 1.2 - 0.05 F \text{ for } 4 < F < 12 \\ 0.6 \text{ for } F \geq 12 \end{cases}$

F = roof slope factor

= number of inches of rise per foot for a sloped roof = $12 \tan \theta$

= rise/span ratio multiplied by 32 for an arch or dome roof

θ = roof slope, degrees

TA = tributary area, ft^2

See Table 2.2 later in this section (IBC Table 1607.1 or ASCE 7 Table 4-1) for the live loads for special-purpose roofs such as roofs used for promenades, roof gardens, or assembly purposes. For a landscaped roof, the weight of the landscaped material must also be included in the dead-load calculations, assuming the soil to be fully saturated. Thus, the saturated density of the soil should be used in calculating the dead loads.

Snow Load

The 50-year ground snow loads P_g are specified in IBC Figure 1608.2 or ASCE 7 Figure 7-1, but for certain areas these snow load maps call for site-specific snow load studies to establish the ground snow loads. The ground snow loads are specified in greater detail in state building codes such as the *New York State Building Code* [3], which is based on the IBC, or the *Ohio State Building Code* [4], and because relatively large variations in snow loads can even occur over small geographic areas, the local or state building codes appear to have better snow load data for the various localities within their jurisdiction compared to the snow load map given in the IBC or ASCE 7. The roof snow load is a function of the ground snow load and is typically smaller than the corresponding ground snow load because of the effect of wind at the roof level.

The roof snow load is dependent on the ground snow load P_g, the roof exposure, the roof slope, the use of the building (e.g., whether or not it is heated), and the terrain conditions at the building site, but it is unaffected by tributary area. The steeper the roof slope, the smaller the snow load because steep roofs are more likely to shed snow, and conversely, the flatter the roof slope, the larger the snow load. Depending on the type of roof, the roof snow load can be either a uniform balanced load or an unbalanced load. A *balanced snow load* is a full uniform snow

EXAMPLE 2.3

Calculation of Tributary Widths and Areas

Determine the tributary widths and tributary areas of the joists, beams, girders, and columns in the framing plan shown in Figure 2.9.

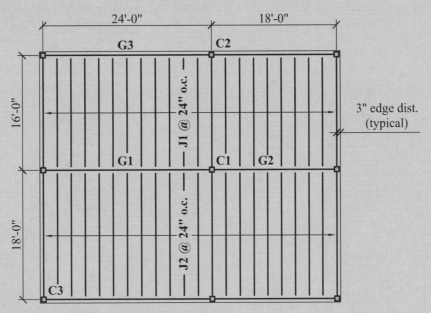

FIGURE 2.9 Roof and floor framing plan.

Solution: The solution is presented in Table 2.1.

TABLE 2.1 Tributary Widths and Areas of Joists, Beams, and Columns

Structural Member	Tributary Width	Tributary Area
Joist J1	$\frac{24 \text{ in.}}{2} + \frac{24 \text{ in.}}{2} = 24 \text{ in.} = 2 \text{ ft}$	$(2 \text{ ft}) \times (16 \text{ ft}) = 32 \text{ ft}^2$
Joist J2	$\frac{24 \text{ in.}}{2} + \frac{24 \text{ in.}}{2} = 24 \text{ in.} = 2 \text{ ft}$	$(2 \text{ ft}) \times (18 \text{ ft}) = 36 \text{ ft}^2$
Beam or girder G1	$\frac{16 \text{ ft}}{2} + \frac{18 \text{ ft}}{2} = 17 \text{ ft}$	$(17 \text{ ft}) \times (24 \text{ ft}) = 408 \text{ ft}^2$
Girder G2	$\frac{16 \text{ ft}}{2} + \frac{18 \text{ ft}}{2} = 17 \text{ ft}$	$(17 \text{ ft}) \times (18 \text{ ft}) = 306 \text{ ft}^2$
Girder G3	$\frac{16 \text{ ft}}{2} + 0.25 \text{ ft} = 8.25 \text{ ft}$	$(8.25 \text{ ft}) \times (24 \text{ ft}) = 198 \text{ ft}^2$
Column C1	—	$\left(\frac{18 \text{ ft}}{2} + \frac{16 \text{ ft}}{2}\right)\left(\frac{24 \text{ ft}}{2} + \frac{18 \text{ ft}}{2}\right) = 357 \text{ ft}^2$
Column C2	—	$\left(\frac{16 \text{ ft}}{2} + 0.25 \text{ ft}\right)\left(\frac{24 \text{ ft}}{2} + \frac{18 \text{ ft}}{2}\right) = 173.3 \text{ ft}^2$
Column C3	—	$\left(\frac{18 \text{ ft}}{2} + 0.25 \text{ ft}\right)\left(\frac{24 \text{ ft}}{2} + 0.25 \text{ ft}\right) = 113.4 \text{ ft}^2$

EXAMPLE 2.4

Roof Load Calculations

For the roof plan shown in Example 2.3, assuming a roof dead load of 10 psf and an essentially flat roof with only a roof slope of $\frac{1}{4}$ in. per foot for drainage, determine the following loads using the IBC load combinations. (a) the uniform total load on joists J2 in lb/ft; (b) the uniform total load on girder G1 in lb/ft; (c) the total axial load on column C1 in lbs. Neglect the rain load R and assume that the snow load S is zero.

Solution: Since the snow and rain load are both zero, the roof live load L_r will be critical. With a roof slope of $\frac{1}{4}$ in. per foot, the number of inches of rise per foot, $F = \frac{1}{4} = 0.25$.

(a) *Joist J2.* The tributary width (TW) = 2 ft and the tributary area (TA) = 36 ft^2 < 200 ft^2. From Section 2.4, we obtain $R_1 = 1.0$ and $R_2 = 1.0$. Using equation (2.4) gives the roof live load

$$L_r = (20)(1)(1) = 20 \text{ psf}$$

The total loads are calculated as follows:

$$w_{TL} \text{ (psf)} = D + L_r = 10 + 20 = 30 \text{ psf}$$

$$w_{TL} \text{ (lb/ft)} = w_{TL}(\text{psf}) \times \text{TW} = (30 \text{ psf})(2 \text{ ft}) = 60 \text{ lb/ft}$$

(b) *Girder G1.* The tributary width = 17 ft and the tributary area = 408 ft^2. Thus, 200 < TA < 600, and from Section 2.4, we obtain $R_1 = 1.2 - (0.001)(408) = 0.792$ and $R_2 = 1.0$. Using equation (2.4) gives the roof live load

$$L_r = (20)(0.792)(1) = 16 \text{ psf}$$

The total loads are calculated as follows:

$$w_{TL}(\text{psf}) = D + L_r = 10 + 16 = 26 \text{ psf}$$

$$w_{TL}(\text{lb/ft}) = w_{TL}(\text{psf}) \times \text{TW} = (26 \text{ psf})(17 \text{ ft}) = 442 \text{ lb/ft}$$

(c) *Column C1.* The tributary area of column C1, TA = 357 ft^2. Thus, 200 < TA < 600, and from Section 2.4 we obtain $R_1 = 1.2 - (0.001)(357) = 0.843$ and $R_2 = 1.0$. Using equation (2.4) gives the roof live load

$$L_r = (20)(0.843)(1) = 17 \text{ psf}$$

so

$$w_{TL}(\text{psf}) = D + L_r = 10 + 17 = 27 \text{ psf}$$

and the column axial load

$$P = (27 \text{ psf})(357 \text{ ft}^2) = 9639 \text{ lb} = 9.7 \text{ kips}$$

---

load over an entire roof surface, and an *unbalanced snow load* is a nonuniform distribution of snow load over a roof surface. Only balanced snow loads are considered in this book. The various possible configurations of snow load are beyond the scope of this book, and the reader is referred to the ASCE 7 load specifications.

The design snow load on a sloped roof, P_s, is obtained using the ASCE 7 load specifications and is given as

$$P_s = C_s P_f \qquad \text{psf} \tag{2.5}$$

and the flat roof snow load

$$P_f = 0.7 C_e C_t I P_g \qquad \text{psf} \tag{2.6}$$

where C_e = exposure factor from ASCE 7 Table 7-2
C_t = thermal factor from ASCE 7 Table 7-3
C_s = roof slope factor from ASCE 7 Figure 7-2
I = importance factor from ASCE Table 7-4
P_g = ground snow load from ASCE Figure 7-1 (or governing state code)

In calculating the roof slope factor C_s, the following should be noted:

1. Slippery surface values can be used only where the roof surface is free of obstruction and sufficient space is available below the eaves to accept all the sliding snow.
2. Examples of slippery surfaces include metal, slate, and glass, and bituminous, rubber, and plastic membranes with a smooth surface.
3. Membranes with embedded aggregates, asphalt shingles, and wood shingles must not be considered slippery.

Snow Drift and Sliding Snow Loads

Snow accumulation or drifting snow could occur on the lower roof where a high roof is adjacent to a low roof or where there are large rooftop units or high parapets or other roof obstructions (Figure 2.12). These snow drift loads, which are caused by wind and are triangular in shape, must be superimposed on the uniform balanced roof snow loads. Sliding snow load occurs on low roofs where a high sloped roof is adjacent to a lower roof. The sliding snow load, which is uniform in shape, must be superimposed on the uniform balanced roof snow loads. However, the snow drift load and the sliding snow load must not be combined but must be treated separately, with the larger load used in the design of the roof members. The calculation of snow drift and sliding snow loads is beyond the scope of this book, and the reader is referred to the ASCE 7 load specifications.

The reader should refer to Sections 7-6 and 7-7 of ASCE 7 [2] for the calculation procedure for unbalanced snow loads (due to wind), snow drifts on lower roofs adjacent to high roofs, snow drifts due to roof projections and roof top units, and sliding snow loads on lower roofs adjacent to sloped high roofs. The snow drift load and sliding snow load are superimposed separately on the balanced snow load to determine the most critical snow load case, but the snow drift loads must not be combined with sliding snow loads.

Floor Live Load

Floor live loads depend on the use and occupancy of the structure and the tributary area of the structural member, and are usually available in IBC Table 1607.1 or ASCE 7 Table 4-1. These gravity forces are either uniform loads specified in units of pounds per square foot of horizontal plan area or concentrated loads specified in pounds. The uniform and concentrated live loads are not to be applied simultaneously and, in design practice, the uniform loads govern most of the time, except in a few cases such as thin slabs, where the concentrated load may result in a more critical situation, due to the possibility of punching shear. Concentrated live loads also govern in the design of stair treads.

Examples of tabulated floor live loads are 40 psf for residential buildings, 50 psf for office spaces in office buildings, and 250 psf for heavy storage buildings. These tabulated live loads consist of transient and sustained load components. The sustained live load is the portion of the total live load that remains virtually permanent on the structure, such as furnishings, and is much smaller than the tabulated live loads. Typical sustained load values are 4 to 8 psf for residential buildings and 11 psf for office buildings [6]. For a more complete listing of the minimum live loads recommended, the reader is referred to Table 2.2.

Floor Live-Load Reduction

To allow for the low probability that floor elements with large tributary areas will have the entire tributary area loaded with the live load at the same time, the IBC and ASCE 7 load specifications allow for floor live loads to be reduced if certain conditions are met. Thus, the floor live-load reduction factor accounts for the low probability that a structural member with a large tributary area will be fully loaded over the entire area at the same time.

The reduced design live load of a floor L in psf is given as

EXAMPLE 2.5

Roof Snow Load Calculation

The building shown in Figure 2.10 has sloped roof rafters (6:12 slope) spaced 4 ft 0 in. o.c. and is located in Lowville, New York. The roof dead load is 10 psf of sloped area. Assume a fully exposed roof and terrain category C, and use the ground snow load from the state snow map to obtain more accurate ground snow load values. Calculate (a) the total uniform load in lb/ft on a horizontal plane using the IBC, and (b) the maximum shear and moment in the roof rafter.

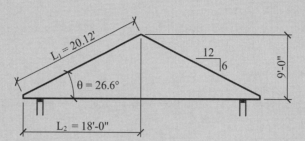

FIGURE 2.10 Sloped roof.

Solution: (a) The roof slope θ for this building is 26.6°.

Roof live load L_r. From Section 2.4, the roof slope factor is obtained as $F = 6$; therefore, $R_2 = 1.2 - (0.05)(6) = 0.9$. Since the tributary area of the rafter = 4 ft × 18 ft = 72 ft^2 < 200 ft^2, $R_1 = 1.0$, the roof live load will be

$$L_r = 20R_1R_2 = (20)(1.0)(0.9) = 18 \text{ psf}$$

Snow load. Using IBC Figure 1608.2 or ASCE 7 Figure 7-1, we find that Lowville, New York falls within the CS areas, where site-specific snow case studies are required. It is common practice in such cases to obtain the ground snow load from the snow map in the local or state building code. From Figure 1608.2 of the *New York State Building Code* [3], the ground snow load P_g for Lowville, New York is found to be 85 psf.

Assuming a building with a warm roof that is fully exposed and a building site with terrain category C, we obtain the coefficients as follows:

exposure coefficient $C_e = 0.9$ (ASCE 7 Table 7-2)

thermal factor $C_t = 1.0$ (ASCE 7 Table 7-3)

importance factor $I = 1.0$ (ASCE Table 7-4)

slope factor $C_s = 1.0$ (ASCE Figure 7-2 with roof slope $\theta = 26.6°$ and a warm roof)

The flat roof snow load

$$P_f = 0.7C_eC_tIP_g = (0.7)(0.9)(1.0)(1.0)(85) = 54 \text{ psf}$$

Thus, the design roof snow load

$$P_s = C_sP_f = (1.0)(54) = 54 \text{ psf}$$

Therefore, the snow load $S = 54$ psf.

The total load of the horizontal plan area is given as

$$w_{TL} = D\frac{L_1}{L_2} + (L_r \text{ or } S \text{ or } R) \qquad \text{psf of horizontal plan area}$$

Since the roof live load L_r (18 psf) is less that the snow load S (54 psf), the snow load is more critical and will be used in calculating the total roof load.

$$w_{TL} = (10 \text{ psf})\left(\frac{20.12 \text{ ft}}{18 \text{ ft}}\right) + 54 \text{ psf}$$

$$= \textbf{65 psf} \text{ of horizontal plan area}$$

The total load in pounds per horizontal linear foot is given as

$$w_{TL}(\text{lb/ft}) = w_{TL}(\text{psf}) \times \text{tributary width or spacing of rafters}$$
$$= (65 \text{ psf})(4 \text{ ft}) = \textbf{260 lb/ft}$$

The horizontal thrust is

$$H = \frac{w_{TL}(L_2)(L_2/2)}{h} = \frac{(260 \text{ lb/ft})(18 \text{ ft})(18 \text{ ft}/2)}{9 \text{ ft}} = 4680 \text{ lb}$$

The collar or ceiling ties must be designed to resist this horizontal thrust (see Figure 2.11).

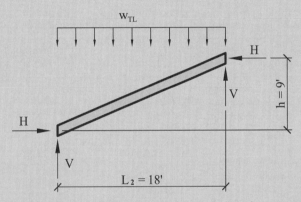

FIGURE 2.11 Free-body diagram of a sloped roof.

(b) $L_2 = 18$ ft. The maximum shear force in the rafter is

$$V_{max} = w_{TL}\frac{L_2}{2} = (260)\left(\frac{18 \text{ ft}}{2}\right) = 23409 \text{ lb} = 2.34 \text{ kips}$$

The maximum moment in the rafter is

$$M_{max} = \frac{w_{TL}(L_2)^2}{8} = \frac{(260)(18 \text{ ft})^2}{8} = 10{,}530 \text{ ft-lb} = 10.53 \text{ ft-kips}$$

$$L = L_0\left(0.25 + \frac{15}{\sqrt{K_{LL}A_T}}\right) \qquad (2.7)$$

$\geq 0.50L_0$ for members supporting *one floor* (e.g., slabs, beams, and girders)
$\geq 0.40L_0$ for members supporting *two or more floors* (e.g., columns)

where L_0 = unreduced design live load from Table 2.2 (IBC Table 1607.1 or ASCE 7 Table 4-1)
K_{LL} = live-load element factor (see Table 4-2 of Ref. 2)
= 4 (interior and exterior columns without cantilever slabs)
= 3 (edge columns with cantilever slabs)

FIGURE 2.12 Snow drift and sliding snow loads.

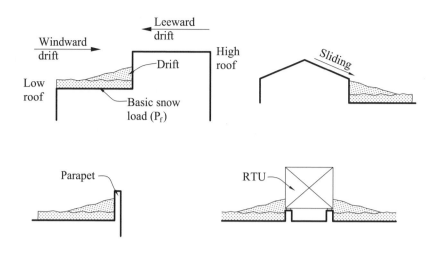

TABLE 2.2 Minimum Uniformly Distributed and Concentrated Floor Live Loads

Occupancy	Uniform Load (psf)	Concentrated Load (lb)
Apartments, hotels, and multifamily houses		
Private rooms and corridors serving them	40	—
Public rooms and corridors serving them	100	—
Balconies		
Exterior	100	—
One- and two-family residences only, and not exceeding 100 ft^2	60	—
Dining rooms and restaurants	100	—
Office buildings		
Lobbies and first-floor corridors	100	2000
Offices	50	2000
Corridors above first floor	80	2000
Residential (one- and two-family dwellings)	40	—
Roofs		
Ordinary flat, pitched, and curved roofs	20	—
Promenades	60	—
Gardens or assembly	100	—
Schools		
Classrooms	40	2000
Corridors above first floor	80	2000
First-floor corridors	100	2000
Stairs and exitways	100	—
One- and two-family residences only	40	—
Storage		
Light	125	—
Heavy	250	—
Stores		
Retail		
First floor	100	1000
Upper floors	75	1000
Wholesale	125	1000

Source: Adapted from Ref 1, Table 1607.1

= 2 (corner columns with cantilever slabs, edge beams without cantilever slabs, interior beams)
= 1 (all other conditions)

A_T = summation of the floor tributary area in ft² supported by the member, excluding the roof area

- For beams and girders (including continuous beams or girders), A_T is the tributary width of the beam or girder multiplied by its center-to-center span between supports.
- For *one-way slabs*, A_T must be less than or equal to 1.5(span of one-way slab)².
- For a member supporting more than one floor area in multistory buildings, A_T will be the summation of all the applicable floor areas supported by that member.

The IBC and ASCE 7 load specifications *do not* permit floor live-load reductions for floors satisfying any one of the following conditions:

- $K_{LL} A_T \leq 400$ ft².
- Floor live load $L_0 > 100$ psf.
- Floors with occupancies used for assembly purposes, such as auditoriums, stadiums, and passenger car garages, because of the high probability of overloading in such occupancies during an emergency.
- For passenger car garage floors, the live load is allowed to be reduced by 20% for members supporting two or more floors.

The following should be noted regarding the tributary area A_T used in calculating the reduced floor live load:

1. Beams are usually supported by girders, which in turn are supported by columns, as indicated in our previous discussions on load paths. The tributary area A_T for beams will usually be smaller than those for girders, and thus beams will have less floor live-load reduction than for girders. The question arises as to which A_T to use for calculating the loads on the girders.
2. For the design of the beams, use the A_T value of the beam to calculate the reduced live load that is used to calculate the moments, shears, and reactions. These load effects are used for the design of the beam and the beam–girder or beam–column connections.
3. For the girders, recalculate the beam reactions using the A_T value of the girder. These smaller beam reactions are used for the design of girders only.
4. For columns, A_T is the summation of the tributary areas of all the floors with *reducible live loads* above the level at which the column load is being determined, and it excludes the roof areas.

2.5 DEFLECTION CRITERIA

The limits on deflections due to gravity loads are intended to ensure user comfort and to prevent excessive cracking of plaster ceilings and architectural partitions. These deflection limits are usually specified in terms of the joist, beam, or girder span, and the deflections are calculated based on elastic analysis of the structural member. The maximum allowable deflections recommended in IBC Table 1604.3 [1] are as follows:

$$\text{Maximum allowable deflection due to live load } \delta_{LL} \leq \frac{L}{360}$$

$$\text{Maximum allowable deflection due to total dead plus live load } \delta_{kD+LL} \leq \frac{L}{240}$$

where k = creep factor
= 0.5 for seasoned lumber used in dry service conditions
= 1.0 for unseasoned or green lumber or seasoned lumber used in wet service conditions

EXAMPLE 2.6

Column Load Calculation

A three-story building has columns spaced at 20 ft in both orthogonal directions and is subjected to the roof and floor loads shown below. Using a column load summation table, calculate the cumulative axial loads on a typical interior column with and without live-load reduction. Assume a roof slope of $\frac{1}{4}$ in. per foot for drainage.

Roof loads:

$$D_{roof} = 25 \text{ psf}$$

$$\text{Snow load } S = 35 \text{ psf}$$

Second- and third-floor loads:

$$D_{floor} = 50 \text{ psf}$$

$$\text{Floor live load } L = 40 \text{ psf}$$

Solution: At each level, the tributary area supported by a typical interior column is (20 ft)(20 ft) = 400 ft.²

Roof live load L_T. From Section 2.4, the roof slope factor is obtained as $F = \frac{1}{4} = 0.25$. Therefore, $R_2 = 1.0$. Since the tributary area of the column = 400 ft², $R_1 = 1.2 - (0.001)(400) = 0.8$. The roof live load will be

$$L_T = 20 R_1 R_2 = (20)(0.8)(1.0) = 16 \text{ psf} < 20 \text{ psf}$$

The reduced or design floor live loads for the second and third floors are calculated using Table 2.3.

TABLE 2.3 Reduced or Design Floor Live-Load Calculations

Member	Levels Supported	A_T Summation of Floor Tributary Area	K_{LL}	Unreduced Floor Live Load, L_0 (psf)	Live-Load Reduction Factor, $0.25 + \dfrac{15}{\sqrt{K_{LL} A_T}}$	Design Floor Live Load, L
Third floor column (i.e., column below roof)	Roof only	Floor live-load reduction *not* applicable to roofs!	—	—	—	35 psf (snow load)
Second-floor column (i.e., column below third floor)	1 floor + roof	1 floor × 400 ft² = **400 ft²**	4 $K_{LL}A_T$ = 1600 ft² > 400 ft²; live load reduction allowed	40	$0.25 + \dfrac{15}{\sqrt{(4)(400)}} = 0.625$	(0.625)(40) = **25 psf** ≥ $0.50 L_0$ = 20 psf
Ground or first-floor column (i.e., column below second floor)	2 floors + roof	2 floors × 400 ft² = **800 ft²**	4 $K_{LL}A_T$ = 3200 ft² > 400 ft²; therefore live load reduction allowed	40	$0.25 + \dfrac{15}{\sqrt{(4)(800)}} = 0.52$	(0.52)(40) = **21 psf** ≥ $0.40 L_0$ = 16 psf

TABLE 2.4 Column Load Summation

Level	Tributary Area, (TA) (ft²)	Dead Load, D (psf)	Live Load, L_0 (S or L_r or R on the Roof) (psf)	Design Live Load Roof: S or L_r or R Floor: L (psf)	Unfactored Total Load at Each Level, w_{s1} Roof: D Floor: D + L (psf)	Unfactored Total Load at Each Level, w_{s2} Roof: D+0.75S Floor: D+0.75L (psf)	Unfactored Column Axial Load at Each Level, P = (TA)(w_{s1}) or (TA)(w_{s2}) (kips)	Cumulative Unfactored Axial Load, ΣP_{D+L} (kips)	Cumulative Unfactored Axial Load, $\Sigma P_{D+0.75L+0.75S}$ (kips)	Maximum Cumulative Unfactored Axial Load, ΣP (kips)
					With Floor Live Load Reduction					
Roof	400	25	35	35	25	51.3	10 or 20.5	10	20.5	20.5
3rd Flr	400	50	40	25	75	68.8	30 or 27.5	40	48	48
2nd Flr	400	50	40	21	71	65.8	28.4 or 26.3	68.4	74.3	74.3
					Without Floor Live Load Reduction					
Roof	400	25	35	35	25	51.3	10 or 20.5	10	20.5	20.5
Third floor	400	50	40	40	90	80	36 or 32	46	52.5	52.5
Second floor	400	50	40	40	90	80	36 or 32	82	84.5	84.5

For this building, all other loads except D, S, L_r and L are zero. Thus,

$$W = H = T = F = R = E = 0$$

Substituting these in the ASD load combinations in Section 2.1, and noting that the snow load being greater than the roof live load and thus becomes the design live load for the roof, yields the following modified load combinations for calculating the unfactored column axial load for this building:

1: D
2: $D + L$
3: $D + S$
4: $D + 0.75L + 0.75S$
5: D
6: $D + 0.75L + 0.75S$
7: $0.6D$
8: $0.6D$

Therefore, the two most critical load combinations for calculating the column axial loads are load combination 2 ($D+L$) and load combinations 4 or 6 ($D+0.75L+0.75S$). By inspection, load combination 3 is not critical, and using the applicable ASD load combinations, $D+L$ and $D+0.75L+0.75S$, the unfactored column axial loads are calculated using the column load summation table (Table 2.4).

Therefore, the ground floor column will be designed for a cumulative axial compression load of 74.3 kips, the second floor column for a load of 48 kips, and the third floor column for a compression load of 20.5 kips. The corresponding values without floor live load reduction are 84.5 kips, 52.5 kips, and 20.5 kips, respectively. For the ground floor column, there is only a 12% reduction in axial load when live load reduction is considered. Thus, the effect of floor live load reduction on columns and column footings is not as critical for low rise buildings as it is for high rise buildings.

D = dead load
LL = live load (i.e., floor live load, roof live load, snow load, or rain load)
L = span of joist, beam or girder

For members that support masonry wall or glazing, the allowable total load deflection should be limited to $L/600$ or ⅜ in., whichever is smaller, to reduce the likelihood of cracking of the masonry wall or glazing. For prefabricated members it is common to camber the beams to help control the total deflection. The amount of the camber is usually some percentage of the dead load to prevent overcambering.

2.6 LATERAL LOADS

The two main types of lateral loads that act on wood buildings are wind and seismic or earthquake loads. These loads are both random and dynamic in nature, but simplified procedures using equivalent static load approaches have been established in the IBC and ASCE 7 for calculating the total lateral forces acting on a building due to these loads. The wind loads specified in the ASCE 7 load specifications have a 50-year recurrence interval, and the seismic loads have a 2500-year recurrence interval. For any building design, both seismic and wind loads have to be checked to determine which one governs. It should be noted that even in cases where the seismic load controls the overall design of the lateral force–resisting system in a building, the wind load may still be critical for the the uplift forces on the roof. We present two examples in this section to illustrate the calculation of wind and seismic lateral forces using the simplified calculation procedures of the ASCE 7 load specifications [2].

Wind Load

There are two types of systems in a building structure for which wind loads are calculated: the main wind force–resisting system (MWFRS) and the components and cladding (C&C). The MWFRS transfers safely to the ground the overall building lateral wind loads from the various levels of the building. Examples are shear walls, braced frames, moment frames, and roof and floor diaphragms, and these elements are usually *parallel* to the direction of the wind force.

The components and cladding (C&C) are members that are loaded as individual components, with the wind load acting perpendicular to these elements. Examples include walls, cladding, roofs, uplift force on a column, and a roof deck fastener. The wind pressures on C&C are usually higher than for the MWFRS because of local spikes in the wind pressures over small areas in components and cladding. The wind pressure is a function of the effective wind area A_e, which is given in the IBC as

$$A_e = \text{(span of member)(tributary width)} \geq \frac{\text{(span of member)}^2}{3} \qquad (2.8)$$

For cladding and deck fasteners, the effective wind area A_e must not exceed the area that is tributary to each fastener. In calculating wind pressures, positive pressures are indicated by a force "pushing" into the wall or roof surface, and negative pressures are shown "pulling" away from the wall or roof surface. The *minimum* design wind pressure for the MWFRS and the components and cladding (C&C) is 10 psf (IBC Section 1609.1.2).

Wind Load Calculation

The three methods for determining the design wind loads on buildings and other structures presented in the ASCE 7 load specifications are as follows:

1. *Simplified method* (method 1): uses projected areas with the net horizontal and vertical wind pressures on the *exterior projected area* of the building. The wind pressures are obtained from ASCE 7 Figure 6-2 for the MWFRS and ASCE Figure 6-3 for the components and cladding. The simplified procedure is applicable only if *all* of the following conditions are satisfied:
    - Simple diaphragm buildings (i.e., wind load is transferred through the roof and floor diaphragms to the vertical main wind force–resisting system
    - Enclosed buildings (i.e., no large opening on any side of the building)
    - Mean roof height of building $\leq$ 60 ft
    - Symmetrical buildings
    - Buildings with mean roof height $\leq$ least horizontal dimension of building
    - Buildings without expansion joints
    - If the building has moment frames, and roof slope $\leq 30°$
2. *Analytical method* (method 2): wind pressures act perpendicular to the exterior wall and roof surfaces.
    - The analytical method is applicable to buildings with a regular shape (see ASCE 7 Section 6-2) that are not subjected to unusual wind forces, such as across-wind loading, vortex shedding, galloping, or flutter.
    - In the analytical method positive pressures are applied to the windward walls, and negative pressures or suction is applied to the leeward walls.
3. *Wind tunnel method* (method 3)
    - If method 1 or 2 cannot be used, the wind tunnel procedure must be used.
    - The wind tunnel procedure is used for very tall and wind-sensitive buildings, and buildings with height/least plan dimension greater than 5.0.

The simplified procedure (method 1) is used in this book.

Simplified Wind-Load Calculation Method (Method 1 for Low-Rise Buildings)

In the simplified method, the wind forces are applied perpendicular to the vertical projected area (VPA) and the horizontal projected area (HPA) of the building. The wind pressure diagram for the MWFRS is shown in Figure 2.13. The horizontal pressures represent the combined windward and leeward pressures, with the internal pressures canceling out; the vertical pressures include the combined effect of the external and internal pressures. The different wind pressure zones in Figure 2.13 are defined in Table 2.5. Where zone E or G falls on a roof overhang, use the windward roof *overhang* wind pressures from ASCE 7 Figure 6-2.

The simplified procedure (method 1) for calculating wind loads involves the following steps:

1. Determine the applicable wind speed for the building location from ASCE 7 Figure 6-1.
2. Calculate the mean roof height and determine the wind exposure category.
3. Determine the applicable *horizontal* and *vertical* wind pressures as a function of the wind speed, roof slope, zones, and effective wind area from ASCE 7 Figure 6-2 for the main wind force–resisting system and ASCE 7 Figure 6-3 for components and cladding. The tabulated values are based on an assumed exposure category of B, a mean roof height of 30 ft, and an importance factor of 1.0. These tabulated wind pressures are to be applied to the horizontal projected area (HPA) and the vertical projected area (VPA) of the building. The horizontal wind pressure on the vertical projected area of the building is the sum of the external windward and external leeward pressures since the internal pressures cancel out. The resulting wind pressures are applied to one side of the building for each wind direction. Note that the wind pressures on *roof overhangs* are much higher than at other locations on the roof because of the external wind pressures acting on both the bottom and top exposed surfaces of the overhang.
4. Obtain the design wind pressures (P_{s30} for MWFRS and P_{net30} for C&C), as a function of the tabulated wind pressures obtained in step 3, the applicable height and exposure adjustment factor (from ASCE 7-2 Figures 6-2 or 6-3), the topography factor K_{zt} (ASCE-7 Section 6.5.7), and the importance factor I_w from IBC table 1604.5 or ASCE-7 Tables 1-1 and 6-1. (For site conditions with flat topography, $K_{zt} = 1.0$)

$$P_{s30} = \lambda\ K_{zt} I_w P_{s30} \geq 10 \text{ psf}$$
$$P_{net30} = \lambda\ K_{zt} I_w P_{net30} \geq 10 \text{ psf}$$

5. Apply the calculated wind pressures to the building as shown in ASCE 7 Figure 6-2 for MWFRS. Note that for MWFRS, the end zone width is equal to *2a*.

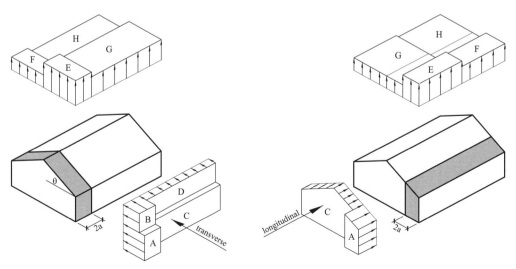

FIGURE 2.13 Wind pressure diagram for a main wind force–resisting system. (Adapted from Ref. 5, Figure 1609.6.2.1.)

TABLE 2.5 Definition of Wind Pressure Zones, MWFRS

Zone	Definition
A	End zone *horizontal* wind pressure on the *vertical* projected *wall* surface
B	End zone *horizontal* wind pressure on the *vertical* projected *roof* surface
C	Interior zone *horizontal* wind pressure on the *vertical* projected *wall* surface
D	Interior zone *horizontal* wind pressure on the *vertical* projected *roof* surface
E	End zone *vertical* wind pressure on the *windward* side of the *horizontal* projected *roof* surface
F	End zone *vertical* wind pressure on the *leeward* side of the *horizontal* projected *roof* surface
G	Interior zone *vertical* wind pressure on the *windward* side of the *horizontal* projected *roof* surface
H	Interior zone *vertical* wind pressure on the *leeward* side of the *horizontal* projected *roof* surface
E_{OH}	End zone *vertical* wind pressure on the *windward* side of the *horizontal* projected *roof overhang* surface
G_{OH}	Interior zone *vertical* wind pressure on the *windward* side of the *horizontal* projected *roof overhang* surface

6. Apply the calculated wind pressures to the building walls and roof as shown in ASCE 7 Figure 6-3 for components and cladding. Note that for C&C, the end zone width is equal to a, where

$$a \leq 0.1 \times \text{least horizontal dimension of building}$$
$$\leq 0.4 \times \text{mean roof height of the building}$$
$$\geq 0.04 \times \text{least horizontal dimension of building}$$
$$\geq 3 \text{ ft}$$

For a roof slope of less than 10°, the eave height should be used in lieu of the mean roof height for calculating the edge zone width $2a$.

Sample calculations of wind loads for the MWFRS using the simplified method (method 1) is presented in Example 2.7, and the calculation of the wind load on C&C is shown in Chapter 9. The more rigorous analytical method (method 2) is beyond the scope of this book.

Seismic Load

Earth's crust consists of several tectonic plates, many of which have fault lines, especially where the various tectonic plates interface with each other. When there is slippage along these fault lines, the energy generated creates vibrations that travel to Earth's surface and create an earthquake or seismic event. The effect of an earthquake on a structure depends on the location of that structure with regard to the point, the *epicenter*, at which the earthquake originated. The supporting soil for a structure plays a significant role in the magnitude of the seismic accelerations to which the structure will be subjected. The lateral force on the structure due to an earthquake is proportional to the mass and acceleration of the structure, in accordance with Newton's second law. Thus, $F = ma = W(a/g)$, where W is the weight of the structure and any components that will be accelerated at the same rate as the structure and a/g is the acceleration coefficient. The IBC and ASCE 7 equations used for calculating seismic lateral forces were derived from this basic equation.

The magnitude of the seismic loads acting on a building structure is a function of the type of lateral load–resisting system used in the building, the weight of the building, and the soil conditions at the building site. A seismic load calculation example for a wood building is presented in Example 2.8. The solution presented here will be in accordance with ASCE 7-05 Section 12.14, which is a simplified analysis procedure. This procedure is permitted for wood-framed structures in seismic use group or occupancy category I that are three stories or less in height (ASCE 7-05 Section 12-14.1.1), which is the category into which most wood structures fall. In the calculations, the building is assumed to be fixed at the base. The more in-depth equivalent lateral force analysis procedure presented in ASCE 7 is also permitted but is beyond the scope of this book.

EXAMPLE 2.7

Simplified Wind Load Calculation for MWFRS

A one-story wood building 50 × 75 ft in plan is shown in Figure 2.14. The truss bearing (or roof datum) elevation is at 15 ft and the truss ridge is 23 ft 4 in. above the ground-floor level. Assume that the building is enclosed and located in Rochester, New York on a site with a category C exposure. Determine the total horizontal wind force on the main wind force–resisting system (MWFRS) in both the transverse and longitudinal directions and the gross and net vertical wind uplift pressures on the roof (MWFRS) in both the transverse and longitudinal directions assuming a roof dead load of 15 psf.

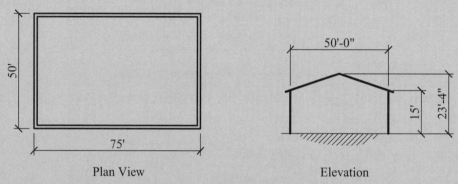

FIGURE 2.14 Building plan and elevation, wind load example.

Solution:

1. Determine the applicable 3-second gust wind speed for the building location from ASCE 7 Figure 6-1. For Rochester, New York, the wind speed is 90 mph.
2. The wind exposure category is C by assumption.
3. Determine the applicable horizontal and vertical wind pressures as a function of the wind speed, roof slope, effective wind area, and wind load direction for MWFRS and C&C from ASCE 7 Figure 6-2 for MWFRS and ASCE 7 Figure 6-3 for C&C:

$$\text{roof slope } \theta = \tan^{-1}\frac{23.33 \text{ ft} - 15 \text{ ft}}{50 \text{ ft}/2} = 18.43°$$

The mean roof height = $(15 \text{ ft} + 23.33 \text{ ft})/2 = 19.2$ ft.

MWFRS

For the building with a roof slope θ of 18.43° and a wind speed of 90 mph, the tabulated horizontal wind pressures on the vertical projected area of the building are obtained by linear interpolation between wind pressure values for θ of 0° and 20° from ASCE 7 Figure 6-2 as follows.

Horizontal wind pressures on MWFRS—transverse wind:

$$\text{End zone pressure on } \textit{wall} \text{ (zone A)} = 12.8 \text{ psf} + \left(\frac{18.43° - 0°}{20°}\right)(17.8 \text{ psf} - 12.8 \text{ psf})$$

$$= 17.4 \text{ psf}$$

$$\text{End zone pressure on } \textit{roof} \text{ (zone B)} = -6.7 \text{ psf} + \left(\frac{18.43° - 0°}{20°}\right)[(-4.7 \text{ psf}) - (-6.7 \text{ psf})]$$

$$= -4.9 \text{ psf} = 0 \text{ psf (see footnote 7 in ASCE 7 Figure 6-2)}$$

Interior zone pressure on *wall* (zone C) = 8.5 psf + $\left(\dfrac{18.43° - 0°}{20°}\right)$(11.9 psf − 8.5 psf)

$$= 11.6 \text{ psf}$$

Interior zone pressure on *roof* (zone D) = −4.0 psf + $\left(\dfrac{18.43° - 0°}{20°}\right)$[(−2.6 psf) − (−4.0 psf)] = −2.7 psf

$$= 0 \text{ psf (see footnote 7 in ASCE 7 Figure 6-2)}$$

Horizontal wind pressures on MWFRS—longitudinal wind:

End zone pressure on *wall* (zone A) = 12.8 psf

Interior zone pressure on *wall* (zone C) = 8.5 psf

If the horizontal wind pressure on roofs in zones B and D (see Figure 2.13) is less than zero, zero horizontal wind pressure is used for these zones. In the longitudinal wind direction, for roofs with sloped rafters but no hip rafters, the triangular area between the roof datum level and the ridge is a wall surface, so the horizontal wall pressures for zones A and C apply on this surface.

The resulting wind pressure on the MWFRS is *not symmetrical,* due to the nonsymmetrical location of the end zones and the higher wind pressures acting on the end zones in ASCE 7 Figure 6-2. In this book, we adopt a simplified approach to finding the wind force, using the average horizontal wind pressure. However, the asymmetrical nature of the wind loading on the building and the resulting torsional effect of such loading should also be investigated. For this building, the tabulated vertical wind pressures on the horizontal projected area of the building are obtained by linear interpolation between wind pressure values for θ of 0° and 20° from the tables in ASCE 7 Figure 6-2 as follows:

Vertical wind pressures on roof—transverse wind:

End zone on *windward* roof (zone E) = −15.4 psf

End zone on *leeward* roof (zone F) = −10.6 psf

Interior zone on *windward* roof (zone G) = −10.7 psf

Interior zone on *leeward* roof (zone H) = −8.0 psf

Vertical wind pressures on roof—longitudinal wind:

End zone on *windward* roof (zone E) = −15.4 psf

End zone on *leeward* roof (zone F) = −8.8 psf

Interior zone on *windward* roof (zone G) = −10.7 psf

Interior zone on *leeward* roof (zone H) = −6.8 psf

The wind pressures are summarized in Figure 2.15. For the MWFRS, the end zone width = *2a* according to ASCE 7 Figure 6-2, where

$a \leq 0.1 \times$ least horizontal dimension of building
$\leq 0.4 \times$ mean roof height of the building and
$\geq 0.04 \times$ least horizontal dimension of building
≥ 3 ft

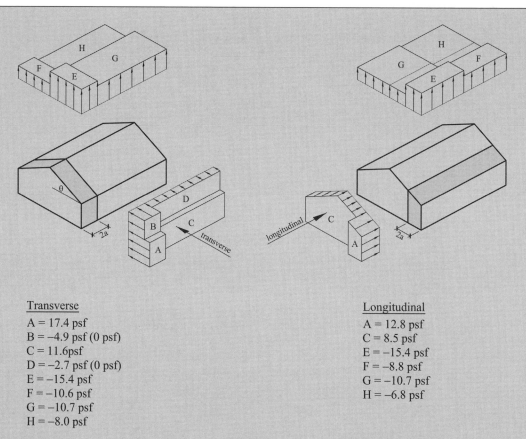

FIGURE 2.15 Summary of wind pressures. (Adapted from Ref. 5, Figure 1609.6.2.1.)

Transverse
A = 17.4 psf
B = −4.9 psf (0 psf)
C = 11.6 psf
D = −2.7 psf (0 psf)
E = −15.4 psf
F = −10.6 psf
G = −10.7 psf
H = −8.0 psf

Longitudinal
A = 12.8 psf
C = 8.5 psf
E = −15.4 psf
F = −8.8 psf
G = −10.7 psf
H = −6.8 psf

4. Multiply the tabulated wind pressures obtained in step 3 by the applicable height and exposure coefficients in ASCE 7 Figure 6-2 and the importance factor from ASCE 7 Table 6-1. Tabulate the MWFRS wind loads. For a category I building, importance factor $I_w = 1.0$ (IBC Table 1604.5).

Longitudinal wind. The end zone width is *2a* according to ASCE 7 Figure 6-2, where

$$a \leq (0.1)(50 \text{ ft}) = \mathbf{5 \text{ ft}} \quad \text{(governs)}$$

$$\leq (0.4)\left(\frac{15 \text{ ft} + 23.33 \text{ ft}}{2}\right) = 7.7 \text{ ft}$$

$$\geq (0.04)(50 \text{ ft}) = 2 \text{ ft}$$

$$\geq 3 \text{ ft}$$

Therefore, the edge zone = $2a = (2)(5 \text{ ft}) = 10$ ft.

For simplicity, as discussed previously, we use the average horizontal wind pressure acting on the entire building, calculated as follows:

average *horizontal* wind pressure in the longitudinal direction, q

$$= \frac{(\text{edge zone})(\text{edge zone pressures}) + (\text{total width of building} - \text{edge zone})(\text{interior zone pressures})}{\text{total width of building}}$$

where the total width of the building perpendicular to the longitudinal wind is 50 ft.

$$q_{\text{average}}(\text{wall}) = \frac{(10 \text{ ft})(12.8 \text{ psf}) + (50 \text{ ft} - 10 \text{ ft})(8.5 \text{ psf})}{50 \text{ ft}} = \mathbf{9.36 \text{ psf}}$$

$$q_{\text{average}}(\text{roof}) = \mathbf{0 \text{ psf}} \quad \text{(per footnote 7 in ASCE 7 Figure 6-2)}$$

Using exposure C as given (refer to ASCE 7-05 Section 6.5.6.3 for a description of the various exposure categories), the horizontal forces on the MWFRS are as given in Table 2.6. The total unfactored wind load on the building at the roof datum level is calculated as

TABLE 2.6 Longitudinal Wind Load, MWFRS

Level	Height (ft)	Exposure/Height Coefficient at Mean Roof Height, λ [ASCE 7 Figure 6-2]	Average Horizontal Wind Pressure, q (psf)	Design Horizontal Wind Pressure, $\lambda I_w q$	Total Unfactored Wind Load on the Building at Each Level (kips)
Ridge	23.33	1.29	**9.36**	(1.29)(1.0)(9.36) = 12.1 psf	—
Roof datum (i.e., at truss bearing elevation)	15	1.29	**9.36**	(1.29)(1.0)(9.36) = 12.1 psf	7.1
Base shear					**7.1**

Note: Mean roof height = 19.2 ft ≈ 20 ft.

$$\left[\left(\frac{1}{2}\right)(12.1 \text{ psf})(23.33 \text{ ft} - 15 \text{ ft}) + 12.1 \text{ psf}\left(\frac{15 \text{ ft}}{2}\right)\right](50 \text{ ft}) = 7058 \text{ lb} = 7.1 \text{ kips}$$

[The $\frac{1}{2}$ term in the equation accounts for the triangular shape of the vertical projected wall area above the roof datum (or truss bearing) level.]

Transverse wind. The end zone width is $2a$, where

$$a \leq (0.1)(50 \text{ ft}) = \mathbf{5 \text{ ft}} \quad \text{(governs)}$$
$$\leq (0.4)\left(\frac{15 \text{ ft} + 23.33 \text{ ft}}{2}\right) = 7.7 \text{ ft}$$
$$\geq (0.04)(50 \text{ ft}) = 2 \text{ ft}$$
$$\geq 3 \text{ ft}$$

Therefore, the edge zone = $2a$ = (2)(5 ft) = 10 ft.

For simplicity, as discussed previously, we use the average horizontal wind pressure acting on the entire building, calculated as follows:

average horizontal wind pressure in the transverse direction, q
$$= \frac{(\text{edge zone})(\text{edge zone pressures}) + (\text{total width of building} - \text{edge zone})(\text{interior zone pressures})}{\text{total width of building}}$$

where the total width of the building perpendicular to the transverse wind is 75 ft.

$$q_{\text{average}}(\text{wall}) = \frac{(10 \text{ ft})(17.4 \text{ psf}) + (75 \text{ ft} - 10 \text{ ft})(11.6 \text{ psf})}{75 \text{ ft}} = \mathbf{12.4 \text{ psf}}$$

$$q_{\text{average}}(\text{roof}) = \mathbf{0 \text{ psf}} \quad \text{(per footnote 7 in ASCE 7 Figure 6-2)}$$

Using exposure C as given (refer to ASCE 7-05 Section 6.5.6.3 for the description of the different exposure categories), the horizontal wind forces on the MWFRS are as given in Table 2.7. The total unfactored wind load on the building at the roof datum level is calculated as

$$\left[(0 \text{ psf})(23.33 \text{ ft} - 15 \text{ ft}) + (16 \text{ psf})\left(\frac{15 \text{ ft}}{2}\right)\right](75 \text{ ft}) = 9000 \text{ lb} = 9.0 \text{ kips}$$

Note that for the transverse wind direction, the vertical projected area of the roof surface above the roof datum (or truss bearing) level is rectangular in area.

TABLE 2.7 Transverse Wind Load, MWFRS

Level	Height (ft)	Exposure/Height Coefficient at Mean Roof Height, λ [ASCE 7 Figure 6-2]	Average Horizontal Wind Pressure, q (psf)	Design Horizontal Wind Pressure, $\lambda I_w q$	Total Unfactored Wind Load on the Building at Each Level (kips)
Ridge	23.33	1.29	0[a]	(1.29)(1.0)(0) = 0 psf	
Roof datum (i.e., at truss bearing elevation)	15	1.29	12.4	(1.29)(1.0)(12.4) = 16.0 psf	9.0
Base shear					9.0

[a] The pressure on the vertical projected area of the roof surface is taken as zero if the wind pressure obtained is negative. (See wind pressures for zones B and D for MWFRS).

5. Calculate the net vertical wind pressures on the roof for the main wind force–resisting system (MWFRS).

The roof wind uplift diagram (Figure 2.16) shows the calculated vertical uplift wind pressures on various zones of the roof. These are obtained by multiplying the tabulated vertical wind pressures by the height/exposure adjustment factor and the importance factor. The exposure category is C and

$$\text{mean roof height} = \frac{15 \text{ ft} + 23.33 \text{ ft}}{2} = 19.2 \text{ ft} \approx 20 \text{ ft}$$

The height/exposure adjustment coefficient, from ASCE 7 Figure 6-2 is 1.29

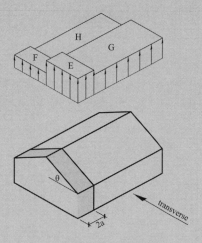

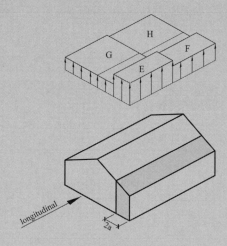

Transverse
E = −19.9 psf
F = −13.7 psf
G = −13.8 psf
H = −10.4 psf

Longitudinal
E = −19.9 psf
F = −11.4 psf
G = −13.8 psf
H = −8.8 psf

FIGURE 2.16 Roof uplift diagram.

The importance factor $I_w = 1.0$.

Vertical wind pressures on roof—longitudinal wind:

$$\text{End zone on } \textit{windward} \text{ roof (zone E)} = (1.29)(1.0)(-15.4 \text{ psf}) = -19.9 \text{ psf}$$

$$\text{End zone on } \textit{leeward} \text{ roof (zone F)} = (1.29)(1.0)(-8.8 \text{ psf}) = -11.4 \text{ psf}$$

$$\text{Interior zone on } \textit{windward} \text{ roof (zone G)} = (1.29)(1.0)(-10.7 \text{ psf}) = -13.8 \text{ psf}$$

$$\text{Interior zone on } \textit{leeward} \text{ roof (zone H)} = (1.29)(1.0)(-6.8 \text{ psf}) = -8.8 \text{ psf}$$

Vertical wind pressures on roof—transverse wind:

$$\text{End zone on } \textit{windward} \text{ roof (zone E)} = (1.29)(1.0)(-15.4 \text{ psf}) = -19.9 \text{ psf}$$

$$\text{End zone on } \textit{leeward} \text{ roof (zone F)} = (1.29)(1.0)(-10.6 \text{ psf}) = -13.7 \text{ psf}$$

$$\text{Interior zone on } \textit{windward} \text{ roof (zone G)} = (1.29)(1.0)(-10.7 \text{ psf}) = -13.8 \text{ psf}$$

$$\text{Interior zone on } \textit{leeward} \text{ roof (zone H)} = (1.29)(1.0)(-8.0 \text{ psf}) = -10.4 \text{ psf}$$

(Negative vertical wind pressures indicate uplift or suction on the roof; and positive values indicate downward pressure on the roof.) Using the given roof dead load of 15 psf, the net wind uplift pressures on the roof are calculated using the controlling IBC load combination 16-14 (i.e., $0.6D + W$) from Section 2.1 as shown below.

Net vertical wind pressures on roof—longitudinal wind:

$$\text{End zone on } \textit{windward} \text{ roof (zone E)} = (0.6)(15 \text{ psf}) - 19.9 \text{ psf} = -10.9 \text{ psf}$$

$$\text{End zone on } \textit{leeward} \text{ roof (zone F)} = (0.6)(15 \text{ psf}) - 11.4 \text{ psf} = -2.4 \text{ psf}$$

$$\text{Interior zone on } \textit{windward} \text{ roof (zone G)} = (0.6)(15 \text{ psf}) - 13.8 \text{ psf} = -4.8 \text{ psf}$$

$$\text{Interior zone on } \textit{leeward} \text{ roof (zone H)} = (0.6)(15 \text{ psf}) - 8.8 \text{ psf} = +0.2 \text{ psf}$$

Net vertical wind pressures on roof—transverse wind:

$$\text{End zone on } \textit{windward} \text{ roof (zone E)} = (0.6)(15 \text{ psf}) - 19.9 \text{ psf} = -10.9 \text{ psf}$$

$$\text{End zone on } \textit{leeward} \text{ roof (zone F)} = (0.6)(15 \text{ psf}) - 13.7 \text{ psf} = -4.7 \text{ psf}$$

$$\text{Interior zone on } \textit{windward} \text{ roof (zone G)} = (0.6)(15 \text{ psf}) - 13.8 \text{ psf} = -4.8 \text{ psf}$$

$$\text{Interior zone on } \textit{leeward} \text{ roof (zone H)} = (0.6)(15 \text{ psf}) - 10.4 \text{ psf} = -1.4 \text{ psf}$$

(Negative net vertical wind pressures indicate uplift or suction or upward pressure on the roof surface, and positive values indicate downward pressure on the roof.) As an alternate, the height/exposure coefficient, λ could have been obtained by linear interpolation of ASCE Figure 6-2 as follows:

$$\lambda = 1.21 + \frac{1.29 - 1.21}{20 \text{ ft} - 15 \text{ ft}} (19.2 \text{ ft} - 15 \text{ ft}) = 1.28$$

The difference between the λ value obtained here and the 1.29 value used earlier is less than 1%.

EXAMPLE 2.8

Seismic Load Calculation

For the two-story wood-framed structure shown in Figure 2.17, calculate the seismic base shear V, in kips, and the lateral seismic load at each level, in kips, given the following information:

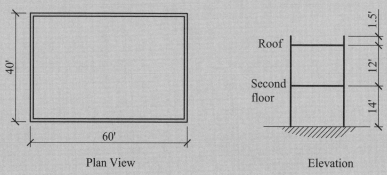

FIGURE 2.17 Building plan and elevation, seismic load example.

floor dead load = 30 psf

roof dead load = 20 psf

exterior walls = 15 psf

snow load P_f = 35 psf

site class = D (ASCE 7 Chapter 20)

importance I_e = 1.0 (ASCE 7 Table 11.5-1)

S_S = 0.25% (ASCE 7 Figure 22-1)

S_1 = 0.07% (ASCE 7 Figure 22-2)

R = 6.5 (ASCE 7 Table 12.14-1)

F = 1.1 for a two-story building (ASCE 7 Section 12.14.8.1)

Solution: First calculate the weight tributary W for each level including the weight of the floor structure, the weight of the walls, and a portion of the snow load. Where the flat roof snow load P_f is greater than 30 psf, 20% of the flat roof snow load is to be included in W for the roof (ASCE 7 Section 12.14.8.1). See Table 2.8 for the calulations.

TABLE 2.8 Building Weights for Seismic Load Calculation

Level	Area	Tributary Height	Weight of Floor	Weight of Walls	W_{total}
Roof	(60 ft)(40 ft) = **2400 ft²**	(12 ft/2) + 1.5 ft = **7.5 ft**	(2400 ft²)[20 psf + (0.2 × 35 psf)] = **64.8 kips**	(7.5 ft)(15 psf) × (2)(60 ft + 40 ft) = **22.5 kips**	64.8 kips + 22.5 kips = **87.3 kips**
Second floor	(60 ft)(40 ft) = **2400 ft²**	(12 ft/2) + (14 ft/2) = **13 ft**	(2400 ft²)(30 psf) = **72.0 kips**	(13 ft)(15 psf) × (2)(60 ft + 40 ft) = **39.0 kips**	72.0 kips + 39.0 kips = **111 kips**
					ΣW = 87.3 kips + 111 kips = **198.3 kips**

Seismic variables:

$$F_a = 1.6 \text{ (ASCE 7 Table 11.4-1)}$$

$$F_v = 2.4 \text{ (ASCE 7 Table 11.4-2))}$$

$$S_{MS} = F_a S_S = (1.6)(0.25) = 0.40 \text{ (ASCE 7 Equation 11.4-1)}$$

$$S_{M1} = F_a S_1 = (2.4)(0.07) = 0.168 \text{ (ASCE 7 Equation 11.4-2)}$$

$$S_{DS} = \tfrac{2}{3} S_{MS} = (\tfrac{2}{3})(0.40) = \mathbf{0.267} \text{ (ASCE 7 Equation 11.4-3)}$$

$$S_{D1} = \tfrac{2}{3} S_{M1} = (\tfrac{2}{3})(0.168) = \mathbf{0.112} \text{ (ASCE 7 Eqquation 11.4-4)}$$

Base shear:

$$V = \frac{F S_{DS} W}{R} \quad \text{(ASCE Equation 12.14-11)}$$

$$= \frac{(1.1)(0.267)(198.3)}{6.5} = \mathbf{8.96 \text{ kips}}$$

Force at each level. The lateral force at each level of the building is calculated using IBC Equation 16-57 as follows:

$$F_x = \frac{F S_{DS} W_x}{R} \quad \text{(ASCE 7 Equations 12.14-11 and 12.14-12)}$$

$$F_R = \frac{(1.1)(0.267)(87.3)}{6.5} = \mathbf{3.94 \text{ kips}}$$

$$F_2 = \frac{(1.1)(0.267)(111)}{6.5} = \mathbf{5.02 \text{ kips}}$$

The seismic loads are summarized in Figure 2.18.

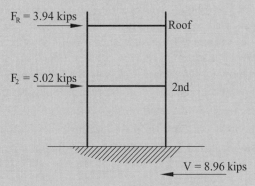

FIGURE 2.18 Summary of seismic loads.

Note: A flat roof was assumed in this example, but for buildings with a pitched roof, the weight of the roof and walls that are assigned to the roof level are assumed to act at the roof datum level, which is equivalent to the top elevation of the exterior walls.

The structural system factor, R used in this example corresponds to that of a plywood panel. Where sheathing of different materials are used on both sides of the studs in a shear wall, the R value for the wall system will be the smaller of the R values of the two sheathing materials, if sheathing on both wall faces are used to resist the lateral load, or the R value of the stronger sheathing material, if the weaker sheathing is neglected and only the stronger sheathing material is used to resist the lateral load. One example of this case is the use of plywood sheathing ($R = 6.5$) on the exterior face of a building and gypsum wallboard ($R = 2$) on the interior face of the building.

REFERENCES

1. ICC (2006), *International Building Code,* International Code Council, Washington, DC.
2. ASCE (2005), *Minimum Design Loads for Buildings and Other Structures,* American Society of Civil Engineers, Reston, VA.
3. State of New York (2002), *New York State Building Code,* Albany, NY.
4. State of Ohio (2005), *Ohio State Building Code,* Columbus, OH.
5. ICC (2003), *International Building Code,* International Code Council, Washington, DC.
6. NAHB (2000), *Residential Structural Design Guide-2000,* National Association of Home Builders Research Center, Upper Marlboro, MD.

PROBLEMS

2.1 Calculate the total uniformly distributed roof dead load in psf of the horizontal plan area for a sloped roof with the design parameters given below.

2×8 rafters at 24 in. o.c.

Asphalt shingles on $\frac{1}{2}$-in. plywood sheathing

6-in. insulation (fiberglass)

Suspended ceiling

Roof slope: 6:12

Mechanical & electrical (i.e., ducts, plumbing, etc.) = 5 psf

2.2 Given the following design parameters for a sloped roof, calculate the uniform total load and the maximum shear and moment on the rafter. Calculate the horizontal thrust on the exterior wall if rafters are used.

Roof dead load D = 20 psf (of sloped roof area)

Roof snow load S = 40 psf (of horizontal plan area)

Horizontal projected length of rafter L_2 = 14 ft

Roof slope: 4:12

Rafter or truss spacing = 4 ft 0 in.

2.3 Determine the tributary widths and tributary areas of the joists, beams, girders, and columns in the panelized roof framing plan shown in Figure 2.19. Assuming a roof dead load of 20 psf and an essentially flat roof with a roof slope of $\frac{1}{4}$ in. per foot for drainage, determine the following loads using the IBC load combinations. Neglect the rain load R and assume that the snow load S is zero. Calculate the following:

(a) The uniform total load on the typical roof joist, in lb/ft

(b) The uniform total load on the typical roof girder, in lb/ft

(c) The total axial load on the typical interior column, in lb

(d) The total axial load on the typical perimeter column, in lb

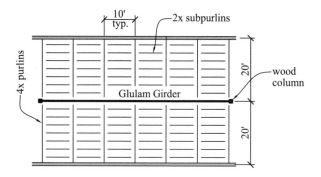

FIGURE 2.19 Floor framing plan.

2.4 A building has sloped roof rafters (5:12 slope) spaced at 2 ft 0 in. on centers and is located in Hartford, Connecticut. The roof dead load is 22 psf of the sloped area. Assume a fully exposed roof with terrain category C, and use the ground snow load from the IBC or ASCE 7 snow map. The rafter horizontal span is 14 ft.
  (a) Calculate the total uniform load in lb/ft on a horizontal plane using the IBC.
  (b) Calculate the maximum shear and moment in the roof rafter.

2.5 A three-story building has columns spaced at 18 ft in both orthogonal directions and is subjected to the roof and floor loads shown below. Using a column load summation table, calculate the cumulative axial loads on a typical interior column with and without a live-load reduction. Assume a roof slope of $\frac{1}{4}$ in. per foot for drainage.

Roof loads:
$$\text{Dead load } D_{\text{roof}} = 20 \text{ psf}$$
$$\text{Snow load } S = 40 \text{ psf}$$

Second- and third-floor loads:
$$\text{Dead load, } D_{\text{floor}} = 40 \text{ psf}$$
$$\text{Floor live load } L = 50 \text{ psf}$$

2.6 A two-story wood-framed structure 36 ft × 75 ft in plan is shown in Figure 2.20 with the following information given. The floor-to-floor height is 10 ft, the truss bearing (or roof datum) elevation is at 20 ft, and the truss ridge is 28 ft 4 in. above the ground-floor level. The building is enclosed and is located in Rochester, New York on a site with a category C exposure. Assume the following additional design parameters:

$$\text{Floor dead load} = 30 \text{ psf}$$
$$\text{Roof dead load} = 20 \text{ psf}$$
$$\text{Exterior walls} = 10 \text{ psf}$$
$$\text{Snow load } P_f = 40 \text{ psf}$$
$$\text{Site class} = D$$
$$\text{Importance } I_e = 1.0$$
$$S_S = 0.25\%$$
$$S_1 = 0.07\%$$
$$R = 6.5$$

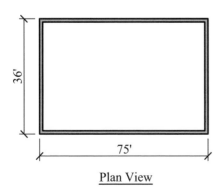
Plan View

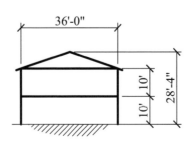
Elevation

FIGURE 2.20 Building plan and elevation.

Calculate the following:
(a) The total horizontal wind force on the main wind force–resisting system (MWFRS) in both the transverse and longitudinal directions
(b) The gross and net vertical wind uplift pressures on the roof (MWFRS) in both the transverse and longitudinal directions
(c) The seismic base shear V, in kips
(d) The lateral seismic load at each level, in kips

chapter *three*

Allowable Stress Design Method for Sawn Lumber and Glued Laminated Timber

3.1 ALLOWABLE STRESS DESIGN METHOD

The *allowable stress design* (ASD) *method* is used widely for the structural design of wood structures in the United States and is adopted in this book. The method is based on stipulating a maximum allowable stress under service conditions that incorporates a factor of safety. The factor of safety is the ratio of the maximum load or stress a structural member can support before it fails to the allowable load permitted by the Code, and it varies between 1.5 and 2.0 for most building structures. This allowable stress must be greater than or equal to the applied stress which is calculated on the basis of an elastic structural analysis. The *load and resistance factor design* (LRFD) *method* was introduced in the 2005 NDS code [1]. In this method, the load effects are multiplied by various load factors, which are usually greater than 1.0, to give the loads and load effects at the ultimate limit state. The load effects are usually determined from an elastic structural analysis. The factored resistance of the structural member or element is determined by multiplying the nominal resistance or theoretical strength of the member by a strength reduction or resistance factor, which is usually less than 1.0. The factored load effect must be less than or equal to the factored resistance for the structural member to be adequate at the ultimate limit state. The member is also checked at the serviceability limit state using the unfactored loads. The LRFD method accounts better than does the ASD method for the variability of loads in assigning the various load factors.

As stated earlier, the ASD method follows the approach that the stress applied must be less than or equal to the code-specified allowable stress. The stresses applied are denoted by a lower case f and the allowable stresses are denoted by an uppercase F. For example, the tension stress applied parallel to the grain f_t $(= P/A)$ must be less than or equal to the allowable tension stress parallel to the grain F'_t, where

$$F'_t = F_{t,\text{ NDS-S}} \text{ (product of all applicable adjustment or } C \text{ factors)}$$

Similarly, the bending stress applied f_b $(= M/S_X)$ must be less than or equal to the allowable bending stress F'_b, where

$$F'_b = F_{b,\text{ NDS-S}} \text{ (product of all applicable adjustment or } C \text{ factors)}$$

In general, the allowable stress is given as

$$F' = F_{\text{NDS-S}} \text{ (product of all applicable adjustment or } C \text{ factors)} \quad (3.1)$$

The applicable adjustment or C factors are given from Tables 3.1 and 3.2. The tabulated design values $F_{\text{NDS-S}}$ given in the NDS Supplement (NDS-S) [2] assume normal load duration and dry-use conditions.

TABLE 3.1 Applicability of Adjustment Factors for Sawn Lumber

	Load Duration Factor	Wet Service Factor	Temperature Factor	Beam Stability Factor	Size Factor	Flat Use Factor	Incising Factor	Repetitive Member Factor	Column Stability Factor	Buckling Stiffness Factor	Bearing Area Factor
$F_b' = F_b \times$	C_D	C_M	C_t	C_L	C_F	C_{fu}	C_i	C_r	—	—	—
$F_t' = F_t \times$	C_D	C_M	C_t	—	C_F	—	C_i	—	—	—	—
$F_v' = F_v \times$	C_D	C_M	C_t	—	—	—	C_i	—	—	—	—
$F_{c\perp}' = F_{c\perp} \times$	—	C_M	C_t	—	—	—	C_i	—	—	—	C_b
$F_c' = F_c \times$	C_D	C_M	C_t	—	C_F	—	C_i	—	C_P	—	—
$E' = E \times$	—	C_M	C_t	—	—	—	C_i	—	—	C_T	—
$E_{min}' = E_{min} \times$	—	C_M	C_t	—	—	—	C_i	—	—	C_T	—

Source: Ref. 1, Table 4.3.1. Courtesy of the American Forest & Paper Association, Washington, DC.

NDS Tabulated Design Stresses

The NDS tabulated design values are based on a 5% exclusion value of the clear-wood strength values with a reduction factor or strength ratio of less than 1.0 to allow for the effects of strength-reducing characteristics such as knot size, slope of grain, splits, checks, and shakes. A 5% exclusion value implies that five of every 100 pieces of wood are likely to have lower strength than the values tabulated in NDS-S. Only the elastic modulus and compression stress perpendicular to grain are based on average values. The combined effect of the strength ratio and other adjustment factors on the 5% exclusion clear-wood strength yield factors of safety for wood that range from 1.25 to 5.0, with average values of approximately 2.5. Table 3.3 shows the strength reduction factors applied to the 5% exclusion *clear-wood test* values. The clear-wood strength properties were established from laboratory tests on small straight-grained wood specimens where the wood members are loaded from zero to failure within a 5-minute duration. New tests known as *in-grade tests* were conducted more recently in which more than 70,000 full-size dimension lumber specimens were tested to failure in bending, tension, and compression parallel to the grain. The in-grade tests were used to verify the design values from the clear-wood tests.

TABLE 3.2 Applicability of Adjustment Factors for Glued Laminated Timber

	Load Duration Factor	Wet Service Factor	Temperature Factor	Beam Stability Factor[a]	Volume Factor[a]	Flat Use Factor	Curvature Factor	Column Stability Factor	Bearing Area Factor
$F_b' = F_b \times$	C_D	C_M	C_t	C_L	C_V	C_{fu}	C_c	—	—
$F_t' = F_t \times$	C_D	C_M	C_t	—	—	—	—	—	—
$F_v' = F_v \times$	C_D	C_M	C_t	—	—	—	—	—	—
$F_{c\perp}' = F_{c\perp} \times$	—	C_M	C_t	—	—	—	—	—	C_b
$F_c' = F_c \times$	C_D	C_M	C_t	—	—	—	—	C_P	—
$E' = E \times$	—	C_M	C_t	—	—	—	—	—	—
$E_{min}' = E_{min} \times$	—	C_M	C_t	—	—	—	—	—	—

Source: Ref. 1, Table 5.3.1. Courtesy of the American Forest & Paper Association, Washington, DC.
[a]The beam stability factor C_L should not be applied simultaneously with the volume factor C_V; the lesser of these two values applies (see NDS Section 5.3.6).

TABLE 3.3 Strength Reduction Factors

Stress Type	Strength Reduction Factors[a]	Limit State
Bending stress, F_b	0.48	Ultimate strength
Tension parallel to the grain, F_t	0.48	Ultimate strength
Compression parallel to the grain, F_c	0.533	Ultimate strength
Shear parallel to the grain, F_v	0.247	Ultimate strength
Compression perpendicular to the grain,[b] $F_{c\perp}$	0.667	0.04 in. deformation
Bending modulus of elasticity,[b] E	1.0	Elastic or proportional limit

Source: Refs. 8 and 10.
[a] The strength reduction factor accounts for strength-reducing defects such as knots, splits, wanes, and decay and normalizes the stress design values with respect to load duration, stress concentrations, and so on.
[b] F_c and E are based on the average values from the tests.

The allowable stress perpendicular to grain is controlled by a deformation limit state and not by crushing failure. Under load perpendicular to the grain, the member continues to support increasing load because of the densification of the member. The NDS code sets a 0.04-in. deformation limit for the determination of $F'_{c\perp}$.

Stress Adjustment Factors

The various stress adjustment factors that are used in wood design and discussed in this chapter are listed in Table 3.4.

Load Duration Factor C_D

Wood is susceptible to creep and continues to deflect under constant load due to moisture loss from the cell walls and cavities of the wood member. This creep property of wood is accounted for in determining the strength of a wood member through use of the load duration factor. The term *duration of load* designates the total accumulated length of time that a load is applied during the life of the structure, and the shorter the load duration, the stronger the wood member. The load duration factor C_D converts stress values for normal load duration to design stress values for other durations of loading. The NDS code defines the normal load duration as 10 years, which means that the load is applied on the structure for a cumulative maximum duration of 10 years, and the load duration factor for this normal load duration is 1.0. The load duration factors for other types of loads are given in Table 3.5.

For any load combination, the governing load duration factor C_D will be the value that corresponds to the shortest-duration load in that load combination. Thus, the governing C_D

TABLE 3.4 Adjustment Factors

Adjustment Factor	Description
C_D	Load duration factor
C_M	Wet service or moisture factor
C_F	Size factor
C_{fu}	Flat-use factor
C_t	Temperature factor
C_r	Repetitive member factor
C_i	Incising factor
C_P	Column stability factor
C_L	Beam stability factor
C_V	Volume factor (applies only to glulam)
C_b	Bearing area factor
C_c	Curvature factor

TABLE 3.5 Load Duration Factors

Type of Load	Cumulative Load Duration	Load Duration Factor, C_D
Dead load	Permanent	0.9
Floor live load	10 years or normal duration	1.0
Snow load	2 months	1.15
Roof live load	7 days	1.25
Construction load	7 days	1.25
Wind	10 minutes	1.6
Seismic (earthquake)	10 minutes	1.6
Impact	1 second or less	2.0

Source: Refs. 1 and 2.

value is equal to the largest C_D value in that load combination. It should be noted that the NDS code recommends a load duration factor of 1.6 for wind and seismic loads, rather than the 1.33 from the older codes, to reflect the very small duration of these loads. For example, 3-second wind gust pressures are now used in modern codes such as the IBC and ASCE 7 for wind load calculations, reflecting a lower load duration than the wind pressures used in older codes such as the *Uniform Building Code* (UBC). However, some jurisdictions may still insist that a load duration factor of 1.33 be used in that locality, so the reader is advised to consult with the local building plans' examiner in their jurisdiction for the appropriate load duration factor for wind load in that area.

EXAMPLE 3.1

Load Duration Factors in Load Combinations

Determine the load duration factor C_D for the following load combinations: (a) dead load + floor live load; (b) dead load + floor live load + snow load; (c) dead load + floor live load + roof live load.

Solution:

(a) *Dead load + floor live load:*

$$C_D \text{ for dead load} = 0.9$$

$$C_D \text{ for floor live load} = 1.0$$

Since the largest C_D value governs, $C_{D(D+L)} = \mathbf{1.0.}$

(b) *Dead load + floor live load + snow load:*

$$C_D \text{ for dead load} = 0.9$$

$$C_D \text{ for floor live load} = 1.0$$

$$C_D \text{ for snow load} = 1.15$$

Since the largest C_D value governs, $C_{D(D+L+S)} = \mathbf{1.15.}$

(c) *Dead load + floor live load + roof live load:*

$$C_D \text{ for dead load} = 0.9$$

$$C_D \text{ for floor live load} = 1.0$$

$$C_D \text{ for roof live load} = 1.25$$

Since the largest C_D value governs, $C_{D(D+L+Lr)} = \mathbf{1.25.}$

TABLE 3.6 Moisture Factor C_M for Sawn Lumber and Glulam

	Equilibrium Moisture Content, EMC (%)	Moisture Factor, C_M
Sawn lumber	≤ 19 (i.e. S-dry)	1.0
	> 19	Use C_M values from NDS-S Tables 4A to 4F, as applicable
Glulam	< 16 (i.e., S-dry)	1.0
	≥ 16	Use C_M values from NDS-S Tables 5A to 5D, as applicable

Moisture Factor C_M

The strength of a wood member is affected by its moisture content. The higher the moisture content, the more susceptible to creep the wood member is, and therefore the lower the wood strength. The tabulated design stresses in NDS-S apply to surface-dried (S-dry) wood used in dry service conditions. Examples will include S-dry interior wood members used in covered and insulated buildings. S-dry condition indicates a maximum moisture content of 19% for sawn lumber and less than 16% for glued laminated timber (glulam). The moisture factor C_M for sawn lumber and glulam is shown in Table 3.6.

Size Factor C_F

The size factor accounts for the effect of the structural member size on the strength of the wood member. As the depth of a wood member increases, the deviation of the stress distribution from the assumed linear stress distribution becomes more pronounced, leading to some reduction in the strength of the wood member. The size factor C_F applies only to sawn lumber and is not applicable to southern pine (or southern yellow pine), glulam, or machine stress–rated (MSR) lumber. Table 3.7 gives the size factors for the various size classifications of sawn lumber.

Repetitive Member Factor C_r

The repetitive factor applies only to flexural members placed in series and takes into account the redundancy in a roof, floor, or wall framing. It accounts for the fact that if certain conditions are satisfied, the failure or reduction in strength of one flexural member, in series with other adjacent members, will not necessarily lead to failure of the entire floor or wall system because of the ability of the structure to redistribute load from a failed member to the adjacent members. Examples of repetitive members include roof or floor joists and rafters, built-up joists, beams, and columns. For a wood structural member to be classified as repetitive, all the following conditions must be satisfied:

- There are at least three parallel members of dimension lumber.
- The member spacing is not greater than 24 in. on centers.
- The members are connected or tied together by roof, floor, or wall sheathing (e.g., plywood).

If a member is repetitive, the repetitive member factor C_r is 1.15, and for all other cases, C_r is 1.0. The repetitive member factor C_r does not apply to timbers (i.e., beam and stringers and post and timbers) or glulam. The repetitive member factor applies to the tabulated bending stresses in NDS-S Tables 4A, 4B, 4C, and 4F only and has already been incorporated into the tabulated bending stress $(F_b C_r)$ for decking given in NDS-S Table 4E.

TABLE 3.7 Size Factor C_F for Sawn Lumber

Size Classification	Size Factor, C_F
Dimension lumber, decking	Use C_F from NDS-S Tables 4A, 4B, 4E, and 4F
Timbers: beam and stringer (B&S), post and timbers (P&T)	$C_F = \left(\dfrac{12}{d}\right)^{1/9} \leq 1.0$ where d is the actual member depth, in inches. For loads applied to B&S, see NDS-S Table 4D

Flat Use Factor C_{fu}

Except for roof or floor decking, the NDS-S tabulated design stresses for dimension lumber apply to wood members bending about their stronger (or x–x) axis, as indicated in Figure 3.1a.

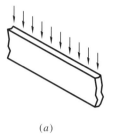 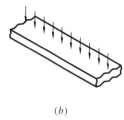

FIGURE 3.1 Direction of loading with respect to orthogonal axes: (a) bending about the strong axis; (b) bending about the weak axis.

For most cases, the flat use factor C_{fu} is 1.0, but for the few situations where wood members are stressed about their weaker (y–y) axis as indicated in Figure 3.1b, the flat use factor C_{fu} is obtained from NDS-S Tables 4A and 4B for dimension lumber, and from NDS-S Tables 5A to 5D for glulam. An example of a situation where dimension lumber may be bent about the weaker (y–y) axis is in stair treads. The flat use factor does not apply to decking because this factor has already been incorporated in the tabulated NDS-S design values for decking.

For glulam members, the flat use factor is usually greater than 1.0 and is applicable to beams subjected to weak axis bending and when the width (or smaller dimension) of the glulam member is less than 12 in. The C_{fu} values for glulam are tabulated in the adjustment factors section of NDS-S Tables 5A to 5D and are to be applied to the F_{by} (or weak axis bending) stress only. Although glulam beams are sometimes subjected to combined bending about the strong and weak axes, it is practical and conservative to assume a C_{fu} value of 1.0 in design.

Incising Factor C_i

Some species of lumber do not easily accept pressure preservative treatment, and small incisions parallel to the grain may be cut in the wood to increase the depth of penetration of preservatives in the wood member. The incising factor applies only to dimension lumber and has a value of 1.0 for lumber that is not incised. Design values are multiplied by the factors shown in Table 3.8 when dimension lumber is incised parallel to the grain, as follows:

$$\text{maximum depth} = 0.4 \text{ in.}$$

$$\text{maximum length} = 0.375 \text{ in.}$$

$$\text{maximum incision density} = 1100 \text{ per ft}^2$$

Temperature Factor C_t

Beyond a temperature of 100°F, an increase in temperature leads to a reduction in the strength of a wood member. The NDS code Table 2.3.3 lists the temperature factors C_t for various temperatures. For temperatures not greater than 100°F, the temperature factor C_t is 1.0 for wood used in dry or wet service conditions, and this applies to most insulated wood building structures. When the temperature in a wood structure is greater than 100°F, as may occur in some industrial structures, the C_t values can be obtained from NDS code Table 2.3.3.

Beam Stability Factor C_L

The beam stability factor accounts for the effect of lateral torsional buckling in a flexural member. When a beam is subjected to bending due to gravity loads, the top edge is usually in compression while the bottom edge is in tension. As a result of the compression stresses, the top edge of the beam will be susceptible to out-of-plane or sideway buckling if it is not braced

TABLE 3.8 Incising Factor

Design Value	Incising Factor, C_i
E, E_{min}	0.95
F_b, F_t, F_c, F_v	0.80
$F_{c\perp}$	1.00

Source: Ref. 1.

adequately. As the top edge moves sideways, the bottom edge, which is in tension, tends to move sideways in the opposite direction, and this causes twisting in the beam if the beam ends are restrained. This tendency to twist is called *lateral-torsional buckling,* and it should be noted that if the beam ends are unrestrained, the beam will become unstable and could roll over. However, lateral restraints are usually always provided, at least at the beam supports. Generally, if the compression edge of a flexural member is continuously braced by decking or plywood sheathing, the beam will not be susceptible to lateral-torsional buckling and the beam stability factor will be 1.0 in this case.

Two methods are prescribed in the NDS code that could be used to determine the beam stability factor C_L: rule-of-thumb method and the calculation method.

Rule-of-Thumb Method (NDS Code Section 3.3.3)   The beam stability factor C_L is 1.0 for each of the following cases provided that the conditions described for each case are satisfied:

1. If $d/b \leq 2$, no lateral support and no blocking are required.
   Examples: 2×4 or 4×8 wood member
2. If $2 < d/b \leq 4$, the beam ends are held in position laterally using solid end blocking.
   Examples: 2×6, 2×8, or 4×16 wood member
3. If $4 < d/b \leq 5$, the compression edge of a beam is laterally supported throughout its length. Solid end blocking is not required by the NDS code, but is recommended.
   Example: 2×10 wood member with plywood sheathing nailed to the top face of the wood member at fairly regular intervals
4. If $5 < d/b \leq 6$, bridging or full-depth solid blocking is provided at 8-ft intervals or less, or plywood sheathing is nailed to the top face of the wood member and blocking provided at the ends of the joist and at midspan where the joist span exceeds 12 ft or plywood sheathing nailed to the top and bottom faces of the joist.
   Example: 2×12
5. If $6 < d/b \leq 7$, both edges of the beam are held in line for the entire length.
   Examples: 2×14 with plywood sheathing nailed to the top face; a ceiling nailed to the bottom face of the wood member

For all five cases above, the member dimensions d and b are the nominal width (larger dimension) and thickness (smaller dimension) of the wood member, respectively.

Calculation Method for C_L   When the nominal width (or larger dimension) d is greater than the nominal thickness (or smaller dimension) b, the beam stability factor is calculated using the equation

$$C_L = \frac{1 + F_{bE}/F_b^*}{1.9} - \sqrt{\left(\frac{1 + F_{bE}/F_b^*}{1.9}\right)^2 - \frac{F_{bE}/F_b^*}{0.95}} \qquad (3.2)$$

where $F_{bE} = 1.20 E'_{\min}/R_B^2$ = Euler critical buckling stress for *bending* members
$\quad E'_{\min} = E_{\min} C_M C_t C_i C_T$ = reduced modulus of elasticity for buckling calculations
$\quad R_B = \sqrt{l_e d/b^2} \leq 50$
   (when R_B is greater than 50, the beam size has to be increased and/or lateral bracing provided to the compression edge of the beam to bring R_B within the limit of 50)
$\quad l_e$ = effective length obtained from Table 3.9 as a function of the unbraced length l_u
$\quad l_u$ = the unsupported length of the compression edge of the beam, or the distance between points of lateral support preventing rotation and/or lateral displacement of the compression edge of the beam
   [l_u is the unbraced length of the compression edge of the beam in the plane (i.e., y–y axis) in which lateral torsional buckling will occur; without continuous lateral support, there is a reduction in the allowable bending stress, but in many cases, continuous lateral support is provided by the plywood roof or floor sheathing, which is nailed and glued to the compression face of the joists and beams, and for these cases, the unbraced length is taken as zero]

$F_{bx}^* = F_{bx, \text{NDS-S}}$ (product of all applicable adjustable factors except C_{fu}, C_V, and C_L)

Column Stability Factor C_p

The column stability factor accounts for buckling stability in columns and is discussed more fully in Chapter 5. It is a function of the column end fixity, the column effective length, and the column cross-sectional dimensions, and it applies only to the compression stress parallel to the grain, F_c'. When a column is fully or completely braced about an axis of buckling such as by an intersecting stud wall or by plywood sheathing, the column stability factor C_p about that axis will be equal to 1.0.

Volume Factor C_V

The volume factor, which accounts for size effects in glulam members, is applicable only for bending stress calculations and is be covered in Section 3.2. When designing glulam members for bending, the smaller of the beam buckling stability factor C_L and the volume factor C_V is used to calculate the allowable bending stress. The volume factor is given as

$$C_V = \left(\frac{21}{L}\right)^{1/x}\left(\frac{12}{d}\right)^{1/x}\left(\frac{5.125}{b}\right)^{1/x} = \left(\frac{1291.5}{bdL}\right)^{1/x} \leq 1.0 \quad (3.3)$$

where L = length of beam between points of zero moment, ft
(Conservatively, assume that L = span of beam)
d = depth of beam, in.
b = width of beam, in. (≤ 10.75 in.)

TABLE 3.9 Effective Length l_e for Bending Members

Span Condition	$l_u/d < 7$	$7 \leq l_u/d \leq 14.3$	$14.3 < l_u/d$
		$7 \leq l_u/d$	
Cantilever			
Uniformly distributed load	$l_e = 1.33l_u$	$l_e = 0.90l_u + 3d$	
Concentrated load at unsupported end	$l_e = 1.87l_u$	$l_e = 1.44l_u + 3d$	
Single span beam			
Uniformly distributed load	$l_e = 2.06l_u$	$l_e = 1.63l_u + 3d$	
Concentrated load at center with no intermediate lateral support	$l_e = 1.80l_u$	$l_e = 1.37l_u + 3d$	
Concentrated load at center with lateral support at center		$l_e = 1.11l_u$	
Two equal concentrated loads at $\frac{1}{3}$ points with lateral support at $\frac{1}{3}$ points		$l_e = 1.68l_u$	
Three equal concentrated loads at $\frac{1}{4}$ points with lateral support at $\frac{1}{4}$ points		$l_e = 1.54l_u$	
Four equal concentrated loads at $\frac{1}{5}$ points with lateral support at $\frac{1}{5}$ points		$l_e = 1.68l_u$	
Five equal concentrated loads at $\frac{1}{6}$ points with lateral support at $\frac{1}{6}$ points		$l_e = 1.73l_u$	
Six equal concentrated loads at $\frac{1}{7}$ points with lateral support at $\frac{1}{7}$ points		$l_e = 1.78l_u$	
Seven or more concentrated loads, evenly spaced, with lateral support at points of load application		$l_e = 1.84l_u$	
Equal end moments		$l_e = 1.84l_u$	
Single span or cantilever conditions not specified above	$l_e = 2.06l_u$	$l_e = 1.63l_u + 3d$	$l_e = 1.84l_u$

Source: Ref. 1, Table 3.3.3. Courtesy of the American Forest & Paper Association, Washington, DC.

$x = 20$ for southern pine

$x = 10$ for all other species

Bearing Area Factor C_b

The bearing area factor accounts for the increase in allowable stresses at interior supports. NDS code Section 3.10.4 indicates that for bearing supports not nearer than 3 in. to the end of a member and where the bearing length l_b measured parallel to the grain of the member is less than 6 in., a bearing area adjustment factor C_b could be applied, where

$$C_b = \frac{l_b + 0.375}{l_b} \quad \text{for } l_b \leq 6 \text{ in.}$$

$$= 1.0 \quad \text{for } l_b > 6 \text{ in.} \quad (3.4)$$

$$= 1.0 \quad \text{for bearings at the ends of a member}$$

This equation accounts for the increase in bearing capacity at interior reactions or interior concentrated loads (see Figure 3.2) where there are unstressed areas surrounding the stress bearing area.

Curvature Factor C_c

The curvature factor is used to adjust the allowable bending stresses for curved glulam members such as arches. The curvature factor C_c applies only to the curved section of a wood member, and for straight wood members, C_c is 1.0. The curvature factor for a curved wood member is given as

$$C_c = 1 - (2000)\left(\frac{t}{R}\right)^2 \quad (3.5)$$

where t = thickness of laminations, in.

R = radius of curvature of the inside face of the lamination, in.

$\dfrac{t}{R} \leq \dfrac{1}{100}$ for hardwoods and southern pine

$\leq \dfrac{1}{125}$ for other softwoods

Buckling Stiffness Factor C_T

When a 2 × 4 or smaller sawn lumber is used as a compression chord in a roof truss with a minimum of $\frac{3}{8}$-in.-thick plywood sheathing attached to its narrow face, the stiffness against buckling is increased by the bucking stiffness factor C_T. This is accounted for by multiplying the reference modulus of elasticity for beam and column stability, $E_{\min}$, by the buckling stiffness factor C_T, which is calculated as

$$C_T = 1 + \frac{K_m l_e}{K_T E} \quad (3.6)$$

where l_e = effective column length of truss compression chord ($= K_e l$) $\leq$ 96 in.

(for $l_e > 96$ in., use $l_e = 96$ in.)

K_e = effective column length factor, usually taken as 1.0 for members supported at both ends

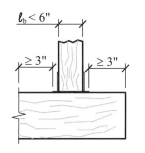

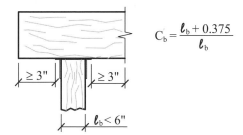

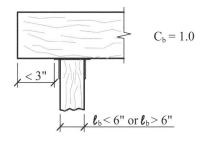

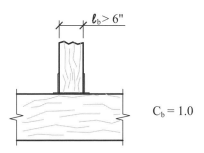

FIGURE 3.2 Bearing stress factors for different bearing conditions.

l = unbraced length of the truss top chord between connections or panel points
K_m = 2300 for wood with a moisture content not greater than 19% at the time of plywood attachment
= 1200 for unseasoned wood or partially seasoned wood at the time of plywood attachment (i.e., wood with moisture content > 19%)
K_T = 0.59 for visually graded lumber
= 0.75 for machine-evaluated lumber (MEL)
= 0.82 for glulams and machine stress-rated (MSR) lumber

Procedure for Calculating Allowable Stress

The procedure for calculating the allowable stresses from the tabulated NDS design stresses is as follows:

1. Identify the size classification of the wood member: that is, whether it is dimension lumber, beam and stringer, or post and timber or decking.
2. Obtain the tabulated design stress values from the appropriate table in NDS-S based on the size classification of the member, the wood species, the stress grade, and the structural function or stress. Refer to Table 1.4 for the applicable NDS-S tables. Note that these design values already includes a factor of safety.
3. Obtain the applicable adjustment or C factors from the NDS code adjustment factors applicability tables (Tables 3.1 and 3.2).
4. Calculate the allowable stress:

$$F' = F_{\text{NDS-S}} \text{ (product of all applicable adjustment factors from Table 3.1 or 3.2)}$$

Moduli of Elasticity for Sawn Lumber

There are two moduli of elasticity design values tabulated in NDS-S for sawn lumber. The reference modulus of elasticity, E, is used for beam deflection calculations, while the reduced modulus of elasticity, $E_{\min}$, is used for beam and column stability calculations. $E_{\min}$ is derived from the reference modulus of elasticity E with a factor safety of 1.66 applied, in addition to accounting for the coefficient of variation in the modulus of elasticity and an adjustment factor to convert reference E values to a pure bending basis.

3.2 GLUED LAMINATED TIMBER

Glulam is the acronym for glued laminated timber, which is made from thin laminations of kiln dried sawn lumber. They are used for very long spans (up to 60 ft or more) and to support very heavy loads where sawn lumber becomes impractical, and are also widely used in architecturally exposed structures because of their aesthetic appeal. Microlams and Parallams are proprietary glulams that are also used for floor beams and girders. Other advantages of glulam compared to sawn lumber include higher strength, better fire rating because of their relatively large size compared to sawn lumber framing, better aesthetics, better quality control because of the small size of the laminations, greater dispersion of defects in the laminations of the glulam, ready formation to curved shape for arches, and economical for long spans and large loads. Glulams are widely used in the construction of arch and long-span wood structures and are fabricated from very thin laminations of sawn lumber as shown in Figure 3.3. They are usually end-jointed and the laminations are stacked and glued together with water-resistant adhesives to produce wood members of practically any size and length. The lamination thickness varies from $1\frac{1}{2}$ or $1\frac{3}{8}$ in. for straight/slightly curved members and is $\frac{3}{4}$ in. for sharply curved members.

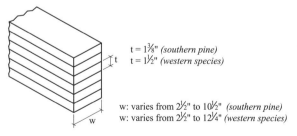

$t = 1\frac{3}{8}"$ *(southern pine)*
$t = 1\frac{1}{2}"$ *(western species)*

w: varies from $2\frac{1}{2}"$ to $10\frac{1}{2}"$ *(southern pine)*
w: varies from $2\frac{1}{2}"$ to $12\frac{1}{4}"$ *(western species)*

FIGURE 3.3 Glulam cross section.

Glulams are available for depths up to 72 inches and the standard glulam widths are $3\frac{1}{8}$, $5\frac{1}{8}$, $6\frac{3}{4}$, $8\frac{3}{4}$, and $10\frac{3}{4}$ in. for western species glulam (i.e., Douglas fir-larch, hem-fir, and California redwood), and 3, 5, $6\frac{3}{4}$, $8\frac{1}{2}$, and $10\frac{1}{2}$ in. for southern pine glulam. The most commonly used wood species for glulam are Douglas fir-larch and southern pine. As noted earlier, the sizes of glulam are usually called out on plans using their actual dimensions, whereas sawn lumber is called out using nominal dimensions. Table 1C of NDS-S lists the various glulam sizes and section properties. Glulams are usually grade-stamped on the top face for identification. In practice, it is usually more economical for the designer to specify the desired allowable stresses than to specify the glulam combination: hence the introduction of the stress class system in NDS-S Table 5A.

End Joints in Glulam

The two most common types of end jointing system used in glulam members are finger joints and scarf joints, shown in Figure 3.4. The end joints are usually staggered in the laminations of the glulam member and no glue is applied at the end joints. Wet-use adhesives are used under pressure to join glulam laminations together at their interfaces.

Grades of Glulam

Traditionally, the two main structural grades of glulam are bending combination and axial combination glulam. The bending combination glulams are given in NDS-S Table 5A–Expanded, and the axial combinations are given in NDS-S Table 5B. The values in NDS-S Tables 5B and 5A are for softwood glulam, the most frequently used glulam in the United States, and NDS-S Tables 5C and 5D are for hardwood glulam. The four standard appearance grades for glulam are framing, industrial, architectural and premium, with the last two grades being the only suitable grades for architecturally exposed structures [11].

Bending Combination Glulams (NDS-S Tables 5A–Expanded and 5C)

The bending combination glulams are used predominantly to resist bending loads, and for efficiency, their outer laminations are usually made of higher-quality lumber than the inner laminations. They could also be used to resist axial loads but may not be as efficient for resist axial loads as axial combination glulams. There are two types of bending combination lamination layout—balanced and unbalanced layout. In the balanced layout, which is recommended for continuous beams, the tension and compression laminations have the same tabulated design bending stress, whereas, in the unbalanced layout, the tension lamination has a higher tabulated bending stress than the compression lamination. Unbalanced layouts are recommended only for simply supported beams. An example of a bending combination glulam is 24F–V4, the most commonly used unbalanced layout, where the "24" refers to a tabulated bending stress of 2400 psi in the tension lamination, "V" refers to glulam with visually graded laminations, and "4" is the combination symbol. Another example is 22F–E5, where the "22" refers to a tabulated bending stress of 2200 psi, "E" refers to glulam with machine-stress-rated laminations, and "5" is the combination symbol. The most commonly used glulam with a balanced lamination layout is 24F-V8 [11].

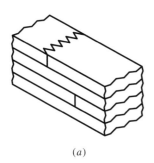
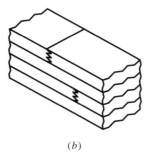
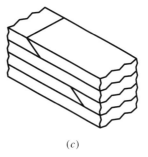

(a) (b) (c)

FIGURE 3.4 End joints in glulam: (*a*) vertical finger joint; (*b*) horizontal finger joint; (*c*) scarf joint.

Axial Combination Glulams (NDS-S Tables 5B and 5D)

Axial combination glulams are used predominantly to resist axial loads, and therefore all the laminations are made of the same quality. They could also be used to resist bending loads but are not as efficient as bending combination glulams. One example of an axial combination glulam is 5DF, where "5" is a combination symbol and "DF" refers to the wood species used in the glulam laminations.

Wood Species Used in Glulam

The wood species or species groups used in glulam laminations are shown in Table 3.10. The outer laminations of glulam are usually of higher-quality lumber than the inner laminations, and often, the tension outer laminations are made of higher-quality lumber than the compression outer laminations. In design practice it is recommended and practical to use glulam that uses the same wood species for both the outer and inner laminations. This would help prevent construction errors that may result from a glulam beam being inadvertently flipped upside down on site such that the lower-strength compression laminations (that should be stressed in compression) become stressed in tension, and the higher-strength tension laminations become stressed in compression.

Stress Class System

In the 2005 NDS-S, Table 5A provides an alternative means for specifying glulam based on the required allowable stress or stress class. This stress class approach is favored by the glulam industry because it gives glulam manufacturers flexibility in providing glulams when the required allowable stresses are specified, rather than specifying a specific glulam combination. The stress class approach greatly reduces the number of glulam combinations that the designer has to choose from, thus increasing design efficiency. In the stress class system of Table 5A, glulam is called out, for example, as 20F–1.5E, where "20" refers to the tabulated bending stress of 2000 psi and "E" refers to a bending modulus of elasticity of 1.5×10^6 psi obtained by mechanical grading. The intent is that future NDS-S editions will use the stress class approach in place of the bending combination glulams. The stress class approach does not apply to axial combination glulams.

NDS-S Tables 5A, 5A–Expanded, and 5B

In NDS-S Tables 5A, 5A–Expanded, and 5B, the allowable bending stress when high-quality tension laminations are stresses in tension is denoted as F_{bx}^+. The allowable bending stress when lower-quality compression laminations are stressed in tension is denoted as F_{bx}^-. As stated previously, it is preferable and recommended to use glulam with equal tensile strengths in the compression and tension laminations (i.e., $F_{bx}^+ = F_{bx}^-$). This is called a *balanced lay-up system*. Five values of elasticity modulus E are given in NDS-S Tables 5A and 5B:

TABLE 3.10 Wood Species Used in Glulam

Wood Species Symbol	Description
DF	Douglas fir-larch
DFS	Douglas fir south
HF	Hem-fir
WW	Western woods
SP	Southern pine
DF/DF	Douglas fir-larch wood species are used in both the outer laminations and the inner laminations
DF/HF	Douglas fir-larch wood species are used in the outer laminations, and hem-fir wood species are used in the inner laminations
AC	Alaska cedar
SW	Softwood species

- E_x is used in deflection calculations for strong (x–x)-axis bending of the glulam member.
- $E_{x,\,min}$ is used in glulam beam and column stability calculations for buckling about the x–x axis. It is used in C_L and C_P calculations for buckling about the x–x axis.
- E_y is used for deflection calculations for weak (y–y axis)-bending of the glulam member.
- $E_{y,\,min}$ is used in glulam beam and column stability calculations for buckling about the y–y axis. It is used in C_L and C_P calculations for buckling about the y–y axis.
- E_{axial} is used to calculate axial deformations in axially loaded glulam members.

Two values of compression stress perpendicular to grain ($F_{c\perp}$) are given in NDS-S Table 5A, and the location of these stresses in a glulam beam is indicated in Figure 3.5.

- $F_{c\perp \text{ tension face}}$ applies to the bearing stress on the outer face of the tension lamination.
- $F_{c\perp \text{ compression face}}$ applies to the bearing on the outer face of the compression lamination.

The allowable bearing stresses on the tension and compression laminations are given as

$$F'_{c\perp\,t} = F_{c\perp \text{ tension face}} C_M C_t$$

$$F'_{c\perp\,c} = F_{c\perp \text{ compression face}} C_M C_t$$

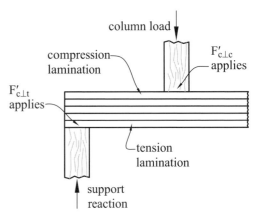

FIGURE 3.5 Bearing stress at supports and at interior loads.

3.3 ALLOWABLE STRESS CALCULATION EXAMPLES

In this section we present several examples to illustrate the calculation of allowable stresses for sawn lumber and glulam members.

3.4 LOAD COMBINATIONS AND THE GOVERNING LOAD DURATION FACTOR

The IBC basic load combinations were presented in Chapter 2. The distinct critical load combinations for which the various structural elements in a typical wood structure need to be designed are given in Table 3.11. For the roof beams, the seismic and floor live loads are assumed to be zero, while for the floor beams, the roof live loads, snow load, seismic load, and wind load are taken as zero since floor beams are not usually subjected to these types of loads. In load combinations 7 and 8 in Chapter 2, the wind and seismic loads oppose the dead load. Therefore, W and E take on negative values only in these two load combinations in order to maximize the uplift or overturning effect. These two load combinations are used to include the effects of uplift or overturning in structural elements and systems due to wind and seismic load effects.

Normalized Load Method

From Table 3.11, it is apparent that except for floor beams, several load combinations need to be considered in the design of wood members, but because of the varying load duration factors C_D, it is not possible or easy to determine by inspection the most critical load combinations for the design of a particular wood member. Thus, it can become quite tedious to design for all these load combinations with differing C_D values, especially when the design is carried out manually.

To help in identifying the most critical or governing load combination quickly, thus reducing the tedium in design, a normalized load method is adopted in this book. In this method the total load for each load combination is normalized with respect to the governing load duration factor C_D for that load combination. The load combination that has the largest normalized load is the most critical, and the wood member is designed for the actual total load corresponding to that governing load combination. The reader is cautioned to design for the actual total load,

EXAMPLE 3.2

Allowable Stresses in Sawn Lumber (Dimension Lumber)

Given a 2 × 10 No. 2 Douglas Fir–Larch wood member that is fully braced laterally, subject to dead load and snow load, and exposed to the weather under normal temperature conditions, determine all the applicable allowable stresses.

Solution:

1. Size classification = dimension lumber. Therefore, use NDS-S Table 4A.
2. From NDS-S Table 4A we obtain the following design values:

$$\text{Bending stress } F_b = 900 \text{ psi}$$

$$\text{Tension stress parallel to the grain } F_t = 575 \text{ psi}$$

$$\text{Horizontal shear stress parallel to the grain } F_v = 180 \text{ psi}$$

$$\text{Compression stress perpendicular to the grain } F_{c\perp} = 625 \text{ psi}$$

$$\text{Compression stress parallel to the grain } F_c = 1350 \text{ psi}$$

$$\text{Pure bending modulus of elasticity } E = 1.6 \times 10^6 \text{ psi}$$

3. Determine the applicable stress adjustment or C factors. From Section 3.1, for a dead load plus snow load ($D + S$) combination, the governing load duration factor C_D is 1.15 (i.e., the largest C_D value in the load combination governs). From the adjustment factors section of NDS-S Table 4A, noting that the wood member is exposed to the weather, we obtain the following C factors:

$$C_M(F_b) = 1.0$$

$$C_M(F_t) = 1.0$$

$$C_M(F_v) = 0.97$$

$$C_M(F_{c\perp}) = 0.67$$

$$C_M(F_c) = 0.8$$

$$C_M(E) = 0.9$$

$$C_F(F_b) = 1.1$$

$$C_F(F_t) = 1.1$$

$$C_F(F_c) = 1.0$$

$$C_P = 1.0 \text{ (column stability factor needs to be calculated, but assume 1.0 for this problem)}$$

$$C_t = 1.0 \text{ (normal temperature condition)}$$

$C_L, C_{fu}, C_i, C_r, C_b = 1.0$ (assumed)

Note: If $F_b C_F \leq 1150$, $C_M(F_b) = 1.0$:

$$F_b C_F = (900)(1.1) = 990 < 1150; \quad \text{therefore, } C_M(F_b) = 1.0$$

If $F_c C_F \leq 750$, $C_M(F_c) = 1.0$

$$F_c C_F = (1350)(1.0) = 1350 > 750; \quad \text{therefore, } C_M(F_c) = 0.8$$

4. Calculate the allowable stresses using the adjustment factors applicability table (Table 3.1).

Allowable bending stress:
$$F'_b = F_b C_D C_M C_t C_L C_F C_{fu} C_i C_r$$
$$= (900)(1.15)(1.0)(1.0)(1.0)(1.1)(1.0)(1.0)(1.0) = 1139 \text{ psi}$$

Allowable tension stress parallel to the grain:
$$F'_t = F_t C_D C_M C_t C_F C_i$$
$$= (575)(1.15)(1.0)(1.0)(1.1)(1.0) = 727 \text{ psi}$$

Allowable horizontal shear stress:
$$F'_v = F_V C_D C_M C_t C_i$$
$$= (180)(1.15)(0.97)(1.0)(1.0) = 200 \text{ psi}$$

Allowable bearing stress or compression stress perpendicular to the grain:
$$F'_{c\perp} = F_{c\perp} C_M C_t C_i C_b$$
$$= (625)(0.67)(1.0)(1.0)(1.0) = 419 \text{ psi}$$

Allowable compression stress parallel to the grain:
$$F'_c = F_c C_D C_M C_t C_F C_i C_P \quad (C_P \text{ needs to be calculated, but assume 1.0 for this Example;}$$
$$\text{this factor is discussed in Chapter 5})$$
$$= (1350)(1.15)(0.8)(1)(1.0)(1.0)(1.0) = 1242 \text{ psi}$$

Allowable pure bending modulus of elasticity:
$$E' = E C_M C_t C_i$$
$$= (1.6 \times 10^6)(0.9)(1.0)(1.0) = 1.44 \times 10^6 \text{ psi}$$

EXAMPLE 3.3

Allowable Stresses in Sawn Lumber (Timbers)

Given an 8 × 20 No. 2 Douglas Fir-Larch roof girder that is fully braced for bending and subject to a dead load, snow load, and wind load combination, calculate the allowable stresses for dry service and normal temperature conditions.

Solution:

1. Size classification is beam & stringer, so use NDS-S Table 4D. For 8 × 20 sawn lumber, $b = 7.5$ in. and $d = 19.5$ in.

2. From NDS-S Table 4D, we obtain the design values:

    Bending stress $F_b = 875$ psi

    Tension stress $F_t = 425$ psi

    Horizontal shear stress $F_v = 170$ psi

    Bearing stress or compression stress perpendicular to the grain $F_{c\perp} = 625$ psi

    Compression stress parallel to the grain $F_c = 600$ psi

    Pure bending modulus of elasticity $E = 1.3 \times 10^6$ psi

3. Determine the applicable stress adjustment or C factors. From Section 3.1, for a dead load plus wind load plus snow load $(D + W + S)$ combination, the governing load duration factor $C_D = 1.6$ (i.e., the largest C_D value in the load combination governs). As discussed previously, the reader should check to ensure that the local code allows a C_D value of 1.6 to be used for wind loads. From the adjustment factors section of NDS-S Table 4D, we obtain the following stress adjustment or C factors:

    $C_M = 1.0$ (dry service)

    $$C_F = \left(\frac{12}{d}\right)^{1/9} \leq 1.0$$

    $$= \left(\frac{12}{19.5}\right)^{1/9} = \mathbf{0.95} < 1.0 \quad \textbf{OK}$$

    $C_P = 1.0$ (column stability factor; need to calculate, but assume 1.0 for now)

    $C_t = 1.0$ (normal temperature conditions apply)

    $C_L = 1.0$ (member is fully braced laterally for bending)

    $C_{fu}, C_i, C_r, C_b = 1.0$ (Assumed)

4. Calculate the allowable stresses using the adjustment factors applicability table (Table 3.1).

    Allowable bending stress:

    $$F'_b = F_b C_D C_M C_t C_L C_F C_{fu} C_i C_r$$

    $$= (875)(1.6)(1.0)(1.0)(1.0)(0.95)(1.0)(1.0)(1.0) = 1330 \text{ psi}$$

    Allowable tension stress parallel to the grain:

    $$F'_t = F_t C_D C_M C_t C_F C_i$$

    $$= (425)(1.6)(1.0)(1.0)(0.95)(1.0) = 646 \text{ psi}$$

    Allowable horizontal shear stress:

    $$F'_v = F_V C_D C_M C_t C_i$$

    $$= (170)(1.6)(1.0)(1.0)(1.0) = 272 \text{ psi}$$

    Allowable bearing stress or compression stress perpendicular to the grain:

    $$F'_{c\perp} = F_{c\perp} C_M C_t C_i C_b$$

    $$= (625)(1.0)(1.0)(1.0)(1.0) = 625 \text{ psi}$$

Allowable compression stress parallel to the grain:

$$F'_c = F_c C_D C_M C_t C_F C_i C_P$$

$$= (600)(1.6)(1)(1)(0.95)(1.0)(1.0) = 912 \text{ psi}$$

Note: C_P needs to be calculated, and this factor is discussed in Chapter 5. For this example, we assume a value of 1.0.

Allowable pure bending modulus of elasticity:

$$E' = E C_M C_t C_i$$

$$= (1.3 \times 10^6)(1.0)(1.0)(1.0) = 1.3 \times 10^6 \text{ psi}$$

EXAMPLE 3.4

Allowable Stresses in Glulam

A $8\frac{3}{4} \times 34\frac{1}{2}$ in. 24F–V4 DF/DF simply supported glulam roof girder spans 32 ft (Figure 3.6), is fully braced for bending, and supports a dead load, snow load, and wind load combination. Calculate the allowable stresses for dry service and normal temperature conditions.

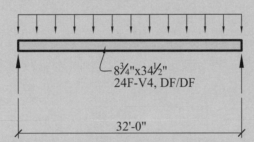

FIGURE 3.6 Simply supported Glulam beam.

Solution:

1. Size classification = glulam in bending (i.e., bending combination glulam). Therefore, use NDS-S Table 5A. For this glulam beam, the width $b = 8.75$ in. and depth $d = 34.5$ in. The distance between points of zero moments $L = 32$ ft (simply supported beam).
2. From NDS-S Table 5A, obtain the design values:

    Bending stress with tension laminations stressed in tension $F^+_{bx} = 2400$ psi

    Bending stress with compression laminations stressed in tension $F^-_{bx} = 1850$ psi

    Bearing stress or compression perpendicular to the grain on tension lamination, $F_{c\perp xx,\, t} = 650$ psi

    Bearing stress or compression perpendicular to the grain on tension lam, $F_{c\perp xx,\, c} = 650$ psi

    Horizontal shear stress parallel to the grain, $F_{v,xx} = 265$ psi

    Pure bending modulus of elasticity $E_{xx} = 1.8 \times 10^6$ psi

3. Determine the applicable stress adjustment or C factors. From Section 3.1, for a dead load plus wind load plus snow load ($D + W + S$) combination, $C_D = 1.6$. (Check if the local code allows a C_D value of 1.6 to be used for wind loads.) From the adjustment factors section of NDS-S Table 5A, we obtain the following C factors:

$$C_M = 1.0 \text{ (dry service)}$$

$$C_t = 1.0 \text{ (normal temperature)}$$

$$C_L = 1.0 \text{ (beam is fully braced)}$$

$$C_b, C_c \text{ and } C_{fu} = 1.0$$

From equation (3.3) we calculate the volume factor as

$$C_v = \left(\frac{21}{L}\right)^{1/x}\left(\frac{12}{d}\right)^{1/x}\left(\frac{5.125}{b}\right)^{1/x} = \left(\frac{1291.5}{bdL}\right)^{1/x} \leq 1.0$$

$$= \left(\frac{1291.5}{(8.75 \text{ in.})(34.5 \text{ in.})(32 \text{ ft})}\right)^{1/10} = 0.82 < 1.0 \quad \textbf{OK}$$

Recall that in calculating the allowable bending stress for glulams, the smaller of the C_V and C_L factors are used. Since $C_V = 0.82$ is less than $C_L = 1.0$, the C_V value of 0.82 will govern for the calculation of the allowable bending stress.

4. Calculate the allowable stresses.

Allowable bending stress with tension laminations stressed in tension:

$$F'^+_{bx} = \text{the smaller of } F^+_{bx}C_D C_M C_t C_L C_{fu} C_c \text{ or } F^+_{bx}C_D C_M C_t C_V C_{fu} C_c$$

$$= (2400)(1.6)(1.0)(1.0)(1.0)(1.0)(1.0) = 3840 \text{ psi}$$

or

$$= (2400)(1.6)(1.0)(1.0)(0.82)(1.0)(1.0) = \textbf{3148 psi} \quad \textbf{(governs)}$$

Therefore,

$$F'^+_{bx} = 3148 \text{ psi}$$

Allowable bending stress with compression laminations stressed in tension:

$$F'^-_{bx} = \text{the smaller of } F^-_{bx}C_D C_M C_t C_L C_{fu} C_c \text{ or } F^-_{bx}C_D C_M C_t C_V C_{fu} C_c$$

$$= (1850)(1.6)(1.0)(1.0)(1.0)(1.0)(1.0) = 2960 \text{ psi}$$

or

$$= (1850)(1.6)(1.0)(1.0)(0.82)(1.0)(1.0) = \textbf{2427 psi} \quad \textbf{(governs)}$$

Therefore,

$$F'^-_{bx} = 2427 \text{ psi}$$

Allowable bearing stress or compression perpendicular to the grain in the tension lamination:

$$F'_{c\perp xx,t} = F_{c\perp xx,t} C_M C_t C_b$$

$$= (650)(1.0)(1.0)(1.0) = 650 \text{ psi}$$

Allowable bearing stress or compression perpendicular to the grain in the compression lamination:

$$F'_{c\perp xx,c} = F_{c\perp xx,c} C_M C_t C_b$$

$$= (650)(1.0)(1.0)(1.0) = 650 \text{ psi}$$

Allowable horizontal shear stress:

$$F'_{v,xx} = F'_{v,xx} C_D C_M C_t$$

$$= (265)(1.6)(1)(1) = 424 \text{ psi}$$

Pure bending modulus of elasticity for bending about the strong x–x axis:

$$E'_{xx} = E_{xx} C_M C_t$$

$$= (1.8 \times 10^6)(1)(1) = 1.8 \times 10^6 \text{ psi}$$

TABLE 3.11 Applicable Design Load Combinations

Roof Beams	Floor Beams	Columns
$E = 0, L = 0, F = 0,$ $H = 0, T = 0^a$	$L_r = 0, S = 0, E = 0, W = 0, H = 0, F = 0, T = 0^a$	
$D + (L_r \text{ or } S \text{ or } R)$	$D + L$	Check all eight load combinations in Section 2.1
$D + 0.75W + 0.75$ $(L_r \text{ or } S \text{ or } R)$	Design for only one load combination	
$0.6D + W$		

^{a}See load combinations in Section 2.1 for definition of the various load symbols.

not the normalized load. The normalized load method is applicable only if all the loads in the load combination are of the same type and pattern. For example, the method is not valid if in the same load combination, some loads are uniform loads (given in psf or ksf) and others are concentrated loads (given in kips or lbs). When a load value is zero in a load combination, that load is neglected when determinating the governing load duration factor C_D for that load combination.

EXAMPLE 3.5

Governing Load Combination (Beams)

For a roof beam subject to the following loads, determine the most critical load combination using the normalized load method: $D = 10$ psf, $L_r = 16$ psf, $S = 20$ psf, $W = 5$ psf (downward). Assume that the wind load was calculated according to the IBC or ASCE 7 load standard.

Solution:

Dead load $D = 10$ psf

Roof live load $L_r = 16$ psf

Snow load $S = 20$ psf

Wind load $W = +5$ psf (since there is no wind uplift, $W = 0$ in load combination 5)

All other loads are neglected.

Note: Loads with zero values are not considered in determination of the load duration factor for that load combination and have been excluded from the load combinations

Since all the loads are distributed uniformly with the same units of pounds per square foot, the load type or patterns are similar. Therefore, the *normalized load method* can be used to determine the most critical load combination. The applicable load combinations are shown in Table 3.12.

TABLE 3.12 Applicable and Governing Load Combination

Load Combination	Value of Load Combination, w	C_D Factor for Load Combination	Normalized Load, w/C_D
D	10 psf	0.9	$\dfrac{10}{0.9} = 11.1$
$D + L_r$	$10 + 16 = 26$ psf	1.25	$\dfrac{26}{1.25} = 20.8$
$D + S$	$10 + 20 =$ **30 psf**	**1.15**	$\dfrac{30}{1.15} = 26.1 \Leftarrow$ (governs)
$D + 0.75W + 0.75L_r$	$10 + (0.75)(5 + 16) = 25.8$ psf	1.6	$\dfrac{25.8}{1.6} = 16.1$
$D + 0.75W + 0.75S$	$10 + (0.75)(5 + 20) = 28.8$ psf	1.6	$\dfrac{28.8}{1.6} = 18.0$
$0.6D + W$	$(0.6)(10) + (5)^a = 11$ psf	1.6	$\dfrac{11}{1.6} = 6.9$

[a] In load combinations 7 and 8 given in Section 2.1, the wind load W and seismic load E are always opposed by the dead load D. Therefore, in these combinations, D takes on a positive number while W and E take on negative values only.

Summary:

- The governing load combination is $D + S$ because it has the highest normalized load.
- The roof beam should be designed for a total uniform load of 30 psf with a load duration factor $C_D = 1.15$.

EXAMPLE 3.6

Governing Load Combination (Columns)

For a two-story column subject to the axial loads given below, determine the most critical load combination using the normalized load method. Assume that the wind loads have been calculated in accordance with the IBC or ASCE 7 load standard.

$$\text{Roof dead load } D_{\text{roof}} = 10 \text{ kips}$$

$$\text{Floor dead load } D_{\text{floor}} = 10 \text{ kips}$$

$$\text{Roof live load } L_r = 12 \text{ kips}$$

$$\text{Floor live load } L = 20 \text{ kips}$$

$$\text{Snow load } S = 18 \text{ kips}$$

$$\text{Wind load } W = \pm 10 \text{ kips}$$

$$\text{Earthquake or seismic load } E = \pm 12 \text{ kips}$$

Solution: In the load combinations, the dead load D is the sum of the dead loads from the all levels of the building. Thus,

$$D = D_{roof} + D_{floor} = 10 \text{ kips} + 10 \text{ kips} = 20 \text{ kips}$$

Since all the loads on this column are concentrated axial loads and therefore similar in pattern or type, the *normalized load method* can be used (Table 3.13)

TABLE 3.13 Applicable and Governing Load Combination

Load Combination	Value of Load Combination, P	C_D Factor for Load Combination	Normalized Load, P/C_D
D	20 kips	0.9	$\frac{20}{0.9} = 22.2$
$D + L$	$20 + 20 = 40$ kips	1.0	$\frac{40}{1.0} = 40$
$D + L_r$	$20 + 12 = 32$ kips	1.25	$\frac{32}{1.25} = 25.6$
$D + S$	$20 + 18 = 38$ kips	1.15	$\frac{38}{1.15} = 33.0$
$D + 0.75L + 0.75L_r$	$20 + (0.75)(20 + 12) = 44$ kips	1.25	$\frac{44}{1.25} = 35.2$
$D + 0.75L + 0.75S$	$20 + (0.75)(20 + 18) = $ **48.5 kips**	**1.15**	$\frac{48.5}{1.15} = 42.2 \Leftarrow$ (governs)
$D + 0.75W + 0.75L + 0.75L_r$	$20 + (0.75)(10 + 20 + 12) = 51.5$ kips	1.6	$\frac{51.5}{1.6} = 32.2$
$D + 0.75W + 0.75L + 0.75S$	$20 + (0.75)(10 + 20 + 18) = 56$ kips	1.6	$\frac{56}{1.6} = 35$
$D + (0.75)(0.7E) + 0.75L + 0.75L_r$	$20 + (0.75)[(0.7)(12) + 20 + 12]$ $= 50.3$ kips	1.6	$\frac{50.3}{1.6} = 31.4$
$D + (0.75)(0.7E) + 0.75L + 0.75S$	$20 + 0.75[(0.7)(12) + 20 + 18] = $ 54.8 kips	1.6	$\frac{54.8}{1.6} = 34.3$
$0.6D + W$	$(0.6)(20) + (-10)^a = 22$ kips	1.6	$\frac{22}{1.6} = 13.8$
$0.6D + 0.7E$	$(0.6)(20) + (0.7)(-12)^a = 20.4$ kips	1.6	$\frac{20.4}{1.6} = 12.8$

[a] In load combinations 7 and 8 given in Section 2.1 the wind load W and seismic load E are always opposed by the dead load D. Therefore, in these combinations, D takes on a positive number while W and E take on negative values only.

Summary:

- The governing load combination is dead load plus floor live load plus snow load ($D + 0.75L + 0.75S$) because it has the highest normalized load.
- The column should be designed for a total axial load $P = 48.5$ kips with a load duration factor $C_D = 1.15$.

REFERENCES

1. ANSI/AF&PA (2005), *National Design Specification for Wood Construction,* American Forest & Paper Association, Washington, DC.
2. ANSI/AF&PA (2005), *National Design Specification Supplement: Design Values for Wood Construction,* American Forest & Paper Association, Washington, DC.
3. Willenbrock, Jack H., Manbeck, Harvey B., and Suchar, Michael G. (1998), *Residential Building Design and Construction,* Prentice Hall, Upper Saddle River, NJ.
4. Ambrose, James (1994), *Simplified Design of Wood Structures,* 5th ed., Wiley, New York.
5. Halperin, Don A., and Bible, G. Thomas (1994), *Principles of Timber Design for Architects and Builders,* Wiley, New York.
6. Stalnaker, Judith J., and Harris, Earnest C. (1997), *Structural Design in Wood,* Chapman & Hall, London.
7. Breyer, Donald et al. (2003), *Design of Wood Structures—ASD,* 5th ed., McGraw-Hill, New York.
8. NAHB (2000), *Residential Structural Design Guide—2000,* National Association of Home Builders Research Center, Upper Marlboro, MD.
9. Kim, Robert H., and Kim, Jai B. (1997), *Timber Design for the Civil and Structural Professional Engineering Exams,* Professional Publications, Belmont, CA.
10. Hoyle, Robert J., Jr. (1978), *Wood Technology in the Design of Structures,* 4th ed., Mountain Press, Missoula, MT.
11. DeStafano, Jim (2003), *Exposed Laminated Timber Roof Structures,* STRUCTURE, pp. 16–17, March.

PROBLEMS

3.1 Explain why and how stress adjustment factors are used in wood design.

3.2 List the various stress adjustment factors used in the design of wood members.

3.3 Determine the load duration factors for the following IBC load combinations.

(a) $D + F$ (IBC Equation 16-8)
(b) $D + H + F + L + T$ (IBC Equation 16-9)
(c) $D + H + F + (L_r$ or S or $R)$ (IBC Equation 16-10)
(d) $D + H + F + 0.75(L + T) + 0.75(L_r$ or S or $R)$ (IBC Equation 16-11)
(e) $D + H + F + (W$ or $0.7E)$ (IBC Equation 16-12)
(f) $D + H + F + 0.75(W$ or $0.7E) + 0.75L + 0.75(L_r$ or S or $R)$ (IBC Equation 16-13)
(g) $0.6D + W + H$ (IBC Equation 16-14)
(h) $0.6D + 0.7E + H$ (IBC Equation 16-15)

3.4 What is glulam, and how is it manufactured?

3.5 Described briefly the various grades of glulam.

3.6 Describe the two ways of specifying glulam.

3.7 What is the advantage of the stress class system for specifying glulam?

3.8 Given a 2 × 12 Douglas fir-larch Select Structural wood member that is fully braced laterally, subject to dead load + wind load + snow load, and exposed to the weather under normal temperature conditions, determine all applicable allowable stresses.

3.9 Given a 10 × 20 Douglas fir-larch Select Structural roof girder that is fully braced for bending and subject to dead load + wind load + snow load. Assuming dry service and normal temperature conditions, calculate the allowable bending stress, shear stress, modulus of elasticity, and bearing stress perpendicular to the grain.

3.10 A $5\frac{1}{8}$ × 36 in. 24F–V8 DF/DF (i.e., 24F–1.8E) simply supported glulam roof girder spans 50 ft, is fully braced for bending, and supports a uniformly distributed dead load + snow load + wind load combination. Assuming dry service and normal temperature conditions, calculate the allowable bending stress, shear stress, modulus of elasticity, and bearing stress perpendicular to the grain.

3.11 A $6\frac{3}{4} \times 36$ in. 24F–V8 DF/DF (i.e., 24F–1.8E) simply supported glulam floor girder spans 64 ft between simple supports and supports a uniformly distributed dead load + floor live load combination. The compression edge of the beam is laterally braced at 8 ft on centers. Assuming dry service and normal temperature conditions, calculate the allowable bending stress, shear stress, modulus of elasticity, and bearing stress perpendicular to the grain.

3.12 For a roof beam subject to the following loads, determine the most critical load combination using the normalized load method: $D = 20$ psf, $L_r = 20$ psf, $S = 35$ psf, and $W = 10$ psf (downward). Assume that the wind load was calculated according to the IBC or ASCE 7 load standard.

3.13 For a column subject to the axial loads given below, determine the most critical load combination using the normalized load method. Assume that the wind loads have been calculated according to the IBC or ASCE 7 load standard.

$$\text{Roof dead load } D_{\text{roof}} = 8 \text{ kips}$$
$$\text{Floor dead load } D_{\text{floor}} = 15 \text{ kips}$$
$$\text{Roof live load } L_r = 15 \text{ kips}$$
$$\text{Floor live load } L = 26 \text{ kips}$$
$$\text{Snow load } S = 20 \text{ kips}$$
$$\text{Wind load } W = \pm 8 \text{ kips}$$
$$\text{Earthquake or seismic load } E = \pm 10 \text{ kips}$$

chapter *four*

DESIGN AND ANALYSIS OF BEAMS AND GIRDERS

4.1 DESIGN OF JOISTS, BEAMS, AND GIRDERS

In this chapter we introduce procedures for the analysis and design of joists, beams, and girders. These horizontal structural members are usually rectangular in cross section with bending about the strong (x–x) axes, and they are used in the framing of roofs and floors of wood buildings. The stresses that need to be checked in the design and analysis of these structural elements include (1) bending stresses (including beam lateral torsional buckling), (2) shear stresses, (3) deflection (live load and long-term deflections), and (4) bearing stresses.

Definition of Beam Span

The *span* of a beam or girder is defined as the clear span face to face of supports, L_n, plus one-half the required bearing length, $l_{b,\text{req'd}}$, at each end of the beam. Thus, the span $L = L_n + l_{b,\text{req'd}}$ (see Figure 4.1). In practice, it is conservative to assume the span of a beam as the distance between the centerline of the beam supports.

Layout of Joists, Beams, and Girders

It is more economical to lay out the joists or beams to span the longer direction of the roof or floor bay with the girders spanning the short direction, since the girders are usually more heavily loaded than are the joists or beams (Figure 4.2). The maximum economical bay size for wood framing using sawn-lumber joists and girders is approximately 14 × 14 ft. Longer girder spans can be achieved using glulams.

Design Procedure

In designing beams and girders, the first step is to calculate all the loads and load effects, and then the beam size is selected to satisfy the bending, shear, deflection, and bearing stress requirements. The design procedure for sawn-lumber joists, beams, or girders and glulam beams and girders is as follows.

Check of Bending Stress

1. Calculate all the loads and load effects, including maximum moment, shear, and reactions. Also calculate the dead and live loads that will be used later for deflection calculations.
    Note: Because of the relatively significant self-weight of girders, an assumption must initially be made for the girder self-weight and included in calculation of the load effects for the girder. This self-weight is checked after the final girder selection is made. As previously discussed, a wood density of 31.2 lb/ft^3 is used in this book to calculate the self-weight of the wood member. Alternatively, the self-weight could be obtained directly

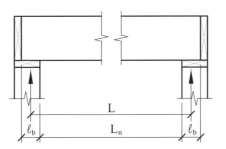

FIGURE 4.1 Definition of beam span.

L = beam span
L_n = clear span
ℓ_b = required bearing length

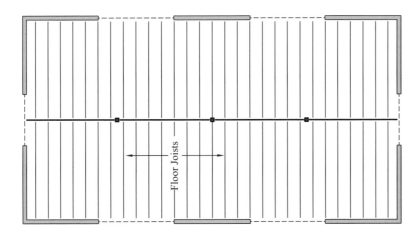

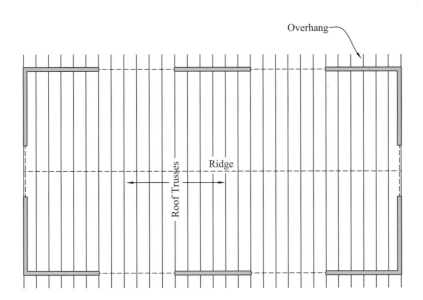

FIGURE 4.2 Typical roof and floor framing layout.

from NDS-S Table 1B, 1C, or 1D by interpolation between the densities given. This not applicable to joists because of their relatively small self-weight, which is usually already included in the pounds per square foot roof or floor dead load.

2. Assume a stress grade (i.e., Select Structural, No. 1 & Better, No. 1, etc.) and size classification (i.e., dimension lumber, timbers, glulam, etc.), and determine the applicable NDS-S table to be used in the design.

3. Assume initially that the allowable bending stress F'_b is equal to the NDS-S tabulated bending stress, F_b. That is, assume that $F'_b = F_{b,\text{NDS-S}}$. Therefore, the required approximate section modulus of the member is given as

$$S_{xx,\text{req'd}} \geq \frac{M_{\max}}{F_b} \qquad (4.1)$$

4. Using the approximate section modulus S_{xx} calculated in step 3, select a trial member size from NDS-S Table 1B (for sawn lumber), NDS-S Table 1C (for western species glulam), or NDS-S Table 1D (for southern pine glulam). For economy, the member size with the least area that has a section modulus at least equal to that required from step 2 should be selected.

5. For the member size selected in step 4, determine all the applicable stress adjustment or C factors, obtain the tabulated NDS-S bending stress, F_b and calculate the actual allowable bending stress,

$$F'_b = F_b \text{ (product of applicable adjustment or } C \text{ factors)} \qquad (4.2)$$

6. Using the size selected in step 4, calculate the applied bending stress f_b and compare this to the actual allowable bending stress calculated in step 5. Check if the applied bending stress $f_b = M_{\max}/S_{xx}$ is less than the actual allowable bending stress F'_b calculated in step 5.
   (a) If $f_b = M_{\max}/S_{xx} \leq F'_b$, the beam is adequate in bending.
   (b) If $f_b = M_{\max}/S_{xx} > F'_b$, the beam is not adequate in bending; increase the member size and repeat the design process.

Check of Shear Stress

Wood is weaker in horizontal shear than in vertical shear because of the horizontal direction of the wood grain relative to the vertical shear force. Thus, horizontal shear is the most critical shear stress for wood members. For vertical shear, the shear force is perpendicular to the grain, whereas for horizontal shear, the shear force is parallel to the grain, and only the lignin that glues the fibers together (see Chapter 1) is available to resist this horizontal shear.

The critical section for shear can conservatively be assumed to be at the face of the beam bearing support. Where the beam support is subjected to confining compressive stresses as shown in Figure 4.3, and where there are no applied concentrated loads within a distance d from the face of the beam support, the critical section for shear can be assumed to be at a distance d from the face of the beam support. Where a beam or joist is supported off the face of a girder on joist or beam hangers, the critical for shear should be taken at the face of the support.

The relationship between the shear stress applied in a rectangular wood beam at the centerline of the beam support f_v and the allowable shear stress F'_v is calculated as

$$f_v = \frac{1.5V}{A} \leq F'_v \qquad (4.3)$$

If applicable, the reduced applied shear stress at a distance d from the face of the beam support is calculated as

$$f'_v = \frac{1.5V'}{A} \leq F'_v \qquad (4.4)$$

where V = maximum shear at the critical section; for simplicity, this critical section can conservatively be assumed to be at the centerline of the joist, beam, or girder bearing support

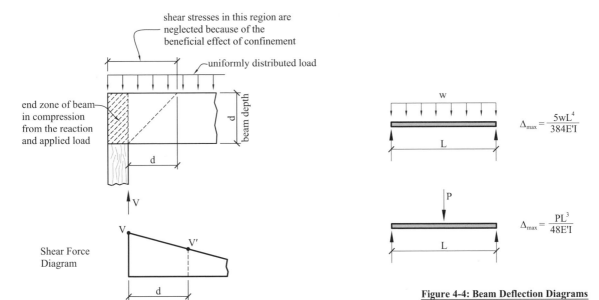

FIGURE 4.3 Critical section for shear. **FIGURE 4.4** Beam deflection diagrams.

V' = applied shear at a distance d from the face of the support
F'_v = allowable shear stress = $F_v C_D C_M C_t C_i$

The C factors are discussed in Chapter 3. Where there is a concentrated load on the beam with full bearing within a distance d from the face of the beam support, the critical section for shear will be at the face of the beam support. For these cases, use equation (4.3). The following shear design checks need to be made:

1. If f_v or $f'_v \leq F'_v$, the beam is adequate in shear.
2. If f_v or $f'_v > F'_v$, the beam is not adequate; therefore, increase the beam size and recheck the shear stress.

Note: If the shear stress f_v at the centerline of the beam support is adequate (i.e., $< F'_v$), there is no need to calculate further the shear stress f'_v at a distance d from the face of the support since it is obvious that f'_v would be less than f_v and thus will be adequate in that case. However, for situations where the calculated shear stress f_v at the centerline of the beam support is not adequate (i.e., $f_v > F'_v$), the shear stress f'_v should then be checked. If f'_v is less than or equal to the allowable shear stress, F'_v, the beam is deemed adequate. If, on the other hand, f'_v is also greater than F'_v, the beam size and/or stress grade would have to be increased until f'_v is just less than or equal to F'_v.

Check of Deflection (Calculating the Required Moment of Inertia)

The beam or girder deflection is calculated using engineering mechanics principles and assuming linear elastic material behavior. The maximum deflection is a function of the type of loading, the span of the beam, and the allowable pure bending modulus of elasticity E' (= $E C_M C_t C_i$). The formulas for the beam deflections under two common loading conditions are shown in Figure 4.4.

The deflection limits from IBC Section 1604.3.6 [1] are given in Table 4.1, and k is a factor that accounts for *creep effects* in wood members, which is the continuous deflection of wood members under constant load due to drying shrinkage and moisture loss. It should be noted that for the dead plus live load deflection, it is the incremental deflection that needs to be calculated and compared to the deflection limits, and this incremental dead plus live load deflection includes the effects of creep, but excludes the instantaneous dead-load deflection, which takes place during construction and before any of the deflection-sensitive elements such as glazing, masonry walls, or marble flooring are installed. If the deflection limits in Table 4.1 are satisfied, the beam is

TABLE 4.1 Deflection Limits for Joists, Beams, and Girders[a]

Deflection	Deflection Limit
Live-load deflection Δ_{LL}	$\dfrac{L}{360}$
Incremental long-term deflection due to dead load plus live load (including creep effects), $\Delta_{TL} = k\Delta_{DL} + \Delta_{LL}$	$\dfrac{L}{240}$

Source: Ref. 1, Section 1604.3.6.
[a] k is 1.0 for green or unseasoned lumber or glulam used in wet service conditions, 1.0 for seasoned lumber used in wet service conditions, and 0.5 for seasoned or dry lumber, glulam, and prefabricated I-joists used in dry service conditions. L is the span of a joist, beam, or girder between two adjacent supports, Δ_{DL} is the deflection due to dead load D, and Δ_{LL} is the deflection due to live load L or (L_r or S).

considered adequate for deflection; otherwise, the beam size has to be increased and the deflection rechecked. When the live load is greater than twice the dead load, the $L/360$ limit will control.

There might be occasions, especially for large-span girders, such as glulam girders, where although the deflection limits are satisfied, the absolute value of these deflections may be sufficiently large (say, greater than 1 in.) as to become unsightly. It can also increase the possibility of ponding in panelized roofs. Ponding is the increase in rain load on flat roofs due to water collecting in the roof depressions caused by the beam and girder deflections. This increase in loading causes more deflections which in turn further increases the roof live load. This cycle can continue and may lead to roof failure. In these cases it may be prudent to camber the girder to negate the effect of the downward deflection. *Cambering* is the fabrication of a beam or girder with a built-in upward deflection, usually equal to *1.5 times the dead-load deflection* for glulam roof beams and *1.0 times the dead load deflection* for floor beams with the intent that when loads are applied, the beam or girder will become essentially flat or horizontal. Cambering of the cantilevered portion of cantilevered beams is not recommended. For beams and girders in flat roofs, further analysis for possible ponding should be carried out when the beam deflection exceeds $\frac{1}{2}$ in. under a uniform load of 5 psf. [24].

The deflection limits provided in this section are intended to limit the cracking of nonstructural elements and finishes, due to large deflections, and are also intended to help control floor vibrations or bouncing floors. However, these deflection limits have been found to be unsatisfactory for controlling floor vibrations for floor joist with spans exceeding about 15 ft. Where floor joist spans exceed about 15 ft, it is recommended [15] that a live-load deflection limit of 0.5 in. be used. It should be noted that where floor joists are supported on floor girders instead of stud walls, the floor vibration will be magnified because of the flexibility of both the joists and the girders. The girders should be designed to a smaller deflection limit than $L/360$ to control the floor vibration. To help control floor vibration, it is also recommended that the plywood sheathing be nailed as well as glued to the floor joists. We discuss the topic of floor vibrations in more detail in Section 4.7.

An alternative approach to calculating the deflections is to calculate the required moment of inertia by equating the deflection to the appropriate deflection limit. For example, for a *uniformly loaded beam or girder,* we obtain the required moment of inertia based on the live load,

$$\Delta_{LL} = \frac{5w_{LL}L^4}{384E'I} = \frac{L}{360}$$

Solving and rearranging the equation above yields the required moment of inertia,

$$I_{req'd} = \frac{5w_{LL}L^3}{384E'}(360) \quad \text{in}^4 \quad (4.5)$$

Similarly, the moment of inertia required based on the total load is obtained for a *uniformly loaded beam or girder* as

$$I_{\text{req'd}} = \frac{5w_{k(\text{DL})+\text{LL}}L^3}{384E'}(240) \quad \text{in}^4 \qquad (4.6)$$

The higher moment of inertia from equations (4.5) and (4.6) will govern.

Check of Bearing Stress (Compression Stress Perpendicular to the Grain)

The NDS-S tabulated design stress perpendicular to the grain, $F_{c\perp}$, is based on a deformation limit of 0.04 in. for a steel plate bearing on a wood member (Figure 4.5). If a lower deformation limit is desired, the NDS code provides an equation for calculating the equivalent tabulated design value for the lower deformation limit of 0.02 in. based on NDS-S tabulated values:

$$F_{c\perp,\,0.02} = 0.73 F_{c\perp,\,\text{NDS-S}} \qquad (4.7)$$

The applied bearing stress or compression stress perpendicular to the grain is calculated as

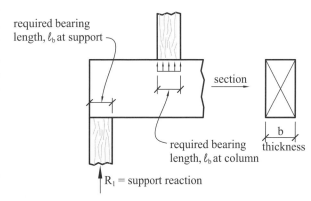

FIGURE 4.5 Bearing at supports and at interior loads.

$$f_{c\perp} = \frac{R_1}{A_{\text{bearing}}} \leq F'_{c\perp} \qquad (4.8)$$

where $F'_{c\perp}$ = allowable bearing stress perpendicular to the grain
R_1 = maximum reaction at the support
A_{bearing} = bearing area = (thickness of beam)(bearing length) = bl_b
$F_{c\perp}$ = NDS-S tabulated design stress perpendicular to the grain
$f_{c\perp}$ = applied bearing stress perpendicular to the grain

The minimum required bearing length l_b is obtained by rearranging equation (4.8). Thus,

$$l_{b,\,\text{req'd}} \geq \frac{R_1}{bF_{c\perp}} \qquad (4.9)$$

If the bearing length l_b required from equation (4.9) is greater than the bearing length available, the designer or engineer can do one of two things: increase beam thickness b and/or use a stress grade that gives a higher value of $F'_{c\perp}$. The practical recommended minimum bearing length is 1.5 in. for wood members bearing on wood and 3 in. for wood members bearing on masonry or concrete walls [15].

Some of the bearing details that could be used for joists, beams, or girders are shown in Figure 4.6. In Figure 4.6*a*, the joists from the adjacent spans butt against each other at the girder support with a tolerance gap of approximately $\frac{1}{2}$ in. provided; thus, the bearing length available for the joist will be less than one-half the thickness b of the supporting girder. In the overlapping joist detail (Figure 4.6*c*), the bearing length available for the joist is equal to the thickness b of the girder. In Figure 4.6*b*, the joist or beam is supported off the face of the girder with proprietary joist or beam hangers. The available bearing length for this detail is given in the manufacturer's catalog for each hanger. Figure 4.6*d* shows the girder support at a wood column using a U-shaped column cap. The available length of bearing for the girder in this detail will depend on the length of the U-shaped column cap.

FIGURE 4.6 Bearing details for joist, beams, and girders.

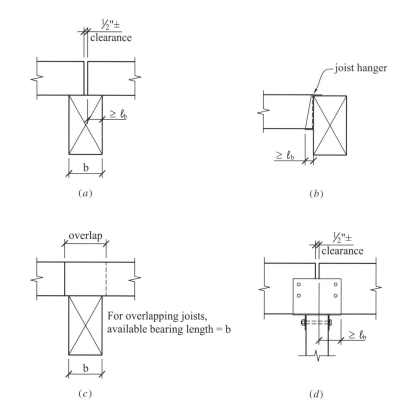

4.2 ANALYSIS OF JOISTS, BEAMS, AND GIRDERS

In the analysis of beams and girders, the beam or girder size, stress grade, and size classification are usually known, and the load is usually known or can be calculated. The intent is to determine if the existing beam or girder is adequate to resist the loads applied or to calculate the load capacity of the beam or girder. The analysis procedure for sawn-lumber joists, beams, or girders and glulam beams and girders is as follows.

Check of Bending Stress

1. Calculate all the loads and load effects, including the maximum moment, shear, and reactions. Also calculate the dead and live loads that will be used later for the deflection calculations. Include the self-weight of the member.
2. From the known stress grade (i.e., Select Structural, No. 1 & Better, No. 1, etc.) and size classification (i.e., dimension lumber, timbers, glulam, etc.), determine the applicable NDS-S table to be used in the design.
3. Determine the cross-sectional area A_{xx}, the section modulus S_{xx}, and the moment of inertia I_{xx} from NDS-S Table 1B (for sawn lumber), NDS-S Table 1C (for western species glulam), or NDS-S Table 1D (for southern pine glulam).
4. Determine the applied bending stress f_b.

$$f_b = \frac{M_{max}}{S_{xx}} \tag{4.10}$$

5. Determine all the applicable stress adjustment or C factors, obtain the tabulated NDS-S bending stress F_b, and calculate the allowable bending stress using equation (4.2):

$$F'_b = F_b \text{ (product of applicable adjustment or } C \text{ factors)}$$

6. Compare the applied stress f_b from step 4 to the allowable bending stress F'_b calculated in step 5.

EXAMPLE 4.1

Design of Joist (or Beams) and Girders

A typical floor plan in the private room areas of a hotel building is shown in Figure 4.7. Design a typical (a) joist B1 and (b) girder G1 assuming the use of Douglas fir-larch.

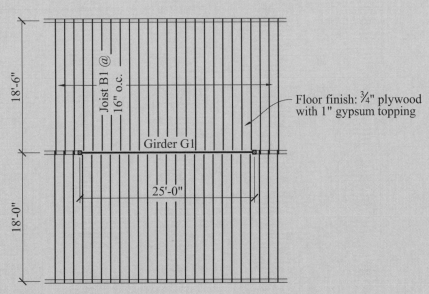

FIGURE 4.7 Typical floor plan.

Solution:

(a) *Design of floor joist B1* (Figure 4.8). Joist span = 18.5 ft. For private room areas and corridors serving them, the floor live load L = 40 psf.

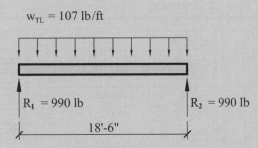

FIGURE 4.8 Free-body diagram of joist B1.

Dead load:

1-in. gypsum	=	6 psf
1-in. plywood sheathing (0.4 psf/$\frac{1}{8}$ in. × 6)	=	2.4 psf (assumed)
Framing: 2 × 14 at 16 in. o.c.	=	3.3 psf (assumed self-weight of floor framing)
Mechanical and electrical	=	8 psf (assumed)
Partitions	=	20 psf (15 psf is the minimum value allowed; see Chapter 2)
Total dead load D	=	39.7 psf ≃ 40 psf

$$\text{Tributary area of joist or beam B1} = \left(\frac{16 \text{ in.}}{12}\right)(18.5 \text{ ft}) = 25 \text{ ft}^2$$

From Chapter 2 and Table 2.2, we obtain the following parameters: A_T, the tributary area = 25 ft² and K_{LL} = 2 (interior beam or girder).

$$K_{LL}A_T = (2)(25 \text{ ft}^2) = 50 \text{ ft}^2 < 400 \text{ ft}^2$$

Therefore, no live-load reduction is permitted. The tributary width (TW) of joist or beam B1 = 16 in. = 1.33 ft. The total uniform load on the joist that will be used to design the joist for bending, shear, and bearing is

$$w_{TL} = (D + L)(TW) = (40 \text{ psf} + 40 \text{ psf})(1.33 \text{ ft}) = 107 \text{ lb/ft}$$

Using the free-body diagram of joist B1, the load effects are calculated as follows:

$$\text{Maximum shear, } V_{max} = R_{max} = \frac{w_{TL}L}{2} = \frac{(107 \text{ lb/ft})(18.5 \text{ ft})}{2} = 990 \text{ lb}$$

$$\text{Maximum moment } M_{max} = \frac{w_{TL}L^2}{8} = \frac{(107 \text{ lb/ft})(18.5 \text{ ft})^2}{8} = 4580 \text{ ft-lb} = 55{,}000 \text{ in.-lb}$$

The following loads will be used for calculating the joist deflections:

$$\text{uniform dead load } w_{DL} = (40 \text{ psf})(1.33 \text{ ft}) = 54 \text{ lb/ft} = 4.5 \text{ lb/in.}$$

$$\text{uniform live load } w_{LL} = (40 \text{ psf})(1.33 \text{ ft}) = 54 \text{ lb/ft} = 4.5 \text{ lb/in.}$$

We now proceed with the design following the steps described previously in this chapter.

Check of Bending Stress

1. Summary of load effects (the self-weight of joist B1 was included in the determination of the floor dead load in psf):

$$\text{maximum shear } V_{max} = 990 \text{ lb}$$

$$\text{maximum reaction } R_{max} = 990 \text{ lb}$$

$$\text{maximum moment } M_{max} = 55{,}000 \text{ in.-lb}$$

$$\text{uniform dead load is } w_{DL} = 4.5 \text{ lb/in.}$$

$$\text{uniform live load is } w_{LL} = 4.5 \text{ lb/in.}$$

2. For the stress grade, assume Select Structural, and assume that the size classification for the joist is dimension lumber. Thus, use NDS-S Table 4A.
3. From NDS-S Table 4A, we find that F_b = 1500 psi. Assume initially that $F_b' = F_b$ = 1500 psi. From equation (4.1) the approximate section modulus of the member required is given as

$$S_{xx, \text{req'd}} \geq \frac{M_{max}}{F_b} = \frac{55{,}000}{1500} = 37 \text{ in}^3$$

4. From NDS-S Table 1B, determine the trial size with the least area that satisfies the section modulus requirement of step 3: Try a 2 × 14 DF-L Select Structural member.

$$b = 1.5 \text{ in.} \quad \text{and} \quad d = 13.25 \text{ in.}$$

Size is still dimension lumber as assumed in step 1.   OK

$$S_{xx} \text{ provided} = 43.9 \text{ in}^3 > 37 \text{ in}^3 \quad \text{OK}$$

$$\text{Area } A \text{ provided} = 19.88 \text{ in}^2$$

$$I_{xx} \text{ provided} = 290.8 \text{ in}^4$$

5. The NDS-S tabulated stresses are

$$F_b = 1500 \text{ psi}$$

$$F_v = 180 \text{ psi}$$

$$F_{c\perp} = 625 \text{ psi}$$

$$E = 1.9 \times 10^6 \text{ psi}$$

The adjustment or C factors are listed in Table 4.2. Using equation (4.2), the allowable bending stress is

$$F_b' = F_b(\text{product of applicable adjustment or } C \text{ factors})$$

$$= F_b C_D C_M C_t C_L C_F C_r C_{fu} C_i$$

$$= (1500)(1.0)(1.0)(1.0)(0.9)(1.15)(1.0)(1.0) = \mathbf{1553 \text{ psi}}$$

TABLE 4.2 Adjustment Factors for Joists

Adjustment Factor	Symbol	Value	Rationale for the Value Chosen
Beam stability factor	C_L	1.0	The compression face is fully braced by the floor sheathing
Size factor	$C_F(F_b)$	0.9	From NDS-S Table 4A
Moisture or wet service factor	C_M	1.0	The equilibrium moisture content is $\leq 19\%$
Load duration factor	C_D	1.0	The largest C_D value in the load combination of dead load plus floor live load (i.e., $D + L$)
Temperature factor	C_t	1.0	Insulated building; therefore, normal temperature conditions apply
Repetitive member factor	C_r	1.15	All the three required conditions are satisfied: • The 2 × 14 trial size is dimension lumber • Spacing = 16 in. $\leq$ 24 in. • Plywood floor sheathing nailed to joists
Bearing stress factor	C_b	1.0	$C_b = \dfrac{l_b + 0.375}{l_b}$ for $l_b \leq 6$ in. = 1.0 for $l_b > 6$ in. = 1.0 *for bearings at the ends of a member* (see Section 3.1)
Flat use factor	C_{fu}	1.0	
Incision factor	C_i	1.0	

6. The actual applied bending stress is

$$f_b = \frac{M_{max}}{S_{xx}} = \frac{55{,}000 \text{ lb-in.}}{43.9 \text{ in}^3} = 1253 \text{ psi}$$

$$< F_b' = 1553 \text{ psi} \quad \text{OK}$$

Therefore, the beam is adequate in bending.

Check of Shear Stress
$V_{max} = 990$ lb. The beam cross-sectional area $A = 19.88$ in^2. The applied shear stress in the wood beam at the centerline of the beam support is

$$f_v = \frac{1.5V}{A} = \frac{(1.5)(990 \text{ lb})}{19.88 \text{ in}^2} = 75 \text{ psi}$$

The allowable shear stress is

$$F_v' = F_v C_D C_M C_t C_i = (180)(1)(1)(1)(1) = 180 \text{ psi}$$

Thus, $f_v < F_v'$; therefore, the beam is adequate in shear.

Note: Since the shear stress f_v at the centerline of the beam support is adequate (i.e., $< F_v'$), there is no need to calculate the shear stress f_v' further at a distance d from the face of the support since it is obvious that f_v' would be less than f_v, and thus will be adequate in that case.

Check of Deflection (see Table 4.3)
The allowable pure bending modulus of elasticity is

$$E' = E C_M C_t C_i = (1.9 \times 10^6)(1.0)(1.0)(1.0) = 1.9 \times 10^6 \text{ psi}$$

The moment of inertia about the strong axis $I_{xx} = 290.8$ in^4, the uniform dead load is $w_{DL} = 4.5$ lb/in., and the uniform live load is $w_{LL} = 4.5$ lb/in.

TABLE 4.3 Joist Deflection Limit

Deflection	Deflection Limit
Live-load deflection Δ_{LL}	$\dfrac{L}{360} = \dfrac{(18.5 \text{ ft})(12)}{360} = 0.62$ in.
Incremental long-term deflection due to dead load plus live load (including creep effects), $\Delta_{TL} = k\Delta_{DL} + \Delta_{LL}$	$\dfrac{L}{240} = \dfrac{(18.5 \text{ ft})(12)}{240} = 0.93$ in.

The dead-load deflection is

$$\Delta_{DL} = \frac{5wL^4}{384E_I'} = \frac{(5)(4.5 \text{ lb/in.})(18.5 \text{ ft} \times 12)^4}{(384)(1.9 \times 10^6 \text{ psi})(290.8 \text{ in}^4)} = 0.26 \text{ in.}$$

The live-load deflection is

$$\Delta_{LL} = \frac{5wL^4}{384E_I'} = \frac{(5)(4.5 \text{ lb/in})(18.5 \text{ ft} \times 12)^4}{(384)(1.9 \times 10^6 \text{ psi})(290.8 \text{ in}^4)} = 0.26 \text{ in.} < \frac{L}{360} \quad \text{OK}$$

Since seasoned wood in dry service conditions is assumed to be used in this building, the creep factor $k = 0.5$. The total incremental dead plus floor live load deflection is

$$\Delta_{TL} = k(\Delta_{DL}) + \Delta_{LL}$$
$$= (0.5)(0.26 \text{ in.}) + 0.26 \text{ in.} = 0.39 \text{ in.} < \frac{L}{240} \quad \text{OK}$$

Alternatively, the required moment of inertia can be calculated using equations (4.5) and (4.6) as follows:

$$I_{\text{req'd}} = \frac{5w_{LL}L^3}{384E'}(360) \quad \text{in}^4 = \frac{(5)(4.5)(18.5 \text{ ft} \times 12)^3}{(384)(1.6 \times 10^6)}(360) = 144.2 \text{ in}^4$$

Similarly, the required moment of inertia based on the total load is obtained for a uniformly loaded beam or girder as

$$I_{\text{req'd}} = \frac{5w_{k(DL)+LL}L^3}{384E'}(240) \quad \text{in}^4 = \frac{(5)(0.5 \times 4.5 + 4.5)(18.5 \text{ ft} \times 12)^3}{(384)(1.6 \times 10^6)}(240) = 144.2 \text{ in}^4$$

Therefore,

the required moment of inertia = 144.2 in^4 < I_{actual} = 290.8 in^4 OK

Check of Bearing Stress (Compression Stress Perpendicular to the Grain)

The maximum reaction at the support R_1 = 990 lb and the thickness of a 2 × 14 sawn-lumber joist, b = 1.5 in. The allowable bearing stress or compression stress parallel to the grain is

$$F'_{c\perp} = F_{c\perp}C_M C_t C_b C_i = (625)(1.0)(1.0)(1.0)(1.0) = 625 \text{ psi}$$

From equation (4.9), the minimum required bearing length l_b is

$$l_{b,\text{req'd}} \geq \frac{R_1}{bF'_{c\perp}} = \frac{990 \text{ lb}}{(1.5 \text{ in.})(625 \text{ psi})} = 1 \text{ in.}$$

The floor joists will be connected to the floor girder using face-mounted joist hangers with the top of the joists at the same level as the top of the girders. From a proprietary joist hanger load table (Figure 4.9) connector catalog, select a face-mounted hanger JH-214 with the following properties:

Joist Size	Model #	Dimensions			Douglas Fir-Larch or So. Pine Allowable Loads (lb)			
		W	H	B	100%	115%	125%	160% (uplift)
2x14	JH-214	1 9/16"	10"	2"	1280	1470	1600	2040
	xx	xx	xx	xx	xx	xx	xx	xx

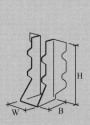
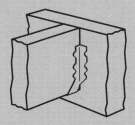

FIGURE 4.9 Generic load table for joist hangers.

Load duration factor, C_D for dead plus floor live load = 1.0 = 100%.

Therefore, the allowable load for JH-214 = 1280 lb > maximum reaction = 990 lb

Available bearing length = 2 in. > $L_{b,\text{req'd}}$ = 1 in.   OK

Use a 2 × 14 DF-L Select Structural for joist B1.

(b) *Design of floor girder G1.* Girder span = 25 ft. Since the joist selected was 2 × 14, we would expect the girder to be larger than a 2 × 14. Thus, the girder is probably going to be a timber (i.e., beam and stringer). Assume the girder self-weight = 90 lb/ft (this will be checked later). For hotels, the live load L = 40 psf.

Dead load:
From the design of the floor joist B1, total dead load, D = 40 psf.

$$\text{Tributary width of girder G1} = \frac{18.5 \text{ ft}}{2} + \frac{18.0 \text{ ft}}{2} = 18.25 \text{ ft}$$

$$\text{Tributary area of girder G1} = (18.25 \text{ ft})(25 \text{ ft}) = 456 \text{ ft}^2$$

From ASCE 7 Table 4-2, A_T, the tributary area = 456 ft^2, and K_{LL} = 2(interior girder).

- $K_{LL}A_T = (2)(456 \text{ ft}^2) = 912 \text{ ft}^2 > 400 \text{ ft}^2$
- Floor live load L = 40 psf < 100 psf
- Floor occupancy is not assembly occupancy

Since all the conditions above are satisfied, live-load reduction is permitted.

Live-load reduction. The reduced or design floor live load for the girder is calculated using equation (2.7) as follows:

$$L = L_0 \left(0.25 + \frac{15}{\sqrt{K_{LL}A_T}}\right) = (40 \text{ psf})\left[0.25 + \frac{15}{\sqrt{(2)(456)}}\right] = \mathbf{30 \text{ psf}}$$

$\geq (0.50)(40 \text{ psf}) = 20$ psf for members supporting one floor (e.g., slabs, beams, and girders)
$\geq (0.40)(40 \text{ psf}) = 16$ psf for members supporting two or more floors (e.g., columns)

The total uniform load on the girder that will be used to design for bending, shear, and bearing is obtained below using the dead load, the reduced live load calculated above, and the assumed self-weight of the girder, which will be checked later.

$$w_{TL} = (D + L)(\text{tributary width}) + \text{girder self-weight}$$

$$= (40 \text{ psf} + 30 \text{ psf})(18.25 \text{ ft}) + 90 \text{ lb/ft} = \mathbf{1368 \text{ lb/ft}}$$

Using the free-body diagram of girder G1 (Figure 4.10), the load effects are calculated as follows:

$$\text{maximum shear } V_{max} = \frac{w_{TL}L}{2} = \frac{(1368 \text{ lb/ft})(25 \text{ ft})}{2} = 17,100 \text{ lb}$$

$$\text{maximum reaction } R_{max} = 17,100 \text{ lb}$$

$$\text{maximum moment } M_{max} = \frac{w_{TL}L^2}{8} = \frac{(1368 \text{ lb/ft})(25 \text{ ft})^2}{8} = 106,875 \text{ ft-lb} = 1,282,500 \text{ in.-lb}$$

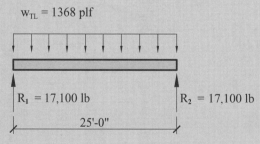

FIGURE 4.10 Free-body diagram of girder G1.

The following loads will be used for calculating the joist deflections:

uniform dead load w_{DL} = (40 psf)(18.25 ft) + 90 lb/ft = 820 lb/ft = 68.33 lb/in.

uniform live load w_{LL} = (30 psf)(18.25 ft) = 547.5 lb/ft = 45.63 lb/in.

We now proceed with the design following the steps described earlier in the chapter.

Check of Bending Stress

1. Summary of load effects (the self-weight of girder G1 was assumed, but this will be checked later in the design process):

   maximum shear V_{max} = 17,100 lb

   maximum reaction R_{max} = 17,100 lb

   maximum moment M_{max} = 1,282,500 in.-lb

   uniform dead load w_{DL} = 68.33 lb/in.

   uniform live load w_{LL} = 45.63 lb/in.

2. For the stress grade, assume Select Structural, and assume that the size classification for the girder is beam and stringer (B&S), that is, timbers. Therefore, use NDS-S Table 4D.

3. From NDS-S Table 4D, we find that F_b = 1600 psi. Assume initially that $F'_b = F_b$ = 1600 psi. From equation (4.1) the required approximate section modulus of the member is given as

$$S_{xx, \text{req'd}} \geq \frac{M_{max}}{F_b} = \frac{1,282,032}{1600} = 801.3 \text{ in}^3$$

4. From NDS-S Table 1B, determine the trial size with the least area that satisfies the section modulus requirement of step 3. Try a 10 × 24 DF-L Select Structural.

   b = 9.5 in. and d = 23.5 in.

   Size is B&S, as assumed in step 1   OK

   S_{xx} provided = 874.4 in^3 > 801.3 in^3 OK

   Area A provided = 223.3 in^2

   I_{xx} provided = 10,270 in^4

5. The NDS-S tabulated stresses are

   F_b = 1600 psi

   F_v = 170 psi

   $F_{c\perp}$ = 625 psi

   E = 1.6 × 10^6 psi

The adjustment or C factors are listed in Table 4.4. Using equation (4.2), the allowable bending stress is

$$F'_b = F_b \text{ (product of applicable adjustment or } C \text{ factors)}$$
$$= F_b C_D C_M C_t C_L C_F C_r C_{fu} C_c$$
$$= (1600)(1.0)(1.0)(1.0)(1.0)(1.0)(1.0)(1.0) = \mathbf{1600 \text{ psi}}$$

TABLE 4.4 Adjustment Factors for Glulam Girder

Adjustment Factor	Symbol	Value	Rationale for the Value Chosen
Beam stability factor	C_L	1.0	The compression face is fully braced by the floor sheathing
Size factor	C_F	0.93	$\left(\dfrac{12}{d}\right)^{1/9} = \left(\dfrac{12}{23.5 \text{ in.}}\right)^{1/9} = 0.93$
Moisture or wet service factor	C_M	1.0	Equilibrium moisture content is $\leq 19\%$
Load duration factor	C_D	1.0	The largest C_D value in the load combination of dead load plus floor live load (i.e., $D + L$)
Temperature factor	C_t	1.0	Insulated building; therefore, normal temperature conditions apply
Bearing stress factor	C_b	1.0	$C_b = \dfrac{l_b + 0.375}{l_b}$ for $l_b \leq 6$ in. $= 1.0$ for $l_b > 6$ in. $= 1.0$ for bearings at the ends of a member
Flat use factor	C_{fu}	1.0	
Curvature factor	C_c	1.0	

6. The actual applied bending stress is

$$f_b = \frac{M_{max}}{S_{xx}} = \frac{1{,}282{,}500 \text{ lb-in.}}{874.4 \text{ in}^3} = 1466 \text{ psi}$$
$$< F'_b = 1600 \text{ psi} \quad \text{OK}$$

Therefore, the beam is adequate in bending.

Check of Shear Stress

$V_{max} = 17{,}100$ lb. The beam cross-sectional area $A = 223.3$ in^2. The applied shear stress in the wood beam at the centerline of the girder support is

$$f_v = \frac{1.5V}{A} = \frac{(1.5)(17{,}100 \text{ lb})}{223.3 \text{ in}^2} = 114.8 \text{ psi}$$

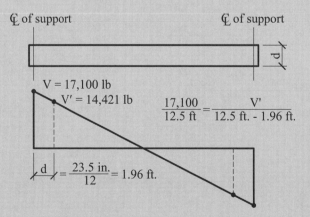

Note: "d" should be measured from the face of the support, but in this example, "d" is conservatively measured from the centerline of support since the bearing dimensions are not yet known

FIGURE 4.11 Girder shear force diagram.

The allowable shear stress is

$$F'_v = F_v C_D C_M C_t C_i = (170)(1)(1)(1)(1) = 170 \text{ psi}$$

Thus, $f_v < F'_v$; therefore, the beam is adequate in shear. See Figure 4.11.

Note: Since the shear stress f_v at the centerline of the girder support is adequate (i.e., $< F'_v$), there is no need to calculate the shear stress f'_v at a distance d from the face of the support since it is obvious that f'_v would be less than f_v and thus will be adequate in that case.

Check of Deflection (see Table 4.5)
The allowable pure bending modulus of elasticity is

$$E' = E C_M C_t C_i = (1.6 \times 10^6)(1.0)(1.0)(1.0) = 1.6 \times 10^6 \text{ psi}$$

The moment of inertia about the strong axis $I_{xx} = 10{,}270 \text{ in}^4$, the uniform dead load is $w_{DL} = 68.33 \text{ lb/in.}$, and the uniform live load is $w_{LL} = 45.63 \text{ lb/in.}$

TABLE 4.5 Girder Deflection Limit

Deflection	Deflection Limit
Live-load deflection Δ_{LL}	$\dfrac{L}{360} = \dfrac{(25 \text{ ft})(12)}{360} = 0.83 \text{ in.}$
Incremental long-term deflection due to dead load plus live load (including creep effects), $\Delta_{TL} = k\Delta_{DL} + \Delta_{LL}$	$\dfrac{L}{240} = \dfrac{(25 \text{ ft})(12)}{240} = 1.25 \text{ in.}$

The dead-load deflection is

$$\Delta_{DL} = \frac{5wL^4}{384E'_t} = \frac{(5)(68.33 \text{ lb/in.})(25 \text{ ft} \times 12)^4}{(384)(1.6 \times 10^6 \text{ psi})(10{,}270 \text{ in}^4)} = 0.44 \text{ in.}$$

The live-load deflection is

$$\Delta_{LL} = \frac{5w_{LL}L^4}{384E'_t} = \frac{(5)(45.63 \text{ lb/in.})(25 \text{ ft} \times 12)^4}{(384)(1.6 \times 10^6 \text{ psi})(10{,}270 \text{ in}^4)} = 0.29 \text{ in.} < \frac{L}{360} = 0.83 \text{ in.} \quad \text{OK}$$

Since seasoned wood in dry service conditions is assumed to be used in this building, the creep factor $k = 0.5$. The total incremental dead plus floor live load deflection is

$$\Delta_{TL} = k\Delta_{DL} + \Delta_{LL}$$
$$= (0.5)(0.44 \text{ in.}) + 0.29 \text{ in.} = 0.51 \text{ in.} < \frac{L}{240} = 1.25 \text{ in.} \quad \text{OK}$$

Alternatively, the required moment of inertia can be calculated using equations (4.5) and (4.6) as follows:

$$I_{\text{req'd}} = \frac{5w_{LL}L^3}{384E'}(360) \text{ in}^4 = \frac{(5)(45.63)(25 \text{ ft} \times 12)^3}{(384)(1.6 \times 10^6)}(360) = 3609 \text{ in}^4$$

Similarly, the required moment of inertia based on total load is obtained for a *uniformly loaded beam or girder* as

$$I_{\text{req'd}} = \frac{5w_{k(DL)+LL}L^3}{384E'}(240) \text{ in}^4 = \frac{(5)(0.5 \times 68.33 + 45.63)(25 \text{ ft} \times 12)^3}{(384)(1.6 \times 10^6)}(240) = 4208 \text{ in}^4$$

Therefore,

$$\text{the required moment of inertia} = 4208 \text{ in}^4 < I_{\text{actual}} = 10{,}270 \text{ in}^4 \quad \text{OK}$$

Check of Bearing Stress (Compression Stress Perpendicular to the Grain)
The maximum reaction at the support $R_1 = 17{,}094$ lb and the thickness of a 10×24 sawn-lumber joist $b = 9.5$ in. The allowable bearing stress or compression stress parallel to the grain is

$$F'_{c\perp} = F_{c\perp} C_M C_t C_b C_i = (625)(1.0)(1.0)(1.0)(1.0) = 625 \text{ psi}$$

From equation (4.7), the minimum required bearing length l_b is

$$l_{b,\text{req'd}} \geq \frac{R_1}{bF'_{c\perp}} = \frac{17{,}100 \text{ lb}}{(9.5 \text{ in.})(625 \text{ psi})} = 2.88 \text{ in.} \quad \text{say, 3 in.}$$

The girders will be connected to the column using a U-shaped column cap detail as shown below, and the minimum length of the column cap required is

$$(3 \text{ in.})(2) + \tfrac{1}{2}\text{-in. clearance between ends of girders} = 6.5 \text{ in.}$$

From a proprietary wood connector catalog, the maximum column cap width available for a U-shaped cap is 9.5 in. Since this is not less than the thickness (smaller dimension) of the girder, the proprietary U-shaped cap can be used. Therefore, use a U-shaped column cap similar to that shown in Figure 4.6d.

Note: The maximum column cap width available in most catalogs is 9.5 in., so where the width of a girder or beam is greater than this value, a special column cap having a minimum width equal to the thickness (smaller dimension) of the girder or beam would have to be designed.

Check of Assumed Girder Self-Weight
As discussed in Chapter 2, the density of wood is assumed in this book to be 31.2 lb/ft³, and this is used to calculate the actual weight of the girder selected.

Assumed girder self-weight = 90 lb/ft

Actual weight of the 10×24 girder selected (from NDS-S Table 1B):

$$\frac{223.3 \text{ in}^2 (31.2 \text{ lb/ft}^3)}{(12 \text{ in.})(12 \text{ in.})} = 49 \text{ lb/ft} \quad \text{which is less than the assumed value of 90 lb/ft} \quad \text{OK}$$

Note: The actual self-weight of the girder in this example is less than the assumed self-weight. However, let us consider a fictitious case where the actual self-weight is greater than the assumed value by, say, 50%. The self-weight calculated should then be used to recalculate the design loads, and the member should be redesigned. However, it should be noted that in many cases, the self-weight of the girder is usually a small percentage of the total load on the girder, and thus the total load is not affected significantly by even large positive variations of up to a maximum of 50% in the girder self-weight.

Assume in our fictitious example that we had selected a girder size that has a self-weight of 135 lb/ft, which is 50% heavier than the 90 lb/ft that was assumed initially in the load calculations. Therefore, the revised total load would become

$$w_{\text{TL}} = (D + L)(\text{tributary width}) + \text{girder self-weight}$$

$$= (40 \text{ psf} + 30 \text{ psf})(18.25 \text{ ft}) + 135 \text{ lb/ft} = 1413 \text{ lb/ft}$$

Compared to the total load of 1368 lb/ft calculated previously using the assumed self-weight of 90 lb/ft, the percentage increase in the total load and load effect for this fictitious example will be $(1413 - 1368)/1368 \times 100\% = 3\%$! This is quite a small and insignificant change for a 50% increase in girder self-weight, and thus its effect can be neglected. So we find that for this girder, even a 50% increase in the self-weight of the girder would not have had a significant effect on the total load and the load effects.

Use a 10×24 DF-L Select Structural for girder G1.

(a) If $f_b = M_{max}/S_{xx} \leq F'_b$, the beam is adequate in bending.
(b) If $f_b = M_{max}/S_{xx} > F'_b$, the beam is not adequate in bending.

7. To determine the maximum allowable load or moment that a beam or girder can support, based on bending stress alone, equate the allowable bending stress F'_b to M_{max}/S_{xx}, and solve for the allowable maximum moment M_{max} and thus the allowable maximum load. Therefore,

$$\text{the allowable maximum moment } M_{max} = F'_b S_{xx} \quad \text{in.-lb}$$
$$= \frac{F'_b S_{xx}}{12} \quad \text{ft-lb}$$

Knowing the allowable maximum moment, we can determine the allowable maximum load based on bending stress alone, using the relationship between maximum moment and the applied load. For example, for a *uniformly loaded beam or girder*, we obtain the applied maximum moment as

$$M_{max} = \frac{w_{TL} L^2}{8}$$

where L is the span of the beam or girder in feet. Solving and rearranging the equation above yields the maximum allowable total load:

$$w_{TL} = \frac{8 M_{max}}{L^2} = \frac{(8)(F'_b S_{xx}/12)}{L^2} \quad \text{lb/ft} \quad (4.11)$$

Knowing the allowable total load w_{TL} and the applied dead load w_{DL} on the member, the allowable live load w_{LL} based on bending stress alone can be obtained using the load combination relationships (see Section 2.1):

$$w_{TL} = w_{DL} + w_{LL}$$

Therefore,

$$\text{the allowable maximum live load } w_{LL} = w_{TL} - w_{DL}$$

The allowable maximum load can be converted to units of pounds per square foot by dividing the load (lb/ft) by the tributary width of the beam or girder in feet.

Check of Shear Stress

8. Determine the applied shear stress f_v or f'_v, if applicable.

$$f_v = \frac{1.5V}{A} \quad (4.12)$$

$$f'_v = \frac{1.5V'}{A} \quad (4.13)$$

where V is the maximum shear at the centerline of the joist, beam, or girder bearing support, and V' is the maximum distance d from the face of the support. Use V' to calculate the applied shear stress *only* for beams or girders with no concentrated loads within a distance d from the face of the support and where the bearing support is subjected only to confining compressive stresses. (i.e., use V' only for beams or girders subject to compression at the support reactions).

9. Determine all the applicable stress adjustment or C factors, obtain the tabulated NDS-S shear stress F_v, and calculate the allowable shear stress using equation (4.14):

$$F'_v = F_v C_D C_M C_t C_i \quad (4.14)$$

10. Compare the applied shear stress f_v or f'_v from step 8 to the allowable shear stress F'_v calculated in step 9.

(a) If f_v or $f'_v \le F'_v$, the beam is adequate in shear.
(b) If f_v or $f'_v > F'_v$, the beam is not adequate in shear.

11. To determine the maximum allowable load or shear that a beam or girder can support, based on shear stress alone, equate the allowable shear stress F'_v to $1.5V/A$ or $1.5V'/A$, and solve for the allowable shear V_{max} and thus the allowable maximum load. That is,

$$\frac{1.5V_{max}}{A_{xx}} = F'_v$$

Therefore,

$$\text{the allowable maximum shear } V_{max} \text{ or } V'_{max} = \frac{F'_v A_{xx}}{1.5} \quad \text{lb}$$

Knowing the allowable maximum shear at the centerline of the support, we can determine the allowable maximum load based on shear stress alone using the relationship between maximum shear and the applied load. However, it should be noted that shear rarely governs the design of joists, beams, and girders.

Check of Deflection

12. Using the appropriate deflection formulas, determine and compare the live load and total load deflections to the appropriate deflection limits. For a *uniformly loaded beam or girder,* we obtain

$$\text{live-load deflection } \Delta_{LL} = \frac{5w_{LL}L^4}{384E'I} \le \frac{L}{360}$$

$$\text{total load deflection (including creep) } \Delta_{k(DL)+LL} = \frac{5w_{k(DL)+LL}L^4}{384E'I} \le \frac{L}{240}$$

where $E' = EC_M C_t C_i$, the stress adjustment factors are as discussed previously, and C_i does not apply to glulam
k = creep effect
= 1.0 for green or unseasoned lumber or glulam used in wet service conditions
= 1.0 for seasoned lumber used in wet service conditions
= 0.5 for seasoned or dry lumber, glulam, and prefabricated I-joists used in dry service conditions
L = span of joist, beam, or girder, in.
$\Delta_{k(DL)+LL}$ = deflection due to total load $k(D) + L$ (or L_r or S), in.
Δ_{LL} = deflection due to live load L (or L_r or S), in.
$w_{k(DL)+LL}$ = total load including $k(D) + L$ (or L_r or S), lb/in.
w_{LL} = maximum live load L (or L_r or S), lb/in.

13. To determine the maximum allowable load that can be supported by a beam or girder based on deflections alone, equate the calculated deflection to the appropriate deflection limit and solve for the allowable maximum load. For example, for a *uniformly loaded beam or girder,* the live-load deflection is calculated as

$$\Delta_{LL} = \frac{5w_{LL}L^4}{384E'I} = \frac{L}{360}$$

Solving and rearranging the equation above yields the allowable maximum live load,

$$w_{LL} = \frac{384E'I}{5L^4}\left(\frac{L}{360}\right) \quad \text{lb/in.}$$

$$= \frac{384E'I}{5L^4}\left(\frac{L}{360}\right)(12) \quad \text{lb/ft} \tag{4.15}$$

Similarly, for a *uniformly loaded beam or girder*, the maximum allowable total load is given as

$$w_{k(DL)+LL} = \frac{384E'I}{5L^4}\left(\frac{L}{240}\right) \quad \text{lb/in.}$$

$$= \frac{384E'I}{5L^4}\left(\frac{L}{240}\right)(12) \quad \text{lb/ft} \qquad (4.16)$$

Knowing that $w_{k(DL)+LL} = w_{k(DL)} + w_{LL} = k(w_{DL}) + w_{LL}$, we can determine the allowable maximum live load w_{LL}. The smaller of this value and the value calculated from equation 4.15 will be the allowable maximum live load. This load can be converted to units of pounds per square foot by dividing this load (lb/ft) by the tributary width of the beam or girder in feet.

Check of Bearing Stress (Compression Stress Perpendicular to the Grain)

14. Determine the applied bearing stress or compression stress perpendicular to the grain and compare to the allowable compression stress perpendicular to the grain.

$$f_{c\perp} = \frac{R_1}{A_{bearing}} = \frac{R_1}{bl_b} \leq F'_{c\perp} \qquad (4.17)$$

where $F'_{c\perp}$ = allowable bearing stress perpendicular to the grain = $F_{c\perp}C_M C_t C_b C_i$
R_1 = maximum reaction at the support
$A_{bearing}$ = bearing area = (thickness of beam)(bearing length) = bL_b
b = thickness (smaller dimension) of the beam or girder
l_b = length of bearing at the beam or girder support
$F_{c\perp}$ = NDS-S tabulated design stress perpendicular to the grain

The minimum required bearing length l_b is obtained by rearranging equation (4.17). Thus,

$$l_{b,\,req'd} \geq \frac{R_1}{bF_{c\perp}} \qquad (4.18)$$

15. To determine the maximum allowable load (or reaction) that can be supported by a beam or girder, based on bearing stress alone, equate the applied bearing stress to the allowable bearing stress and solve for the reaction R_1:

$$\frac{R_1}{A_{bearing}} = F'_{c\perp}$$

Thus,

$$R_1 = A_{bearing} F'_{c\perp} = bl_b F'_{c\perp}$$

Knowing the allowable maximum reaction R_1, we can determine the allowable maximum load based on bearing stress alone using the relationship between the maximum reaction and the applied load. For example, for a *uniformly loaded beam or girder*, we obtain

$$R_{1,max} = \frac{w_{TL} L}{2}$$

where L is the span of the beam or girder in feet. Solving and rearranging the equation above yields the maximum allowable total load,

$$w_{TL} = \frac{2R_{1,max}}{L} = \frac{2A_{bearing} F'_{c\perp}}{L} \quad \text{lb/ft} \qquad (4.19)$$

$$= \frac{2bl_b F'_{c\perp}}{L} \quad \text{lb/ft}$$

Knowing the allowable total load w_{TL} and the applied dead load w_{DL} on the member, the allowable live load w_{LL} based on bearing stress alone can be obtained using the load combinations in Chapter 2:

$$w_{TL} = w_{DL} + w_{LL}$$

Therefore, the allowable maximum live load based on bearing stress alone,

$$w_{LL} = w_{TL} - w_{DL}$$

Note: The actual allowable maximum loads for the beam or girder will be the smallest of the maximum loads calculated for bending stress (step 7), shear stress (step 11), deflection (step 13), and bearing stress (step 15).

Design Examples

In this section we present design examples for glulam girders and the structural elements in stairs.

EXAMPLE 4.2

Analysis of a Wood Beam

Determine the adequacy of a 4 × 14 southern pine No. 1 sawn-lumber beam to support a concentrated moving load of 2000 lb that can be located anywhere on the beam. The beam span is 15 ft, normal temperature and dry service conditions apply, and the beam is laterally braced at the supports only. Assume a wood density of 36 lb/ft³ and a load duration factor C_D of 1.0. The available bearing length l_b is 3 in. Determine the maximum concentrated live load that can be supported by this beam. Assume the creep factor, $k = 1.0$.

Solution: Since the concentrated load is a moving load, the most critical load locations that will result in maximum moment, shear, and reaction in the beam have to be determined. For the maximum moment in this simply supported beam, the critical location of the crane load will be at the midspan of the beam, and for maximum shear the critical location of the concentrated moving load is within a distance d from the face of the support. For maximum reaction, the most critical location for the moving load is when the load is as close to the centerline of the beam support as possible.

Check of Bending Stress

1. Calculate all the loads and load effects, including maximum moment, shear, and reactions. Also calculate the dead and live loads that will be used for the deflection calculations later. Include the self-weight of the member.

   (a) Calculate the self-weight of the beam:

   $$w_{D,\text{beam}} = (36 \text{ lb/ft}^3)(46.38 \text{ in}^2/144) \equiv 12 \text{ lb/ft}$$

   (b) Calculate the maximum moment and shear. Since the concentrated load is a moving load, the maximum moment in the beam will occur when the moving load is at midspan, whereas the maximum shear and reaction will occur when the moving load is at or near the support (Figure 4.12). The maximum moment is calculated as

   $$M_{\max} = \frac{w_{D,\text{beam}} L^2}{8} + \frac{PL}{4} = \frac{(12 \text{ lb/ft})(15 \text{ ft})^2}{8} + \frac{(2000 \text{ lb})(15 \text{ ft})}{4}$$

   $$= 7838 \text{ ft-lb} = 94{,}050 \text{ in.-lb}$$

   The maximum shear at the centerline of the beam support when the concentrated moving load is assumed to be at or very near the face of the beam support (i.e., within a distance d from the face of the support) is calculated as

   $$V_{\max} = \frac{w_{D,\text{beam}} L}{2} + P = \frac{(12 \text{ lb/ft})(15 \text{ ft})}{2} + (2000 \text{ lb})\left(\frac{15 \text{ ft} - 1.23 \text{ ft}}{15 \text{ ft}}\right) = 1926 \text{ lb}$$

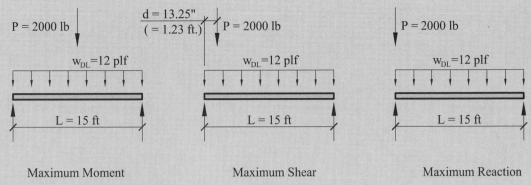

FIGURE 4.12 Free-body diagram of a beam for different loading conditions.

The maximum reaction will occur when the concentrated load is at or very near the centerline of the beam support and is calculated as

$$R_{max} = \frac{w_{D,beam}L}{2} + P = \frac{(12 \text{ lb/ft})(15 \text{ ft})}{2} + 2000 \text{ lb} = 2090 \text{ lb}$$

2. From the known stress grade (i.e., Select Structural, No. 1 & Better, No. 1, etc.) and size classification (i.e., dimension lumber, timbers, glulam, etc.), determine the applicable NDS-S table to be used in the design. For 4 × 14 southern pine No. 1 dimension lumber, use NDS-S Table 4B.

3. Determine the cross-sectional area A_{xx}, the section modulus S_{xx}, and the moment of inertia I_{xx} from NDS-S Table 1B (for sawn lumber), NDS-S Table 1C (for western species glulam), or NDS-S Table 1D (for southern pine glulam). From NDS-S Table 1B, we obtain the section properties for 4 × 14 sawn lumber:

$$A_{xx} = 46.38 \text{ in}^2 \ (b = 3.5 \text{ in. and } d = 13.25 \text{ in.})$$

$$I_{xx} = 678.5 \text{ in}^4$$

$$S_{xx} = 102.4 \text{ in}^4$$

From NDS-S Table 4B, we obtain the tabulated stress values:

$$F_b = 2850 \text{ psi}$$

$$F_v = 175 \text{ psi}$$

$$E = 1.8 \times 10^6 \text{ psi}$$

$$E_{min} = 0.66 \times 10^6 \text{ psi}$$

$$F_{c\perp} = 565 \text{ psi}$$

4. Determine the applied bending stress f_b from equation (4.10):

$$f_b = \frac{M_{max}}{S_{xx}} = \frac{94,050 \text{ in.-lb}}{102.4 \text{ in}^3} = 918.5 \text{ psi}$$

5. Determine all the applicable stress adjustment or C factors, obtain the tabulated NDS-S bending stress F_b, and calculate the allowable bending stress using equation (4.2). The stress adjustment or C factors from NDS-S Table 4B are given in Table 4.6. The allowable bending stress of the beam if the beam stability coefficient C_L is 1.0 is given as

$$F_b^* = F_b C_D C_M C_t C_F C_r C_c C_{fu} C_i = (2850)(1.0)(1.0)(1.0)(0.9)(1.0)(1.0)(1.0) = \mathbf{2565 \text{ psi}}$$

The allowable pure bending modulus of elasticity E' and the bending stability modulus of elasticity E'_{min} are calculated as

$$E' = E C_M C_t C_i = (1.8 \times 10^6)(1.0)(1.0)(1.0) = 1.8 \times 10^6 \text{ psi}$$

$$E'_{min} = E_{min} C_M C_t C_i C_T = (0.66 \times 10^6)(1.0)(1.0)(1.0)(1.0) = 0.66 \times 10^6 \text{ psi}$$

TABLE 4.6 Beam Adjustment Factors

Adjustment Factor	Symbol	Value	Rationale for the Value Chosen
Beam stability factor	C_L	0.746	See calculations below
Size factor	$C_F(F_b)$	0.90	See adjustment factors section of NDS-S Table 4B
Moisture or wet service factor	C_M	1.0	Equilibrium moisture content is $\leq 19\%$
Load duration factor	C_D	1.0	The largest C_D value in the load combination of dead load plus floor live load (i.e., $D + L$)
Temperature factor	C_t	1.0	Insulated building, therefore, normal temperature conditions apply
Repetitive member factor	C_r	1.0	Does not satisfy all three requirements needed for a member to be classified as repetitive
Incision factor	C_r	1.0	1.0 for wood that is not incised, even if pressure treated
Buckling stiffness factor	C_T	1.0	1.0 except for a 2 × 4 truss top chords with plywood sheathing attached
Bearing area factor	C_b	1.0	$C_b = \dfrac{l_b + 0.375}{l_b}$ for $l_b \leq 6$ in. $= 1.0$ for $l_b > 6$ in. $= 1.0$ for bearings at the ends of a member
Flat use factor	C_{fu}	1.0	
Incising factor	C_i	1.0	

Calculation of the Beam Stability Factor C_L

From equation (3.2) the beam stability factor is calculated as follows. The unsupported length of the compression edge of the beam or distance between points of lateral support preventing rotation and/or lateral displacement of the compression edge of the beam is

$$l_u = 15 \text{ ft} = 180 \text{ in.}$$

$$\frac{l_u}{d} = \frac{180 \text{ in.}}{13.25 \text{ in.}} = 13.6$$

Using the l_u/d value, the effective length of the beam of the beam is obtained from Table 3.9 (or NDS code Table 3.3.3) as

$$l_e = 1.63 l_u + 3d = (1.63)(180 \text{ in.}) + (3)(13.25 \text{ in.}) = 333 \text{ in.}$$

$$R_B = \sqrt{\frac{l_e d}{b^2}} = \sqrt{\frac{(333)(13.25 \text{ in.})}{(3.5 \text{ in.})^2}} = 19 < 50 \quad \text{OK}$$

$$F_{bE} = \frac{1.20 E'_{\min}}{R_B^2} = \frac{(1.20)(0.66 \times 10^6)}{(19)^2} = 2194 \text{ psi}$$

$$\frac{F_{bE}}{F_b^*} = \frac{2194 \text{ psi}}{2565 \text{ psi}} = 0.855$$

From equation (3.2) the beam stability factor is calculated as

$$C_L = \frac{1 + F_{bE}/F_b^*}{1.9} - \sqrt{\left(\frac{1 + F_{bE}/F_b^*}{1.9}\right)^2 - \frac{F_{bE}/F_b^*}{0.95}}$$

$$= \frac{1 + 0.855}{1.9} - \sqrt{\left(\frac{1 + 0.855}{1.9}\right)^2 - \frac{0.855}{0.95}} = 0.746$$

The allowable bending stress in the beam is calculated as

$$F'_b = F_b C_D C_M C_t C_L C_F C_r = F_b^* C_L = (2565)(0.746) = \mathbf{1913.5 \text{ psi}}$$

6. Compare the applied stress f_b from step 4 to the allowable bending stress F'_b calculated in step 5. The applied bending stress (see step 4) $f_b = 918.5$ psi $< F'_b = 1913.5$ psi. Therefore, the beam is adequate in bending.

7. To determine the maximum allowable load or moment the beam or girder can support based on bending stress alone, equate the allowable bending stress F'_b to M_{max}/S_{xx}, and solve for the allowable maximum moment M_{max} and thus the allowable maximum load. The applied maximum moment is given as

$$M_{max} = \frac{w_{D,\,beam}L^2}{8} + \frac{PL}{4}$$

The allowable maximum moment is given as

$$M_{max} = F'_b S_{xx} \quad \text{in.-lb}$$

$$= \frac{F'_b S_{xx}}{12} \text{ ft-lb}$$

Equating the M_{max} equations and solving for the maximum allowable concentrated load yields

$$\frac{w_{D,\,beam}L^2}{8} + \frac{PL}{4} = \frac{F'_b S_{xx}}{12} \quad \text{ft-lb}$$

Therefore,

$$P_{LL} = \left(\frac{4}{L}\right)\left(\frac{F'_b S_{xx}}{12} - \frac{w_{D,\,beam}L^2}{8}\right) = \left(\frac{4}{15\text{ ft}}\right)\left[\frac{(1913.5\text{ psi})(102.4\text{ in}^3)}{12} - \frac{(12\text{ lb/ft})(15\text{ ft})^2}{8}\right] \quad \text{lb}$$

$$= \mathbf{4264\text{ lb}}$$

Since the concentrated load is a live load, the maximum allowable moving load P_{LL} (based on bending stress alone) is 4264 lb. This assumes that the only dead load on the beam is the uniform self-weight of the beam, as specified in the problem statement.

Check of Shear Stress

8. Determine the applied shear stress f_v or f'_v, if applicable. In this problem, since the concentrated load could be located within a distance d from the face of the beam, only the shear stress f_v calculated using the shear at the centerline of the beam support V is applicable. Therefore, from equation (4.12) we obtain the applied shear stress as

$$f_v = \frac{1.5V}{A} = \frac{(1.5)(1926)}{46.38} = 62.3 \text{ psi}$$

9. Determine all the applicable stress adjustment or C factors, obtain the tabulated NDS-S shear stress F_v, and calculate the allowable shear stress using Table 3.1. The allowable shear stress from equation (4.14) is

$$F'_v = F_v C_D C_M C_t C_i = (175\text{ psi})(1.0)(1.0)(1.0)(1.0) = 175 \text{ psi}$$

10. Compare the applied shear stress f_v or f'_v from step 8 to the allowable shear stress F'_v calculated in step 9. Since $f_v = 62.3$ psi $< F'_v = 175$ psi, the beam is adequate in shear.

11. To determine the maximum allowable load or shear that can be supported by the beam or girder based on shear stress alone, equate the allowable shear stress F'_v to $1.5V/A$ or $1.5V'/A$, and solve for the allowable shear V_{max} and thus the allowable maximum load. That is,

$$\frac{1.5V_{max}}{A_{xx}} = F'_v$$

The allowable maximum shear

$$V_{max} = \frac{F'_v A_{xx}}{1.5} \quad \text{lb}$$

Knowing the allowable maximum shear at the centerline of the support, we can determine the allowable maximum load based on shear stress alone, using the relationship between maximum shear and the applied load, while recalling that the concentrated live load is a moving load. For this case, the maximum shear due to applied loads was obtained in step 2 as

$$V_{max} = \frac{w_{D,\,beam}L}{2} + P\left(\frac{15\text{ ft} - 1.23\text{ ft}}{15\text{ ft}}\right) = \frac{F'_v A_{xx}}{1.5}$$

Thus, the maximum allowable total concentrated load (based on shear stress alone) is

$$0.918 P_{LL} = \frac{F'_v A_{xx}}{1.5} - \frac{w_{D,\,beam}L}{2} \quad \text{lb}$$

$$= \frac{(175\text{ psi})(46.38\text{ in}^2)}{1.5} - \frac{(12\text{ lb/ft})(15\text{ ft})}{2}$$

$$P_{LL} = 5796 \text{ lb}$$

Since this is greater than the value obtained from step 7, this value will not govern because the beam will fail first in bending at the lower load $P_{LL} = 4264$ lb before the concentrated load reaches the value of 5796 lb. As mentioned previously, shear rarely governs the design of wood joists, beams, and girders.

Check of Deflection

12. Using the appropriate deflection formula for a beam with a uniformly distributed dead load (due to self-weight) plus a concentrated crane (live) load, determine and compare the live load and total load deflections to the appropriate deflection limits. Since the maximum deflection will take place at midspan when the concentrated live load is at the midspan, we obtain the deflections as

$$\text{live-load deflection } \Delta_{LL} = \frac{P_{LL}L^3}{48E'I} = \frac{(2000\text{ lb})(15\text{ ft} \times 12\text{ in./ft})^3}{(48)(1.8 \times 10^6\text{ psi})(678.5\text{ in}^4)}$$

$$= 0.20 \text{ in.}$$

$$< \frac{L}{360} = 0.50 \text{ in.} \quad \text{OK}$$

Total load deflection (including creep)

$$\Delta_{k(DL)+LL} = \frac{5w_{k(DL)}L^4}{384E'I} + \frac{P_{LL}L^3}{48E'I}$$

$$= \frac{(5)(12/12\text{ lb/in})(15\text{ ft} \times 12\text{ in./ft})^4}{(384)(1.8 \times 10^6\text{ psi})(678.5\text{ in}^4)} + \frac{(2000\text{ lb})(15\text{ ft} \times 12\text{ in./ft})^3}{(48)(1.8 \times 10^6\text{ psi})(678.5\text{ in}^4)}$$

$$= 0.21 \text{ in.}$$

$$\leq \frac{L}{240} = \frac{(15\text{ ft})(12\text{ in./ft})}{240} = 0.75 \text{ in.} \quad \text{OK}$$

13. To determine the maximum allowable load that can be supported by the beam or girder based on deflections alone, equate the calculated deflection to the appropriate deflection limit and solve for the allowable maximum load. That is,

$$\Delta_{LL} = \frac{P_{LL}L^3}{48E'I} = \frac{L}{360}$$

Therefore, the allowable maximum live load

$$P_{LL} = \frac{48E'I}{L^3}\left(\frac{L}{360}\right) = \frac{(48)(1.8 \times 10^6)(678.5 \text{ in}^4)}{(15 \text{ ft} \times 12 \text{ in.}/\text{ft})^3}\left(\frac{15 \text{ ft} \times 12 \text{ in.}/\text{ft}}{360}\right) = 5026 \text{ lb}$$

The maximum total load deflection

$$\Delta_{k(DL)+LL} = \frac{5w_{k(DL)}L^4}{384E'I} + \frac{P_{LL}L^3}{48E'I} = \frac{L}{240}$$

Therefore, the allowable maximum live load

$$P_{LL} = \frac{48E'I}{L^3}\left(\frac{L}{240} - \frac{5w_{k(DL)}L^4}{384E'I}\right)$$

$$= \frac{(48)(1.8 \times 10^6)(678.5 \text{ in}^4)}{(15 \text{ ft} \times 12 \text{ in.}/\text{ft})^3}\left[\frac{(15 \text{ ft} \times 12 \text{ in.}/\text{ft})}{240} - \frac{(5)(12/12 \text{ lb}/\text{in})(15 \text{ ft} \times 12 \text{ in.}/\text{ft})^4}{(384)(1.8 \times 10^6)(678.5 \text{ in}^4)}\right]$$

$$= 7425 \text{ lb}$$

The smaller P_{LL} value of 5026 lb governs as far as deflections are concerned. However, this value is still less than the allowable live load calculated in step 7 based on bending stress, so the value of 4264 lb (calculated in step 7) remains the governing allowable maximum live load for this beam.

Check of Bearing Stress (Compression Stress Perpendicular to the Grain)

14. Determine the bearing stress or compression stress applied perpendicular to the grain and compare to the allowable compression stress perpendicular to the grain. From equation (4.17) the bearing stress applied is

$$f_{c\perp} = \frac{R_1}{A_{\text{bearing}}} = \frac{R_1}{bl_b} = \frac{2090 \text{ lb}}{(3.5 \text{ in.})(3 \text{ in.})} = 199 \text{ psi}$$

where $b = 3.5$ in. and l_b (given in the problem statement) $= 3$ in. The allowable bearing stress or compression stress perpendicular to the grain is

$$F'_{c\perp} = F_{c\perp}C_MC_tC_bC_i = (565 \text{ psi})(1.0)(1.0)(1.0)(1.0) = 565 \text{ psi} > f_{c\perp} = 199 \text{ psi} \quad \text{OK}$$

Similar to shear, the beam reaction rarely controls the design strength of a joist or member, and as such will not be checked. The actual allowable maximum loads for the beam or girder will be the smallest of the maximum loads calculated for bending stress (step 7), shear stress (step 11), and deflection (step 13). Therefore, the allowable maximum live load

$$P_{LL} \text{ (from step 7)} = \textbf{4624 lb}$$

Thus, the bending stress criterion controls the allowable maximum live load in this beam.

A 4 × 14 southern pine No. 1 is adequate; the maximum live load $P_{LL}(\text{max}) = 4624$ lb.

EXAMPLE 4.3

Design of a Glulam Girder

The panelized roof framing system shown in Figure 4.13 consists of 2× subpurlins supported on 4× purlins that are supported on simply supported glulam girders. The total roof dead load, including the weight of the subpurlins and purlins, is 20 psf and the snow load is 35 psf. The wind load on the roof is ±10 psf. Design the glulam girder assuming wood density of 36 lb/ft^3, dry service and normal temperature conditions.

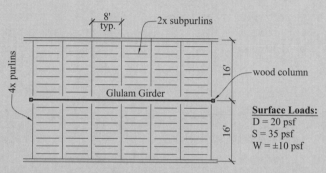

FIGURE 4.13 Framing plan.

Solution: The total roof dead load D is 20 psf (excludes the self-weight of girder). Assume that the girder self-weight is 100 lb/ft (this will be checked later). The snow load S is 35 psf and the wind load W is ±10 psf. All other loads, E, L, R, H, F, and $T = 0$.

$$\text{Tributary width of glulam girder} = \frac{16 \text{ ft}}{2} + \frac{16 \text{ ft}}{2} = 16$$

$$\text{Tributary area of glulam girder} = (16 \text{ ft})(48 \text{ ft}) = 768 \text{ ft}^2 > 600 \text{ ft}^2$$

From Section 2.4, $R_1 = 0.6$. Since we have a flat roof with minimum slope ($\frac{1}{4}$ in./ft) for drainage, the roof slope factor (see Section 2.4) $F = 0.25$; therefore, $R_2 = 1.0$. From equation (2.4), we calculate the roof live load as

TABLE 4.7 Applicable and Governing Load Combinations

Load Combination	Value of Load Combination, w	C_D Factor for Load Combination	Normalized Load, w/C_D
D	20 psf	0.9	$\frac{20}{0.9} = 22.2$
$D + L_r$	20 + 12 = 32 psf	1.25	$\frac{32}{1.25} = 25.6$
$D + S$	20 + 35 = **55 psf**	1.15	$\frac{55}{1.15} = 47.8 \Leftarrow$ (governs)
$D + 0.75W + 0.75L_r$	20 + (0.75)(10 + 12) = 36.5 psf	1.6	$\frac{36.5}{1.6} = 22.8$
$D + 0.75W + 0.75S$	20 + (0.75)(10 + 35) = 53.8 psf	1.6	$\frac{53.8}{1.6} = 33.6$
$0.6D + W$	$(0.6)(20) + (-5)^a = 7$ psf	1.6	$\frac{7}{1.6} = 4.4$

[a] In load combinations 7 and 8 given in Section 2.1, the wind load W and seismic load E are always opposed by the dead load D. Therefore, in these combinations, D takes on a positive number while W and E take on negative values only.

$$L_r = 20R_1R_2 = (20)(1.0)(0.6) = 12 \text{ psf} < 20 \text{ psf}$$

Therefore, from Section 2.1, the governing load combination for this roof girder is determined using the normalized load method. Since the loads E, L, R, H, F, and T are zero, the applicable load combinations are given in Table 4.7.

Summary: The governing load combination is $D + S$ with a load duration factor C_D of 1.15.

We assume that this girder is subjected to uniform loads because there are at least five equally spaced reaction points where the purlins frame into the girder with equal reactions, indicating that the load on the girder is actually close to being uniform. The total uniform load on the girder that will be used to design for bending, shear, and bearing is obtained below using the dead load, the snow load, and the assumed self-weight of the girder, which will be checked later.

$$w_{TL} = (D + S) \text{ (tributary width)} + \text{girder self-weight}$$
$$= (20 \text{ psf} + 35 \text{ psf})(16 \text{ ft}) + 100 \text{ lb/ft} = \mathbf{980 \text{ lb/ft}}$$

Using the free-body diagram of the girder (Figure 4.14), the load effects are calculated as follows:

$$\text{maximum shear } V_{max} = \frac{w_{TL}L}{2} = \frac{(980 \text{ lb/ft})(48 \text{ ft})}{2} = 23{,}520 \text{ lb}$$

maximum reaction $R_{max} = 23{,}520$ lb

$$\text{maximum moment } M_{max} = \frac{w_{TL}L^2}{8} = \frac{(980 \text{ lb/ft})(48 \text{ ft})^2}{8} = 282{,}240 \text{ ft-lb}$$

$$= 3{,}386{,}880 \text{ in.-lb}$$

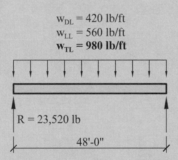

FIGURE 4.14 Free-body diagram of a glulam girder.

The following loads will be used for calculating the girder deflections:

uniform dead load $w_{DL} = (20 \text{ psf})(16 \text{ ft}) + 100 \text{ lb/ft} = 420 \text{ lb/ft} = 35 \text{ lb/in.}$

uniform live load $w_{LL} = (35 \text{ psf})(16 \text{ ft}) = 560 \text{ lb/ft} = 46.7 \text{ lb/in.}$

We now proceed with the design following the steps described earlier in the chapter.

Check of Bending Stress (Girders)

7. Summary of load effects (the self-weight of girder G1 was assumed, but this will be checked later in the design process):

    maximum shear $V_{max} = 23{,}520$ lb

    maximum reaction $R_{max} = 23{,}520$ lb

    maximum moment $M_{max} = 3{,}386{,}880$ in.-lb

    uniform dead load is $w_{DL} = 35$ lb/in.

    uniform live load is $w_{LL} = 46.7$ lb/in.

8. For glulam used primarily in bending, use NDS-S Table 5A–Expanded.
9. Using NDS-S Table 5A–Expanded, assume a 24F–V4 DF/DF glulam; therefore, $F_{bx}^+ = 2400$ psi. (tension lamination stressed in tension). Assume initially that $F'^+_{bx} = F^+_{bx} = 2400$ psi. From equation (4.1), the required approximate section modulus of the member is given as

$$S_{xx, \text{req'd}} \geq \frac{M_{\max}}{F_{bx}^+} = \frac{3{,}386{,}880 \text{ in.-lb}}{2400 \text{ psi}} = 1411.2 \text{ in}^3$$

10. From NDS-S Table 1C (for western species glulam), the trial size with the least area that satisfies the section modulus requirement of step 3 is $6\frac{3}{4}$ in. $\times$ 36 in.:

$$b = 6.75 \text{ in.} \quad \text{and} \quad d = 36 \text{ in.}$$

$$S_{xx} \text{ provided} = 1458 \text{ in}^3 > 1411.2 \text{ in}^3 \quad \text{OK}$$

$$\text{area } A \text{ provided} = 243 \text{ in}^2$$

$$I_{xx} \text{ provided} = 26{,}240 \text{ in}^4$$

11. The NDS-S (Table 5A–Expanded) tabulated stresses are

$$F_{bx}^+ = 2400 \text{ psi} \quad \text{(tension lamination stressed in tension)}$$

$$F_{v,xx} = 265 \text{ psi}$$

$$F_{c\perp xx, \text{ tension lam}} = 650 \text{ psi}$$

$$E_x = 1.8 \times 10^6 \text{ psi}$$

$$E_y = 1.6 \times 10^6 \text{ psi}$$

$$E_{y, \min} = 0.83 \times 10^6 \text{ psi} \quad (E_{y, \min}, \text{ and not } E_{x, \min}, \text{ is used for lateral buckling of the girder about the weak axis})$$

The adjustment or C factors are given in Table 4.8.

TABLE 4.8 Adjustment Factors for Glulam

Adjustment Factor	Symbol	Value	Rationale for the Value Chosen
Beam stability factor	C_L	0.97	See calculation below
Volume factor	C_V	0.802	See calculation below
Curvature factor	C_c	1.0	Glulam girder is straight
Flat use factor	C_{fu}	1.0	Glulam is bending about its strong x–x axis
Moisture or wet service factor	C_M	1.0	Equilibrium moisture content is $< 16\%$
Load duration factor	C_D	1.15	The largest C_D value in the load combination of dead load plus snow load (i.e., $D + S$)
Temperature factor	C_t	1.0	Insulated building; therefore, normal temperature conditions apply
Repetitive member factor	C_r	—	Does not apply to glulams; neglect or enter a value of 1.0
Bearing area factor	C_b	1.0	$C_b = \dfrac{l_b + 0.375}{l_b}$ for $l_b \leq 6$ in. $= 1.0$ for $l_b > 6$ in. $= 1.0$ for bearings at the ends of a member

From the adjustment factor applicability table for glulam (Table 3.2), we obtain the allowable bending stress of the glulam girder with C_V and C_L equal to 1.0 as

$$F_{bx}^*+ = F_{bx}^+ C_D C_M C_t C_{fu} C_c C_{fu} = (2400)(1.15)(1.0)(1.0)(1.0)(1.0) = \mathbf{2760\ psi}$$

The allowable pure bending modulus of elasticity E'_x and the bending stability modulus of elasticity $E'_{y,\min}$ are calculated as

$$E'_x = E_x C_M C_t = (1.8 \times 10^6)(1.0)(1.0) = 1.8 \times 10^6\ psi$$

$$E'_{y,\min} = E_{y,\min} C_M C_t = (0.83 \times 10^6)(1.0)(1.0) = 0.83 \times 10^6\ psi$$

Calculating the Beam Stability Factor C_L

From equation (3.2) the beam stability factor is calculated as follows: The unsupported length of the compression edge of the beam or distance between points of lateral support preventing rotation and/or lateral displacement of the compression edge of the beam is

$$l_u = 8\ ft = 96\ in.\ (i.e., the\ distance\ between\ lateral\ supports\ provided\ by\ the\ purlins)$$

$$\frac{l_u}{d} = \frac{96\ in.}{36\ in.} = 2.67$$

We previously assumed a uniformly loaded girder; therefore, using the l_u/d value, the effective length of the beam of the beam is obtained from Table 3.9 (or NDS code Table 3.3.3) as

$$l_e = 2.06 l_u = (2.06)(96\ in.) = 198\ in.$$

$$R_B = \sqrt{\frac{l_e d}{b^2}} = \sqrt{\frac{(198)(36\ in.)}{(6.75\ in.)^2}} = 12.5 \leq 50 \quad OK$$

$$F_{bE} = \frac{1.20 E'_{\min}}{R_B^2} = \frac{(1.20)(0.83 \times 10^6)}{(12.5)^2} = 6374\ psi$$

$$\frac{F_{bE}}{F_b^*} = \frac{6374\ psi}{2760\ psi} = 2.31$$

From equation (3.2) the beam stability factor is calculated as

$$C_L = \frac{1 + F_{bE}/F_b^*}{1.9} - \sqrt{\left(\frac{1 + F_{bE}/F_b^*}{1.9}\right)^2 - \frac{F_{bE}/F_b^*}{0.95}}$$

$$= \frac{1 + 2.31}{1.9} - \sqrt{\left(\frac{1 + 2.31}{1.9}\right)^2 - \frac{2.31}{0.95}} = 0.97$$

Calculating the Volume Factor C_V

L = length of beam in feet between adjacent points of zero moment = 48 ft

d = depth of beam = 36 in.

b = width of beam, in. ($\leq$ 10.75 in.) = 6.75 in.

x = 10 (for western species glulam)

From equation (3.3),

$$C_V = \left(\frac{21}{L}\right)^{1/x}\left(\frac{12}{d}\right)^{1/x}\left(\frac{5.125}{b}\right)^{1/x} = \left(\frac{1291.5}{bdL}\right)^{1/x} = \left[\frac{1291.5}{(6.75\ in.)(36\ in.)(48\ ft)}\right]^{1/10} = 0.802 \leq 1.0$$

The smaller of C_V and C_L will govern and is used in the allowable bending stress calculation. Since $C_V = 0.802 < C_L = 0.97$, use $C_V = 0.802$. Using Table 3.2 (i.e., adjustment factor applicability table for glulam), we obtain the allowable bending stress as

$$F_{bx}'^+ = F_{bx}^+ C_D C_M C_t C_F C_r \, (C_L \text{ or } C_V) = F_b^*(C_L \text{ or } C_V) = (2760)(0.802) = \mathbf{2213.5 \text{ psi}}$$

12. Using equation (4.10), the actual bending stress applied is

$$f_b = \frac{M_{\max}}{S_{xx}} = \frac{3{,}386{,}880 \text{ lb-in.}}{1458 \text{ in}^3} = 2322 \text{ psi}$$

$$> \text{allowable bending stress } F_b' = 2213.5 \text{ psi} \quad \textbf{not good!}$$

Although this section is only about 5% overstressed, it is prudent to select a larger section that will yield an applied stress that is slightly lower than the allowable stress. To do this we will need to recalculate C_V, C_L, the bending stress applied, and the allowable bending stress.

Try a $6\frac{3}{4} \times 37\frac{1}{2}$ in. 24F–V4 DF/DF glulam girder ($A_{xx} = 253.1$ in^2; $S_{xx} = 1582$ in^3; $I_{xx} = 29{,}660$ in^4), and the reader should verify that the recalculated values of C_V, C_L, and F_b' are as follows:

$$C_V = 0.799$$

$$C_L = 0.963$$

$$F_{bx}'^+ = 2205 \text{ psi}$$

$$f_b = 2141 \text{ psi} < F_{bx}'^+ \quad \text{OK}$$

Therefore, a $6\frac{3}{4} \times 37\frac{1}{2}$ in. 24F–V4 DF/DF glulam girder is adequate in bending.

Check of Shear Stress

$V_{\max} = 23{,}520$ lb. The beam cross-sectional area $A = 253.1$ in^2. The shear stress applied in the wood beam at the centerline of the girder support is

$$f_v = \frac{1.5V}{A} = \frac{(1.5)(23{,}520 \text{ lb})}{253.1 \text{ in}^2} = 139.4 \text{ psi}$$

Using the adjustment factor applicability table for glulam from Table 3.2, we obtain the allowable shear stress as

$$F_v' = F_v C_D C_M C_t = (265)(1)(1)(1) = 265 \text{ psi}$$

Thus, $f_v < F_v'$; therefore, the girder is adequate in shear. See Figure 4.15.

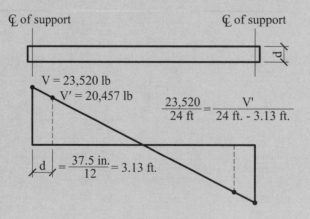

Note: "d" should be measured from the face of the support, but in this example, "d" is conservatively measured from the centerline of support since the bearing dimensions are not yet known

FIGURE 4.15 Glulam girder shear force diagram.

Note: Since the shear stress f_v at the centerline of the girder support is adequate (i.e., $< F_v'$), there is no need to further calculate the shear stress f_v' at a distance d from the face of the support since it is obvious that f_v' would be less than f_v and thus will also be adequate.

Check of Deflection (see Table 4.9)
The allowable pure bending modulus of elasticity for strong (i.e., x–x)-axis bending of the girder was calculated previously:

$$E_x' = E_x C_M C_t = (1.8 \times 10^6)(1.0)(1.0) = 1.8 \times 10^6 \text{ psi}$$

The moment of inertia about the strong axis $I_{xx} = 29{,}660 \text{ in}^4$.

$$\text{uniform dead load } w_{DL} = 35 \text{ lb/in.}$$

$$\text{uniform live load } w_{DL} = 46.7 \text{ lb/in.}$$

TABLE 4.9 Glulam Girder Deflection Limits

Deflection	Deflection Limit
Live-load deflection Δ_{LL}	$\dfrac{L}{360} = \dfrac{(48 \text{ ft})(12)}{360} = 1.6 \text{ in.}$
Incremental long-term deflection due to dead load plus live load (including creep effects), $\Delta_{TL} = k\Delta_{DL} + \Delta_{LL}$	$\dfrac{L}{240} = \dfrac{(48 \text{ ft})(12)}{240} = 2.4 \text{ in.}$

The dead-load deflection is

$$\Delta_{DL} = \frac{5 w_{DL} L^4}{384 EI} = \frac{(5)(35 \text{ lb/in.})(48 \text{ ft} \times 12)^4}{(384)(1.8 \times 10^6 \text{ psi})(29{,}660 \text{ in}^4)} = 0.94 \text{ in.}$$

The live-load deflection is

$$\Delta_{LL} = \frac{5 w_{LL} L^4}{384 EI} = \frac{(5)(46.7 \text{ lb/in.})(48 \text{ ft} \times 12)^4}{(384)(1.8 \times 10^6 \text{ psi})(29{,}660 \text{ in}^4)} = \mathbf{1.25 \text{ in.}} < \frac{L}{360} = 1.6 \text{ in.} \quad \text{OK}$$

Since seasoned wood in dry service conditions is assumed to be used in this building, the creep factor $k = 0.5$ (Table 4.1). The total incremental dead plus floor live load deflection is

$$\Delta_{TL} = k\Delta_{DL} + \Delta_{LL}$$

$$= (0.5)(0.94 \text{ in.}) + 1.25 \text{ in.} = \mathbf{1.72 \text{ in.}} < \frac{L}{240} = 2.4 \text{ in.} \quad \text{OK}$$

Note: Although the deflections calculated are within the IBC deflection limits, the total load deflection and the live-load deflection are on the high side (> 1 in.), and therefore the beam should be cambered. The camber provided for glulam roof beams should be approximately $1.5 \times$ the dead-load deflection (Section 4.1). With the camber provided, the net downward total deflection becomes

$$1.72 \text{ in.} - (1.5)(0.94 \text{ in.}) = 0.31 \text{ in.} = \frac{L}{1858} << \frac{L}{240} \quad \text{OK}$$

Check of Bearing Stress (Compression Stress Perpendicular to the Grain)

Maximum reaction at the support $R_1 = 23{,}520$ lb

Width of **$6\frac{3}{4} \times 37\frac{1}{2}$ in.** glulam girder $b = 6.75$ in.

The allowable bearing stress or compression stress parallel to the grain is

$$F'_{c\perp} = F_{c\perp} C_M C_t C_b = (650)(1)(1)(1) = 650 \text{ psi}$$

From equation (4.9), the minimum required bearing length l_b is

$$l_{b,\,\text{req'd}} \geq \frac{R_1}{bF'_{c\perp}} = \frac{23{,}520 \text{ lb}}{(6.75 \text{ in.})(650 \text{ psi})} = 5.36 \text{ in., say } 5.5 \text{ in.}$$

The girders will be connected to the column using a U-shaped column cap detail similar to that shown in Figure 4.6d, and the minimum length of column cap required is

$$(5.5 \text{ in.})(2) + \tfrac{1}{2}\text{-in. clearance between ends of girders} = 11.5 \text{ in.}$$

Check of Assumed Self-Weight of Girder

A wood density of 36 lb/ft^3 specified in this example is used to calculate the actual weight of the girder selected.

Assumed girder self-weight = 100 lb/ft

Actual weight of a $6\frac{3}{4} \times 37\frac{1}{2}$ in. glulam girder selected from NDS-S Table 1C

$$= \frac{(253.1 \text{ in}^2)(36 \text{ lb/ft}^3)}{(12 \text{ in.})(12 \text{ in.})}$$

$$= 63 \text{ lb/ft} \quad \text{which is less than the assumed value of 100 lb/ft} \quad \text{OK}$$

Note: Even though the assumed self-weight of the girder differs from the actual self-weight of the girder selected, this difference does not affect the results significantly, as demonstrated below, and thus can be neglected. The total load calculated previously using the assumed self-weight of 100 lb/ft was

$$w_{\text{TL}} = (D + S)(\text{tributary width}) + \text{girder self-weight}$$

$$= (20 \text{ psf} + 35 \text{ psf})(16 \text{ ft}) + 100 \text{ lb/ft} = 980 \text{ lb/ft}$$

The corresponding total load using the actual self-weight of the girder selected is

$$w_{\text{TL}} = (D + S)(\text{tributary width}) + \text{girder self-weight}$$

$$= (20 \text{ psf} + 35 \text{ psf})(16 \text{ ft}) + 63 \text{ lb/ft} = 943 \text{ lb/ft}$$

The difference between the assumed and actual self-weight as a percentage of the actual self-weight is

$$\frac{100 - 63}{63} \times 100\% = 59\%$$

The difference between the total load calculated using the assumed self-weight and the actual total load w_{TL} is

$$\frac{980 - 943}{943} \times 100\% = 4\%!$$

Thus, a 59% error in the self-weight of the girder yields only a 4% error in the total load, which is insignificant and can therefore be neglected, especially since we err on the safe side (i.e., the assumed self-weight was greater than the actual self-weight).

Use a $6\frac{3}{4} \times 37\frac{1}{2}$ in. 24F–V4 DF/DF glulam girder.

EXAMPLE 4.4

Design of Stair Treads and Stringers

For the typical floor plan shown in Figure 4.16 the second-floor dead load is 20 psf and the live load is 40 psf. Assuming hem-fir wood species, design the stair treads and stair stringers (Figure 4.17).

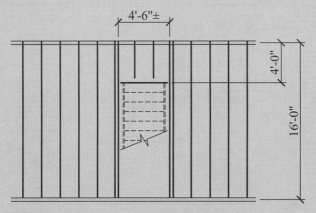

FIGURE 4.16 Second-floor framing plan.

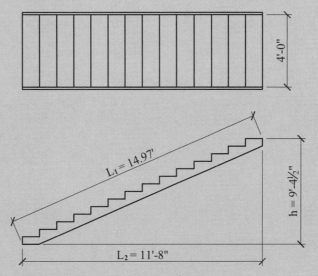

FIGURE 4.17 Plan and elevation of stair.

Solution: Some IBC requirement for stairs (IBC 1009) include the following:

Stair risers must be $\geq$ 4 in. and $\leq$ 7 in. (7.75 in. maximum in groups R-2 and R-3).
Stair treads (excluding nosing) $\geq$ 11 in. (10 in. minimum in groups R-2 and R-3).
Minimum width of stair = 3 ft 8 in.
Minimum head room clearance = 6 ft 8 in.
The floor-to-floor height between stair landings must not exceed 12 ft.

To provide comfort for the users, the following rules of thumb are commonly used:

$$\text{Number of treads} = \text{number of risers} - 1$$

$$2R + T = 25 \text{ (ensures the comfort of people using the stair)}$$

where T is the tread width in inches and R is the riser height in inches.

Risers:

$$\text{Floor-to-floor height} = 9 \text{ ft } 4\tfrac{1}{2} \text{ in.} = 112.5 \text{ in.}$$

$$\text{Assuming 7-in. risers, number of risers} = \frac{112.5 \text{ in.}}{7 \text{ in.}} = 16.07$$

$$\text{Assuming 7.5-in. risers, number of risers} = \frac{112.5 \text{ in.}}{7.5 \text{ in.}} = 15$$

Use 15 risers at 7.5 in. = 112.5 in. floor-to-floor height.

Treads:

$$T + 2 \times R = 25$$

$$T = 25 - (2)(7.5 \text{ in.}) = 10 \text{ in.}$$

$$\text{Number of treads} = \text{number of risers} - 1 = 15 - 1 = 14$$

Use 15 risers at 7.5 in. and 14 treads at 10 in.

$$\text{Horizontal run of stair} = \text{span of stair} = 14 \text{ treads} \times 10 \text{ in.} = 140 \text{ in.} = 11.67 \text{ ft}$$

$$\text{Width of stair} = 4 \text{ ft}$$

Design of tread. The tread is bending about its weaker (y–y) axis, and it should be noted that the uniform stair live load of 100 psf will not usually govern for the design of the treads. Instead, the most critical load for which the tread should be designed is the alternative 300-lb concentrated live load in addition to the uniform tread self-weight and finishes (see Table 2.2) for the alternative concentrated floor live loads.

Assume 2 × 12 for the treads (actual size = 1.5 in. × 11.25 in.). Therefore, the tread width = 11.25 in. (including nosing of 11.25 in. − 10 in. = 1.25 in.)

$$\text{Self-weight of tread} = \frac{(1.5 \text{ in.})(11.25 \text{ in.})}{144}(26.9 \text{ lb/ft}^3) = 3.2 \text{ lb/ft (density of hem-fir} = 26.9 \text{ lb/ft}^3)$$

Span of tread = width of stair = 4 ft

$$M_{\max} = \frac{(3.2 \text{ lb/ft})(4 \text{ ft})^2}{8} + \frac{(300 \text{ lb})(4 \text{ ft})}{4} = 307 \text{ ft-lb (concentrated load at midspan)}$$

$$V_{\max} = \frac{(3.2 \text{ lb/ft})(4 \text{ ft})}{2} + 300 \text{ lb} = 307 \text{ lb (concentrated load at support)}$$

Try a 2 × 12 hem-fir No. 1 tread. Since this is dimension-sawn lumber, use NDS-S Table 4A. From NDS-S Table 1B,

$$S_{yy} = 4.219 \text{ in}^3 \quad \text{(note that } S_{yy} \text{ is used because the tread is bending about the weak axis)}$$

$$F_b = 975 \text{ psi}$$

$$F_v = 150 \text{ psi}$$

From the adjustment or C factors section of NDS-S Table 4A, we obtain

$$C_D(D + L) = 1.0$$

$$C_M = 1.0 \text{ (dry service conditions)}$$

$$C_t = 1.0 \text{ (normal temperature conditions)}$$

$$C_F = 1.0$$

$$C_{fu} = 1.2 \text{ (the flat use factor is used since the 2} \times \text{12 tread is used}$$
$$\text{on the flat and is bending about the weak axis)}$$

Bending stress:

$$\text{Applied bending stress } f_{b,yy} = \frac{M_{\max}}{S_{yy}} = \frac{(307)(12)}{4.219} = 873 \text{ psi}$$

$$\text{Allowable bending stress } F'_b = F_b C_D C_M C_t C_F C_{fu} C_i C_r$$
$$= (975)(1.0)(1.0)(1.0)(1.0)(1.2)(1.0)(1.0) = 1170 \text{ psi} > 873 \text{ psi} \quad \text{OK}$$

Shear stress:

$$\text{Applied shear stress } f_{v,yy} = \frac{1.5V}{A} = \frac{(1.5)(307)}{16.88} = 28 \text{ psi}$$

$$\text{Allowable shear stress } F'_v = F_v C_D C_M C_t C_i = (150)(1.0)(1.0)(1.0)(1.0) = 150 \text{ psi} > 28 \text{ psi} \quad \text{OK}$$

Use a 2 × 12 hem-fir No. 1 tread (nosing = 1.25 in.).

Analysis and Design of Stair Stringer

Width of stair = 4 ft

Actual size of riser = 1.5 in. × 6 in. (see Figure 4.18)

Horizontal run L_2 = 140 in. = 11.67 ft

Floor-to-floor height H = 9.375 ft

Sloped length of stair $L_1 = \sqrt{(H)^2 + (L_2)^2} = \sqrt{(9.375 \text{ ft})^2 + (11.67 \text{ ft})^2} = 15 \text{ ft}$

Stair loads:

Treads = (14 treads × 3.7 lb/ft × 4-ft span)(6.0 in./11.25 in.)/(4 ft × 11.67 ft)	= 4.5 psf
Risers = (15 risers × 3.7 lb/ft × 4-ft span)(6.0 in./11.25 in.)/(4 ft × 11.67 ft)	= 2.6 psf
Two stringers (assume 3 × 16 per stringer) = (2 × 8.3 lb/ft)/4 ft	= 4.2 psf
Handrails (assume 2 psf)	= 2 psf
Total dead load on stair D	= 13.3 psf ≈ 14 psf of sloped area
Live load on stair L	= 100 psf of horizontal plan area

To convert the dead load to units of psf of horizontal plan area, we use the method introduced in Chapter 2:

$$\text{Total load on stair } w_{TL} \text{ (psf on horizontal plane)} = D\frac{L_1}{L_2} + L \quad \text{(see Chapter 2)}$$

$$= (14 \text{ psf})\left(\frac{15 \text{ ft}}{11.67 \text{ ft}}\right) + 100 \text{ psf}$$

$$= 118 \text{ psf}$$

$$\text{Tributary width per stair stringer} = \frac{4 \text{ ft}}{2} = 2 \text{ ft}$$

Total load on stair stringer w_{TL} (lb/ft on horizontal plane) = (118 psf)(2 ft) = **236 lb/ft**

Horizontal span of stair stringer $L_2 = 11.67$ ft

The load effects on the stair stringer are calculated as follows:

$$\text{maximum moment } M_{max} = \frac{w_{TL}(L_2)^2}{8} = \frac{(236 \text{ lb/ft})(11.67 \text{ ft})^2}{8} = 4018 \text{ ft-lb}$$

$$\text{maximum shear } V_{max} = \frac{w_{TL} L_2}{2} = \frac{(236 \text{ lb/ft})(11.67 \text{ ft})}{2} = 1378 \text{ lb}$$

$$\text{maximum reaction } R_{max} = \frac{w_{TL} L_2}{2} = \frac{(236 \text{ lb/ft})(11.67 \text{ ft})}{2} = 1378 \text{ lb}$$

Effective depth of stringer. Typically, the stair stringers are notched to provide support for the treads and risers. As a result of this notch, the effective depth of the stringer is reduced (Figure 4.18). The effective depth of stair stringer to be used in design is

actual depth − depth to allow for the treads/risers "cutting" into the stringer for support

= width of unnotched stair stringer, d − tread width × sin θ

where

$$\theta = \text{angle of slope of stair} = \tan^{-1}\frac{H}{L_2} = \tan^{-1}\frac{9.375 \text{ ft}}{11.67 \text{ ft}} = 39°$$

The tread width (excluding the nosing) is 10 in. If we assume 3 × 16 stair stringers, the effective stringer depth is 15.25 in. − (10 in.)(sin 39°) = **9 in.** Therefore, the effective stringer section = 2.5 × 9 in.

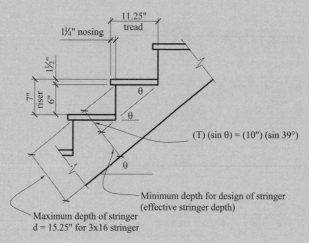

FIGURE 4.18 Effective depth of stair stringer.

The stair stringer should then be designed as a beam (similar to the beam design examples presented previously in this chapter) for the load effects calculated previously using this effective section. The header and trimmer joists supporting the stair stringer at the second-floor level will be subjected to concentrated loads due to reactions from the stair stringers, in addition to the second-floor uniform loads. Once the joists have been analyzed for the maximum shears, moments, and reactions, the design will follow the same approach as was used in Example 4.1. A structural analysis software may be needed for the deflection computations for these nonuniformly loaded joists.

Continuous Beams and Girders

Thus far, we have focused on simply supported single-span beams and girders. For girders supported on three or more columns, it is more cost-effective to analyze the beam or girder as a continuous beam, subject to the limitations on the length of beams or girders that can be transported. However, the analysis of continuous beams is considerably more tedious than the analysis of simply supported single-span beams. For continuous beams, a *pattern load analysis* has to be carried out in which the live load is randomly located on the spans of the beam or girders to produce the maximum load effects. It should be noted that the dead load, on the other hand, is always located over the entire length of the girder. The girder is then analyzed for the different load cases and the absolute maximum negative or positive moment, the absolute maximum positive or negative shear, and the absolute maximum reaction are obtained; the girder is then designed for these absolute maximum values. The maximum deflection at the midspan of the endspan of the continuous beam is calculated and compared to the beam deflection limits discussed earlier. For continuous roof beams in areas with snow, the partial snow load prescribed in Section 7.5 of ASCE 7 should also be considered.

Beams and Girders with Overhangs or Cantilevers

These types of beams or girders are used to frame balconies or overhanging portions of buildings. It consists of the overhang or cantilever and the back span. It is more efficient to have the length of the back span be at least three times the length of the overhang or cantilever span. In these types of beams, the designer needs to, among other things, check the deflection of the tip of the cantilever or overhang and the uplift force at the back-span end, as these could be critical.

Similar to continuous beams, a pattern load analysis is required to obtain the maximum possible moment, shear, reaction, and deflection. The beam would then be designed for these maximum load effects using the procedure discussed previously. The partial snowload prescribed in Section 7.5 of ASCE 7 should also be considered for cantilevered roof beams in areas with snow. The deflection limits for the back span are as presented previously. However, for the cantilever or overhang, the appropriate deflection limits are presented in Table 4.10. The reader should note that the number "2" in the cantilever deflection limits below is a factor that accounts for the fact that the deflected profile of a cantilever beam is one-half that of a simply supported single-span beam (IBC 1604.3).

At the supports of continuous beams or the interior support of a cantilever girder with an overhang, the bottom face of the beam between the point of inflection within the back span (near the interior support) and the tip of the cantilever or overhang (point D) is unbraced because of the negative moment, which causes compression stresses in the bottom fibers. Lateral bracing is required at the bottom of the beam at the interior support (point C) to prevent lateral instability of the column support, and for the same reason, lateral bracing is also required at the supports of continuous beams and girders. The lateral support can be provided by a member of similar

TABLE 4.10 Cantilever Beam Deflection Limits[a]

	Deflection Limit	
Deflection	Back Span	Overhang or Cantilever
Live-load deflection Δ_{LL}	$\dfrac{L_{bs}}{360}$	$\dfrac{2L_{cant}}{360}$
Incremental long-term deflection due to dead load plus live load (including creep effects), $\Delta_{TL} = k\Delta_{DL} + \Delta_{LL}$	$\dfrac{L_{bs}}{240}$	$\dfrac{2L_{cant}}{240}$

[a] $k = 1.0$ for green or unseasoned lumber or glulam used in wet service conditions, 1.0 for seasoned lumber used in wet service conditions, and 0.5 for seasoned or dry lumber, glulam, and prefabricated I-joists used in dry service condition. L_{bs} is the back span of a joist, beam, or girder between two adjacent supports, L_{cant} is the cantilever or overhang span of joist, beam, or girder, Δ_{DL} is the deflection due to dead load D, and Δ_{LL} is the deflection due to live load L or (L_r or S).

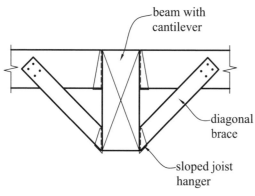

FIGURE 4.19 Beam lateral bracing detail.

depth framing into the continuous beam at the supports or the cantilever beam at point C. Lateral bracing can also be accomplished by using a diagonal kicker as shown in Figure 4.19. The diagonal brace or kicker reduces the unbraced length of the bottom face (i.e., compression zone) of the beam to the distance between the brace and the farthest point of inflection on opposite sides of the interior support.

4.3 SAWN-LUMBER DECKING

In lieu of using plywood sheathing (see Chapter 6) to resist gravity loads on roofs and floors, tongue-and-grooved decking may be used to span longer distances between roof trusses or floor beams. Sawn-lumber decking may also be required on floors where equipment or forklift loading produce concentrated loads for which plywood

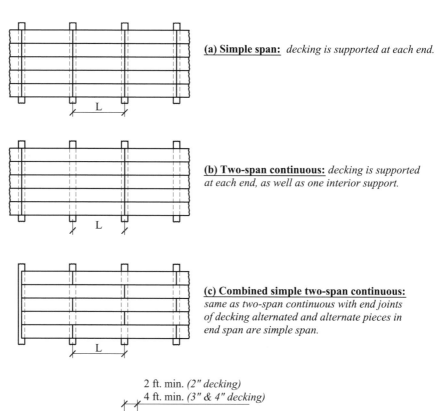

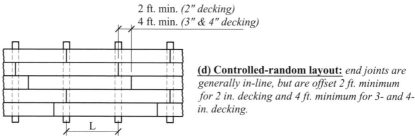

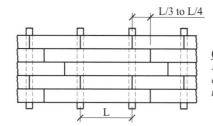

(a) Simple span: *decking is supported at each end.*

(b) Two-span continuous: *decking is supported at each end, as well as one interior support.*

(c) Combined simple two-span continuous: *same as two-span continuous with end joints of decking alternated and alternate pieces in end span are simple span.*

(d) Controlled-random layout: *end joints are generally in-line, but are offset 2 ft. minimum for 2 in. decking and 4 ft. minimum for 3- and 4-in. decking.*

(e) Cantilevered Pieces Intermixed: *decking is simple span in every third course; other pieces are cantilevered over the supports in alternating third or quarter points of the span.*

FIGURE 4.20 Common deck layout profiles.

TABLE 4.11 Maximum Moments and Deflections for Various Decking Layouts

Type of Decking Layout	Maximum Moment, M_{max}	Maximum Deflection, Δ	Remarks
Simple span	$\dfrac{wL^2}{8}$	$\dfrac{5wL^4}{384EI}$	Has the least stiffness of all the deck layout systems
Two-span continuous	$\dfrac{wL^2}{8}$	$\dfrac{wL^4}{185EI}$	
Combined simple and two-span continuous	$\dfrac{wL^2}{8}$	$\dfrac{wL^4}{109EI}$	Average of simple-span and two-span maximum deflections
Controlled random layout	$\dfrac{wL^2}{10}$	$\dfrac{wL^4}{145EI_e}$	Maximum deflection for three-span decking used; $I_e = 0.67(I)$ for 2-in. decking and $0.80I$ for 3 and 4 in. decking
Cantilever pieces intermixed	$\dfrac{wL^2}{10}$	$\dfrac{wL^4}{105EI_e}$	

Source: Refs. 10 and 13.

sheathing is inadequate. Decking is also aesthetically pleasing because the underside of the decking creates an attractive ceiling feature, obviating the need for additional plaster or suspended ceiling. Wood decking are manufactured as 2-in.-thick single tongue-and-groove panels at a moisture content of 15% or as 3- and 4-in. thick double tongue-and-groove panels at a moisture content of 19%. The standard nominal sizes for wood decking are 2×6, 2×8, 3×6, and 4×6. Prior to installation, the lumber decking should be allowed to reach the moisture content of the surrounding atmosphere, to avoid problems that may result from shrinkage.

The tongue and grooves prevent differential deflections of adjacent deck panels, and each piece of deck is usually nailed through the tongue at each support. Decking consists of repetitive sawn-lumber members that are used in a flat orientation. However, the flat-use factor should not be applied in calculating the allowable stresses because this factor is already embedded in the NDS-S tabulated design values for decking in NDS-S Table 4E. Single-member and repetitive-member bending stresses are given in this table. The different stress grades of decking include Select, Commercial Dex and Select Dex.

There are several possible decking layouts as shown in Figure 4.20. The two most popular decking patterns are (1) nonstaggered decking panels supported on only two supports with all the panel joints continuous, and (2) staggered decking panels supported on at least two supports with continuous panel joints parallel to the longer dimension of the panels. It should be noted that tongue-and-groove lumber decking, although efficient for supporting gravity loads, is not very efficient as horizontal diaphragms in resisting lateral loads from wind or seismic forces. It possesses very low diaphragm shear capacity in the direction where the panel joints are continuous. If appreciable diaphragm shear capacity is required in a floor framing with sawn-lumber decking, the horizontal diaphragm shear capacity can be enhanced by using $\frac{5}{16}$-in. plywood sheathing nailed to, and placed on top of, the sawn-lumber decking. Sawn-lumber decking also provides better fire ratings than plywood-sheathed floors. The maximum moments and deflections for the five types of sawn-lumber decking layout are shown in Table 4.11.

EXAMPLE 4.5

Design of Sawn-Lumber Decking

Design sawn-lumber roof decking to span 10 ft 0 in. between roof trusses. The superimposed roof dead load is 16 psf and the snow load is 35 psf. Assume that dry service and normal temperature conditions apply and Douglas fir-larch commercial decking (DF-L Commercial Dex).

Solution:: Assume 3 × 6 in. decking (actual size = 2.5 × 5.5 in.) and simple-span decking layout in Table 4.11. The section properties for 3 × 6 in. decking are

$$A_{xx} = (13.75 \text{ in}^2)(12 \text{ in.}/5.5 \text{ in.}) = 30 \text{ in}^2 \text{ per foot width of deck}$$

$$S_{yy} = (5.729 \text{ in}^3)(12 \text{ in.}/5.5 \text{ in.}) = 12.5 \text{ in}^3 \text{ per foot width of deck}$$

$$I_{yy} = (7.161 \text{ in}^4)(12 \text{ in.}/5.5 \text{ in.}) = 15.62 \text{ in}^4 \text{ per foot width of deck}$$

Loads:

$$\text{Decking self-weight} \left(\frac{2.5 \text{ in.}}{12} \times 31.2 \text{ psf}\right) = 6.5 \text{ psf}$$

$$\underline{\text{Superimposed dead loads} \hspace{2em} = 16 \text{ psf}}$$

$$\text{Total dead load } D = 22.5 \text{ psf}$$

$$\text{Snow load } S = 35 \text{ psf}$$

The governing load combination for this roof deck will be dead load plus snow load ($D + S$):

$$\text{total load} = D + S = 22.5 + 35 = 57.5 \text{ psf}$$

Bending stress:

$$\text{Maximum moment } M_{max} = \frac{wL^2}{8} = \frac{(57.5 \text{ psf})(10 \text{ ft})^2}{8} = 719 \text{ ft-lb per foot width of deck}$$

$$= 8628 \text{ in.-lb per foot width of deck}$$

The tabulated stress values and appropriate adjustment factors from NDS-S Table 4E for 3 × 6 in. decking are:

$$F_b C_r = 1650 \text{ psi (repetitive member)}$$

$$F_{c\perp} = 625 \text{ psi}$$

$$E = 1.7 \times 10^6$$

$$C_F = 1.04 \quad \text{(for 3-in. decking)}$$

$$C_i = 1.0 \quad \text{(lumber is not incised)}$$

The allowable stresses are calculated using the sawn-lumber adjustment factors applicability table (Table 3.1) as follows:

$$F'_b = F_b C_r C_D C_M C_t C_L C_F C_i = (1650)(1)(1)(1)(1)(1)(1.04)(1) = 1716 \text{ psi}$$

$$E' = E C_M C_t C_i = (1.7 \times 10^6)(1)(1)(1) = 1.7 \times 10^6 \text{ psi}$$

The applied bending stress for bending about the weak (y–y) axis is

$$f_{by} = \frac{M_{max}}{S_{yy}} = \frac{8628 \text{ lb-in.}}{12.5 \text{ in}^3} = 690 \text{ psi} \ll \text{allowable stress } F'_b = 1716 \text{ psi} \quad \text{OK}$$

Deflection:

Dead load D = 22.5 psf = 22.5 lb/ft per foot width of deck = 1.88 lb/in. per foot width of deck

Live load (snow) S = 35 psf = 35 lb/ft per foot width of deck = 2.92 lb/in per foot width of deck

The dead-load deflection is

$$\Delta_{DL} = \frac{5wL^4}{384 E' I_{yy}} = \frac{(5)(1.88 \text{ lb/in.})(10 \text{ ft} \times 12)^4}{(384)(1.7 \times 10^6 \text{ psi})(15.62 \text{ in}^4)} = 0.19 \text{ in.}$$

The live-load deflection is

$$\Delta_{LL} = \frac{5wL^4}{384E'I_{yy}} = \frac{(5)(2.92\text{ lb/in.})(10\text{ ft} \times 12)^4}{(384)(1.7 \times 10^6\text{ psi})(15.62\text{ in}^4)} = 0.30\text{ in.}$$

$$< \frac{L}{360} = \frac{(10\text{ ft})(12\text{ in./ft})}{360} = 0.33\text{ in.} \quad \text{OK}$$

Since seasoned wood in dry service conditions is assumed to be used in this building, the creep factor $k = 0.5$. The total incremental dead plus floor live load deflection is

$$\Delta_{TL} = k\Delta_{DL} + \Delta_{LL}$$

$$= (0.5)(0.19\text{ in.}) + 0.30\text{ in.} = 0.40\text{ in.} < \frac{L}{240} = \frac{(10\text{ ft})(12\text{ in./ft})}{240} = 0.5\text{ in.} \quad \text{OK}$$

Use a 3 × 6 in. DF-L Commercial Dex with a simple-span deck layout.

4.4 MISCELLANEOUS STRESSES IN WOOD MEMBERS

In this section, several other types of stresses that can occur in wood members are discussed. Examples include amplification of the shear stress in notched beams, end-to-end or parallel-to-grain bearing stress, and bearing stress at an angle to the grain.

Shear Stress in Notched Beams

Sometimes, it may be necessary to provide notches and holes in wood beams to allow for passage of mechanical ducts or because of headroom limitations as shown in Figure 4.21. The limitations on the size and placement of notches [12] and holes [20] are shown in the figure. The sizes and location of holes in preengineered lumber is specified by the manufacturer. Preengineered light-gage metal plates that are screwed and/or glued to the sides of the joist can be used for reinforcing floor joists with larger notches and holes [21].

The notching of a beam creates stress concentrations, in addition to reducing the allowable shear stress due to a reduction in the depth of the beam at the notch. The NDS code requirements for notched beams are as follows:

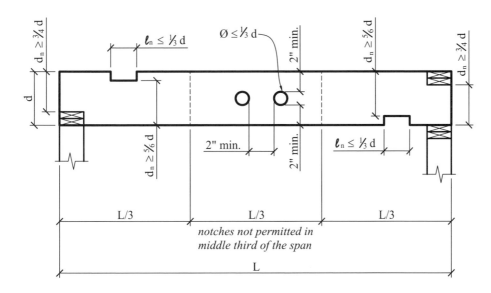

FIGURE 4.21 Notches and holes in wood beams.

- No notches are permitted within the middle third of the beam span.
- No notches are permitted within the outer thirds of the beam span on the tension face (except at beam bearing supports) when the thickness of a beam b is ≥ 3.5 in.
- Within the outer thirds of the beam span, the depth of the notched beam, d_n, must be $\geq \frac{5}{6}d$.
- At the beam supports, the depth of the notched beam, d_n, must be $\geq \frac{3}{4}d$.
- The notch length l_n must be $\leq \frac{1}{3}d$.

The NDS code equation for calculating the adjusted applied shear stress when the notch is on the *tension* face is given as

$$f_v = \frac{\frac{3}{2}V}{bd_n}\left(\frac{d}{d_n}\right)^2$$
$$\leq \text{allowable shear stress } F'_v \qquad (4.20)$$

For a notch on the *compression* face, the adjusted applied shear stress is given as

$$f_v = \frac{\frac{3}{2}V}{b\{d - [(d - d_n/d_n)]e\}}$$
$$\leq \text{allowable shear stress } F'_v \qquad (4.21)$$

where d = total depth of the unnotched beam
d_n = depth of beam remaining at the notch
e = horizontal distance from inside face of support to edge of notch (e must be $\leq d_n$; if $e > d_n$, use $e = d_n$ in calculating the shear stress)

Bearing Stress Parallel to the Grain

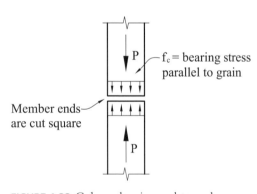

FIGURE 4.22 Column bearing end to end.

Bearing stress parallel to the grain occurs at the ends of two axially loaded compression members bearing on each other end to end as shown in Figure 4.22. The applied bearing stress parallel to the grain is given as

$$f_c = \frac{P}{A}$$
$$\leq F_c^* \qquad (4.22)$$

where A = net bearing area
$F_c^* = F_c C_D C_M C_t C_F C_i$ = allowable compression stress parallel to the grain, excluding the effects of the column stability factor, C_P
P = compression load
F_c = tabulated compression stress parallel to grain (NDS-S Tables 4A to 4D)

The adjustment or C factors are as discussed in Chapter 3. NDS code Section 3.10.1.3 requires that a steel bearing plate be provided whenever the bearing stress parallel to the grain, f_c, is greater than $0.75F_c^*$ to help distribute the load.

Bearing Stress at an Angle to the Grain

This bearing situation occurs at the supports of sloped rafters or where a sloped member frames and bears on a horizontal member as shown in Figure 4.23. The rafter will be subjected to the bearing stress at an angle to the grain, f_θ, and the beam or header or horizontal member will be subjected to bearing stress perpendicular to the grain, $f_{c\perp}$. The bearing stress applied in the rafter at an angle θ to the grain is calculated as

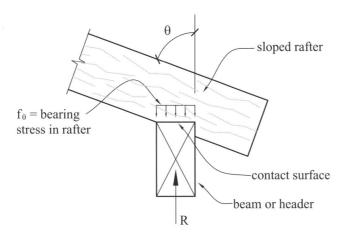

FIGURE 4.23 Bearing at an angle to the grain.

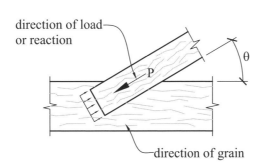

$$f_\theta = \frac{P}{A}$$
$$\leq F'_\theta \qquad (4.23)$$

where P = applied load or reaction
A = bearing area
F'_θ = allowable bearing stress at an angle θ to the grain
θ = angle between the direction of the reaction or bearing stress and the direction of the grain (see Figure 4.23)

The allowable bearing stress at an angle to grain is given in the NDS code by the Hankinson formula as

$$F'_\theta = \frac{F_c^* F'_{c\perp}}{F_c^* \sin^2\theta + F'_{c\perp} \cos^2\theta} \qquad (4.24)$$

where $F_c^* = F_c C_D C_M C_t C_F C_i$ = allowable bearing stress *parallel* to the grain of the rafter or sloped member excluding column stability effects
$F'_{c\perp} = F_{c\perp} C_M C_t C_i C_b$ = allowable bearing stress *perpendicular* to the grain of the beam or header

Note that when θ is zero, the load or reaction will be parallel to the grain, and F'_θ will be equal to F_c^*. When θ is 90°, the load or reaction will be perpendicular to the grain, and F'_θ will be equal to $F'_{c\perp}$.

Sloped Rafter Connection

The design procedure for a sloped rafter connection is as follows:

- If $f_\theta \leq F'_\theta$ in the sloped rafter, the rafter is adequate in bearing.

124 | STRUCTURAL WOOD DESIGN

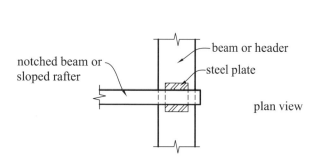

FIGURE 4.24 Bearing stress in a header.

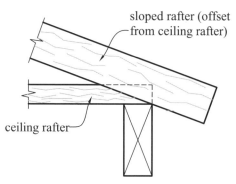

FIGURE 4.25 Alternative sloped rafter detail.

- If $f_\theta > F'_\theta$, increase the bearing area A for the rafter either by increasing the thickness b of the rafter and/or the bearing length provided (i.e., the thickness b of the header).
- If $f_{c\perp} \leq F'_{c\perp}$ in the beam or header, the header is adequate in bearing.
- If $f_{c\perp} > F'_{c\perp}$ in the beam or header, the header is not adequate; therefore, either increase the thickness b of the header or use a steel plate to increase the bearing area A_{bearing} and thus reduce the bearing stress perpendicular to the grain in the beam or header (see Figure 4.24). Note that the steel plate will only help relieve the bearing stress in the beam or header, not the sloped rafter.

The bearing stress at an angle to the grain f_θ can be avoided by using the detail shown in Figure 4.25.

EXAMPLE 4.6

Bearing Stresses at a Sloped Rafter-to-Stud Wall Connection

The sloped rafter shown in Figure 4.26 is supported on a ridge beam at the ridge line and on the exterior stud wall. The reaction from dead load plus the snow load on the rafter at the exterior stud wall is 1800 lb. Assuming No. 1 spruce-pine-fir (SPF) and normal temperature and dry service conditions, determine (a) the allowable bearing stresses at an angle to the grain F'_θ and the allowable bearing stress or allowable stress perpendicular to the grain $F'_{c\perp}$; (b) the bearing stress at an angle to the grain f_θ in the rafter; and (c) the bearing stress perpendicular to the grain $f_{c\perp}$ in the two top plates.

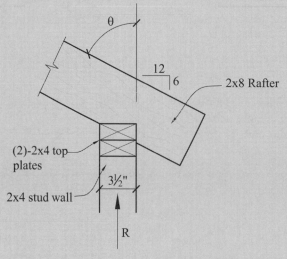

FIGURE 4.26 Sloped rafter-to-stud wall connection.

Solution:

$$2 \times 8 \text{ sloped rafter: } b = 1.5 \text{ in.}; d = 7.25 \text{ in.}$$

$$2 \times 4 \text{ top plates on top of stud wall: } d = 3.5 \text{ in.}$$

Since the 2×8 rafter and 2×4 top plates are dimension lumber, use NDS-S Table 4A. Using the table we obtain the tabulated stresses for spruce-pine-fir (SPF) as

$$\text{bearing stress perpendicular to the grain } F_{c\perp} = 425 \text{ psi}$$

$$\text{compression stress parallel to the grain } F_c = 1150 \text{ psi}$$

The stress adjustment or C factors are

$$C_M = 1.0 \text{ (dry service conditions)}$$

$$C_t = 1.0 \text{ (normal temperature conditions)}$$

$$C_i = 1.0 \text{ (wood is not incised)}$$

$$C_F(F_c) = 1.05$$

$$C_D = 1.15 \text{ (dead load plus snow load)}$$

The bearing length in the 2×4 top plate measured parallel to the grain is the same as the thickness of the sloped rafter. $l_b = b_{\text{rafter}} = 1.5$ in. Since $l_b = 1.5$ in. $<$ 6 in. and the bearing is not nearer than 3 in. from the end of the top plate and the end of the rafter (see Chapter 3), the bearing stress factor is calculated as

$$C_b = \frac{l_b + 0.375 \text{ in.}}{l_b} = \frac{1.5 \text{ in.} + 0.375 \text{ in.}}{1.5 \text{ in.}} = 1.25$$

The bearing area at the rafter–stud wall connection is

$$A_{\text{bearing}} = (\text{thickness } b \text{ of rafter})(\text{width } d \text{ of the } 2 \times 4 \text{ top plates})$$
$$= (1.5 \text{ in.})(3.5 \text{ in.}) = 5.25 \text{ in}^2$$

(a) *Allowable bearing stresses:*

$$F'_{c\perp} = F_{c\perp} C_M C_t C_i C_b = (425)(1)(1)(1)(1.25) = 531 \text{ psi}$$

$$F^*_c = F_c C_D C_M C_t C_F C_i = (1150)(1.15)(1)(1)(1.05)(1) = 1389 \text{ psi}$$

θ is the angle between the direction of the reaction or bearing stress and the direction of grain in the sloped wood member. Therefore, $\tan \theta = 12/6$ and $\theta = 63.4°$. Using equation (4.24), the allowable stress at an angle to grain is given as

$$F'_\theta = \frac{F^*_c F'_{c\perp}}{F^*_c \sin^2 \theta + F'_{c\perp} \cos^2 \theta} = \frac{(1389)(531)}{1389 \sin^2 63.4 + 531 \cos^2 63.4} = 606 \text{ psi}$$

(b) *Bearing stress applied at angle θ to the grain of the rafter:*

$$f_\theta = \frac{P}{A_{\text{bearing}}} = \frac{1800 \text{ lb}}{5.25 \text{ in}^2}$$

$$= 343 \text{ psi} < F'_\theta = 606 \text{ psi} \quad \text{OK}$$

(c) *Bearing stress applied perpendicular to the grain of the top plate:*

$$f_{c\perp} = \frac{P}{A_{\text{bearing}}} = \frac{1800 \text{ lb}}{5.25 \text{ in}^2}$$

$$= 343 \text{ psi} < F'_{c\perp} = 531 \text{ psi} \quad \text{OK}$$

4.5 PREENGINEERED LUMBER HEADERS

In this section we discuss the selection of preengineered lumber such as laminated veneer lumber (LVL) and parallel strand lumber (PSL) using manufacturers' published load tables. These proprietary wood members usually have fewer defects and a lower moisture content than sawn-lumber beams and columns. They are less susceptible to shrinkage and are used for spans and loadings that exceed the capacity of sawn-lumber beams.

EXAMPLE 4.7

Header Beam Design Using Preengineered Lumber

Refer to Figure 4.27 and the following given information:

floor dead load = 15 psf

floor live load = 40 psf

normal temperature and moisture conditions

full lateral stability is provided ($C_L = 1.0$)

Determine the required header size using prefabricated laminated veneer lumber (LVL) with the following material properties: $F_b = 2600$ psi, $F_v = 285$ psi, and $E = 1900$ ksi.

Solution: Calculate the maximum moment and shear in the header:

Span of header beam $L = 15$ ft

Tributary width of the header beam = 8 ft

Dead load $w_{DL} = (15 \text{ psf})(8 \text{ ft}) = 120$ lb/ft

Live load $w_{LL} = (40 \text{ psf})(8 \text{ ft}) = 320$ lb/ft

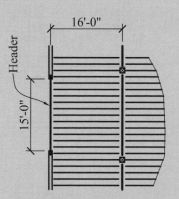

FIGURE 4.27 Partial floor framing plan.

From Section 2.1 the governing load combination is dead plus floor live load; thus, the total load is

$$w_{TL} = w_{DL} + w_{LL} = 120 + 320 = 440 \text{ lb/ft}$$

See Figure 4.28.

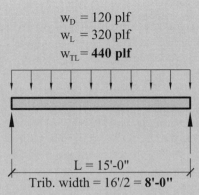

FIGURE 4.28 Header loading diagram.

$$M_{\max} = \frac{w_{TL}L^2}{8} = \frac{(440)(15)^2}{8} = 12{,}375 \text{ ft-lb}$$

$$V_{\max} = \frac{w_{TL}L}{2} = \frac{(440)(15)}{2} = 3300 \text{ lb}$$

Note: After calculating the maximum shear and moment values, the designer could select a specific LVL directly from manufacturers' literature since the manufacturer typically makes this type of information available. After selecting an appropriate product based on shear and moment capacity, the designer would then check the deflection.

Required section properties:

$$S_{\text{req'd}} = \frac{M}{F_b} = \frac{(12{,}375)(12)}{2600} = 57.1 \text{ in}^3$$

$$A_{\text{req'd}} = \frac{1.5V}{F_v} = \frac{(1.5)(3300)}{285} = 17.3 \text{ in}^2$$

Select two $1\frac{3}{4} \times 11\frac{1}{4}$ in. LVLs.
Section properties:

$$A = 2bh = (2)(1.75 \text{ in.})(11.25 \text{ in.}) = 39.3 \text{ in}^2 > 17.3 \text{ in}^2 \quad \text{OK}$$

$$S_x = \frac{2bh^2}{6} = \frac{(2)(1.75 \text{ in.})(11.25 \text{ in.})^2}{6} = 73.8 \text{ in}^3 > 57.1 \text{ in}^3 \quad \text{OK}$$

$$I_x = \frac{2bh^3}{12} = \frac{(2)(1.75 \text{ in.})(11.25 \text{ in.})^3}{12} = 416 \text{ in}^4$$

Check the deflection. The live-load deflection is

$$\Delta_L = \frac{5w_{LL}L^4}{384EI} = \frac{(5)(320/12)(15 \times 12)^4}{(384)(1.9 \times 10^6)(416)} = 0.47 \text{ in.}$$

$$\Delta_{\text{allowable}} = \frac{L}{360} = \frac{(15)(12)}{360} = 0.5 \text{ in.} > 0.47 \text{ in.} \quad \text{OK}$$

Use two $1\frac{3}{4} \times 11\frac{1}{4}$ in. LVLs.

Note: Most manufacturers specify fastening requirements for multiple plies of LVLs or PSLs. The fastening can be either nailed or bolted, but the designer should refer to the published requirements from the specific manufacturer for this information.

4.6 FLITCH BEAMS

A *flitch beam* is a composite section comprised of a steel plate placed between wood members or steel plates or channels attached to opposite sides of an existing wood beam. These are typically used where a solid wood member is not practical, such as depth limitations or heavy loads or where an existing wood beam needs to be strengthened to resist higher loads. Historically, flitch beams have been used as an economical alternative to higher-strength preengineered wood members. Currently, preengineered wood members such as laminated veneer lumber and parallel strand lumber are typically readily available, so the use of flitch beams is limited.

A flitch beam consists of dissimilar materials, so the section properties will have to be transformed in the design process so that the two bonded materials will experience the same strain and deformation. In a transformed section, one material is transformed into an equivalent quantity of the other material. In this case we transform the steel plate into an equivalent amount of wood material. This equivalence is based on the ratio between the modulus of elasticity of steel and the modulus of elasticity of wood.

$$n = \frac{E_s}{E_w}$$

where n = modular ratio
E_s = modulus of elasticity of steel (= 29,000 ksi)
E_w = modulus of elasticity of wood (varies with the wood species)

Note that connection design is covered in Chapter 8. The example here is intended to clarify the full scope of what is required for flitch beam design.

EXAMPLE 4.8

Flitch Beam Design

For this example, the following will be assumed (see Figure 4.29):

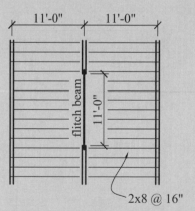

FIGURE 4.29 Framing plan for flitch beam design.

Flitch beam supports 2 × 8's at 16 in. spanning 11 ft on each side.
Maximum depth of the flitch beam is $7\frac{1}{4}$ in. (depth of a 2 × 8).
Total load = 55 psf (dead plus live).
Lumber is Douglas Fir Larch, Select Structural.

A $\frac{3}{8} \times 7\frac{1}{4}$ in. steel plate will be assumed. The approach taken here will be to transform the $\frac{3}{8}$-in. width of steel section to an equivalent wood section while maintaining the same depth.

Solution: From NDS Table 4A:

$$F_b = 1500 \text{ psi}$$

$$F_v = 180 \text{ psi}$$

$$E_w = 1900 \text{ ksi}$$

The modular ratio is

$$n = \frac{E_s}{E_w} = \frac{29,000}{1900} = 15.2$$

The section properties of the steel plates and wood side members are

$$A_p = nbh = (15.2)(0.375 \text{ in.})(7.25 \text{ in.}) = 41.4 \text{ in}^2$$

$$S_p = \frac{nbh^2}{6} = \frac{(15.2)(0.375)(7.25)^2}{6} = 50.1 \text{ in}^3$$

$$I_p = \frac{nbh^3}{12} = \frac{(15.2)(0.375)(7.25)^3}{12} = 181 \text{ in}^4$$

$$A_w = (2)(10.88 \text{ in}^2) = 21.75 \text{ in}^2$$

$$S_w = (2)(13.14 \text{ in}^2) = 26.28 \text{ in}^3$$

$$I_w = (2)(47.6 \text{ in}^2) = 95.3 \text{ in}^4$$

Allowable stress values: (all C-factors are 1.0 except $C_f = 1.2$):

$$F'_b = 1500 \text{ psi}$$

$$(1.2) = 1800 \text{ psi}$$

$$F'_v = 180 \text{ psi}$$

$$E'_w = 1900 \text{ ksi}$$

Combining the composite section properties yields

$$A_c = 41.4 + 21.75 = 63.1 \text{ in}^2$$

$$S_c = 50.1 + 26.28 = 76.3 \text{ in}^3$$

$$I_c = 181 + 95.3 = 276 \text{ in}^4$$

The total load on the beam is

$$W_{TL} = (55 \text{ psf})(11 \text{ ft}) = 605 \text{ plf}$$

Check the bending and shear stresses:

$$M = \frac{wL^2}{8} = \frac{(605)(11)^2}{8} = 9150 \text{ ft-lb}$$

$$V = \frac{wL}{2} = \frac{(605)(11)}{2} = 3328 \text{ lb}$$

$$f_b = \frac{M}{S} = \frac{(9150)(12)}{76.3} = \mathbf{1439 \text{ psi}} < F_b = \mathbf{1500 \text{ psi}} \qquad \text{OK}$$

$$f_v = \frac{1.5V}{A} = \frac{(1.5)(3328)}{63.1} = \mathbf{79.1 \text{ psi}} < F_v = \mathbf{150 \text{ psi}} \qquad \text{OK}$$

Check the deflection: (ignore creep effects because of the presence of the steel plates)

$$\Delta_{DL+LL} = \frac{5wL^4}{385EI} = \frac{(5)(605/12)(11 \text{ ft} \times 12)^4}{(384)(1.9 \times 10^6)(276)}$$

$$= \mathbf{0.38 \text{ in.}} < \frac{L}{240} = \frac{(11 \text{ ft})(12)}{240} = \mathbf{0.55 \text{ in.}} \qquad \text{OK}$$

The composite section is adequate for bending, shear, and deflection. One other item that may need to be designed is the connectors between the wood side members and the plate. If the joists frame on top of the flitch beam, only nominal connectors need to be provided. If the joists frame into the side of the wood members, the size and spacing of the connectors become critical. One possible solution is to provide a top flange type of connector (see Figure 4.30).

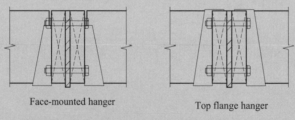

FIGURE 4.30 Various connections for flitch beams.

For this example it will be assumed that the joists frame into the side member with a standard joist hanger connection. The total load to one wood side member is

$$W_{side} = (55 \text{ psf})(5.5 \text{ ft}) = 303 \text{ plf}$$

For practical reasons, the connectors should be spaced in increments consistent with the joist spacing. In this example, the joist spacing is 16 in. Assuming $\frac{1}{2}$-in. connectors, the allowable shear perpendicular to the grain of the wood side member for one connector is

$$Z_\perp = 310 \text{ lb (NDS Table 11B)}$$

The required spacing is then

$$S_{req'd} = \frac{303 \text{ lb}}{310 \text{ plf}} = 0.97 \text{ ft or } 11.7 \text{ in.}$$

Based on the joist spacing, provide $\frac{1}{2}$-in. bolts at 8 in. o.c. The bolts should also be staggered for stability (see Figure 4.31). The minimum spacing for the connectors (from Table 8.3) is as follows:

$$\text{edge distance} = 4D = (4)(\tfrac{1}{2} \text{ in.}) = \textbf{2.0 in.} \text{ (2 in. provided)}$$

$$\text{center-to-center} = 3D = (3)(\tfrac{1}{2} \text{ in.}) = \textbf{1.5 in.} \text{ (16 in. provided)}$$

$$\text{row spacing} = 5D = (5)(\tfrac{1}{2} \text{ in.}) = \textbf{2.5 in.} \text{ (3.25 in. provided)}$$

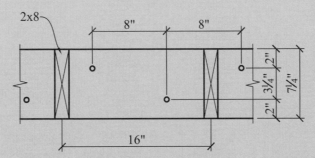

FIGURE 4.31 Flitch beam connection detail.

Note that connection design is covered in Chapter 8. The example here is intended to clarify the full scope of what is required for flitch beam design.

4.7 FLOOR VIBRATIONS

Floor vibrations are usually caused by human activities such as walking or rhythmic activity. Machinery or other external forces, such as vehicular traffic, can also cause floor vibrations. Problems associated with floor vibration are due primarily to resonance, which is a phenomenon that occurs when the forcing frequency or frequency of the human activity is at or near the natural frequency of the structure. When human activity such as aerobics occurs, the cyclical force due to this activity produces acceleration in the floor, and as each cycle of loading adds energy to the system, the vibration increases and the acceleration reaches a maximum. This vibration is mitigated by the presence of damping elements such as partitions or ceiling elements.

The NDS code does not provide any specific guidance regarding floor vibrations because it is a serviceability issue and not safety related. Traditionally, the main serviceability criteria used to control vibrations was to limit deflection under uniform design loads to $L/360$. For shorter spans the $L/360$ limit is typically found to be adequate to control floor vibrations. With the advent of preengineered joists, combined with longer spans and more wide-open areas (and thus no partitions to provide damping), some floors have been found not to have good performance with respect to floor vibrations. Some designers use higher deflection limits (usually between $L/480$ and $L/720$) to control floor vibrations. The acceptability of floor vibrations is a function of the occupant's sensitivity to floor vibrations, which can be quite subjective and variable. In general, occupants are more sensitive to floor vibrations when they are engaged in low amounts of activity. The limits on vibration and floor acceleration are usually expressed as a percentage of the acceleration due to gravity g. Table 4.12 shows the generally accepted limits on floor acceleration.

Floor Vibration Design Criteria

Rhythmic activity on light-frame floors (i.e., wood floors) is not very common. These activities usually occur on combined concrete and steel structures, and preferably should occur on the ground floor of a building, if possible. To control floor vibrations due to rhythmic activity, the

TABLE 4.12 Acceleration Limits for Floor Vibrations (Percent)

Activity	Acceleration Limit, $\frac{a_0}{g} \times 100\%$
Hospital (operating rooms)	0.25
Office, residential, church	0.50
Shopping malls	1.5
Dining, weight lifting	2
Rhythmic activity	5

Source: Ref. 3.

natural frequency of the floor needs to be higher than the forcing frequency of the highest harmonic of the activity in question. Table 4.13 lists typical minimum floor frequencies.

The natural frequency of a floor system can estimated as follows:

$$f_n = 0.18\sqrt{\frac{g}{\Delta_T}} \qquad (4.25)$$

where f_n = natural frequency of the floor system, Hz
g = 386 in./s^2
Δ_T = total floor deflection, in.
   = $\Delta_j + \Delta_g + \Delta_c$
Δ_j = joist deflection, in.
Δ_g = girder deflection, in.
Δ_c = column deflection, in.

For framing systems without girders or columns (such as wood floor framing supported by bearing walls), the girder and column deflection terms would be ignored. A more in-depth coverage of controlling vibrations due to rhythmic activity is given in other references [3].

As stated previously, rhythmic activity on wood-framed floors is not very common. Since wood-framed floors are used primarily for residential, multifamily, and hotel occupancies, the primary design consideration is with walking vibrations. With rhythmic activity, the accumulated energy due to a cyclical load was the primary area of concern. With walking vibrations, the main source of annoyance to the occupants is floor jolts due to a person walking across the floor.

Current research by the ATC, AISC, and others [3–6] is based on response data from occupants on various floor types and is used in this book. To control walking vibrations, floor deflections with respect to beam span should be controlled in accordance with the following equation [4]:

$$\Delta_p \leq 0.024 + 0.1e^{-0.18(L-6.4)} \leq 0.08 \text{ in.} \qquad (4.26)$$

where Δ_p = maximum floor deflection (in.) under a concentrated load of 225 lb
L = beam span, ft

This equation can also be shown graphically (Figure 4.32).

TABLE 4.13 Typical Minimum Floor Frequencies (Hertz)

Activity	Steel/Concrete Floor	Light-Frame Floor
Dancing and dining	5	10
Rhythmic activity	9	13

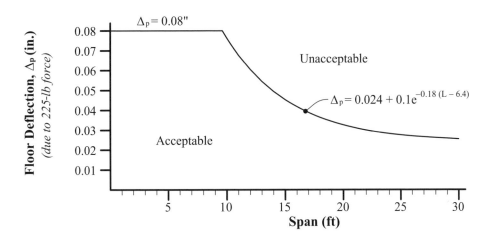

FIGURE 4.32 Floor deflection versus span curve for floor vibrations. (Adapted from Ref. 4, Figure 2-1. Courtesy of the Applied Technology Council, Arlington, VA.)

The formula for calculating the floor deflection under a 225-lb concentrated load is given as

$$\Delta_p = \frac{C_{pd}}{N_{eff}} \frac{PL^3}{48EI_{eff}} \quad (4.27)$$

where Δ_p = maximum floor deflection, in.
C_{pd} = continuity factor for point load
= 0.7 for continuous
= 1.0 for simple span
N_{eff} = number of effective joists, ≥ 1.0
P = 225 lb
L = beam span, in.
EI_{eff} = effective flexural stiffness of the floor panel, lb-in^2

In calculating the maximum floor deflection due to a point load, the deflection of the supports (i.e., girders and columns) can be ignored if the deflection under total load (dead load plus live load) is less than $L/360$ or 0.5 in., whichever is smaller.

The effective flexural stiffness of preengineered products such as I-joists or wood trusses is usually given by the manufacturer of the specific product. For sawn lumber or where product information is not available, the effective flexural stiffness can be calculated as follows:

$$EI_{eff} = \frac{EI}{1 + \gamma EI / C_{fn} EI_m} \quad (4.28)$$

where EI = stiffness of the floor panel, lb-in^2 (includes decking and concrete topping)
C_{fn} = 1.0 for simple spans (see Figure 4.33 for continuous spans)
E = modulus of elasticity, psi
G = modulus of rigidity, psi (for wood, use $G = 0.1 \times 10^6$ psi or $E/G = 20$)

$$\gamma = \frac{14.4}{(L/r)^2} \frac{E}{G} \quad \text{for sawn lumber} \quad (4.29)$$

$$\gamma = \frac{96EI_m}{K_s L^2} \quad \text{for preengineered joists} \quad (4.30)$$

L = joist span, in.
r = radius of gyration of the joist, in.
= $\sqrt{I_m/A_m}$
I_m = moment of inertia of the joist
K_s = shear deflection constant
= $(0.4 \times 10^6)d$ lb (for preengineered I-joists)
= $d \times 10^6$ lb (for preengineered metal web trusses)
= 2×10^6 lb (for preengineered metal plate–connected wood trusses)
d = joist depth, in.

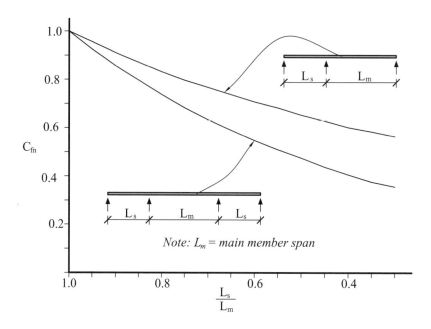

FIGURE 4.33
Continuity factor C_{fn} for continuous spans. (Adapted from Ref. 4, Figure 4-3. Courtesy of the Applied Technology Council, Arlington, VA.)

It should be noted that shear deflection is not critical for solid sawn members, but shear deflection is a factor to be considered for preengineered members because of the thinner webs and shear deformation at the truss joints.

With reference to Figure 4.34, the flexural stiffness of the floor panel, EI, can be determined taking into account the composite action of the decking and concrete topping. For noncomposite floor systems, the flexural stiffness of the floor panel is equal to the flexural stiffness of the joist:

$$EI = EI_m + EI_{top} + EA_m y^2 + EA_{top}(h_{top} - y)^2 \qquad (4.31)$$

where EI_m = flexural stiffness of the joist, lb-in^2
EI_{top} = flexural stiffness of the floor deck parallel to the joist, lb-in^2

$$EI_{top} = EI_{w,\,par} + EI_c + \frac{EA_{w,\,par} EA_c h_{cw}^3}{EA_{w,\,par} + EA_c} \qquad (4.32)$$

EA_{top} = effective axial stiffness of the floor deck parallel to the joist

$$EA_{top} = \frac{EA_{flr}}{1 + 10 EA_{flr}/S_{flr} L_{flr}^2} \qquad (4.33)$$

EA_m = axial stiffness of the joist
EA_{flr} = axial stiffness of the floor deck, lb (see $EA_{w,\,par}$ in Table 4.14 for wood panels)
EA_c = axial stiffness of the concrete deck, lb

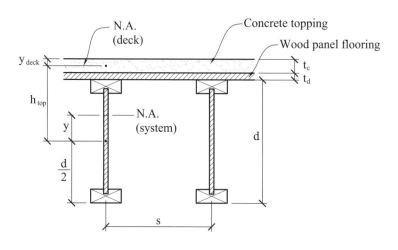

FIGURE 4.34 Typical floor section.

TABLE 4.14 Approximate Stiffness Properties of Wood Subflooring (OSB and Plywood) Per Inch Width[a]

Panel Thickness (in.)	$EI_{w,\,par}$ (lb-in²/in. × 10⁶)	$EI_{w,\,perp}$ (lb-in²/in. × 10⁶)	$EA_{w,\,par}$ (lb/in. × 10⁶)	$EA_{w,\,perp}$ (lb/in. × 10⁶)
$\frac{5}{8}$	0.008	0.027	0.35	0.52
$\frac{3}{4}$	0.016	0.040	0.46	0.65

[a] $EI_{w,\,par}$, flexural stiffness of the wood panel parallel to the member; $EI_{w,\,perp}$, flexural stiffness of the wood panel perpendicular to the member; $EA_{w,\,par}$, axial stiffness of the wood panel parallel to the member; $EA_{w,\,perp}$, axial stiffness of the wood panel perpendicular to the member.

EI_c = flexural stiffness of the concrete deck, lb. in.²
S_{flr} = slip modulus for floor deck connection to the joist (see Table 4.15)
L_{flr} = width of the floor deck (distance between joints in the floor deck), in.
= 48 in. or beam span, L in inches for concrete-topped floors
s = joist spacing, in.
h_{cw} = half of the total deck thickness, in.
= $\frac{1}{2}(t_d + t_c)$
h_{top} = distance from the centroid of the floor deck to the centroid of the joist, in.

$$h_{top} = \frac{d}{2} + t_d + t_c - y_{deck} \qquad (4.34)$$

$$y_{deck} = \frac{EA_{w,\,par}\,[(t_d/2) + t_c] + EA_c\,(t_c/2)}{EA_{w,\,par} + EA_c} \qquad (4.35)$$

y_{deck} = distance from the top of the concrete to the centroid of the floor deck
t_d = thickness of the deck (plywood)
t_c = thickness of the concrete
y = distance from the centroid of the system to the centroid of the joist, in.

$$y = \frac{EA_{top}h_{top}}{EA_m + EA_{top}} \qquad (4.36)$$

When calculating the axial and flexural stiffness of the concrete deck, the modulus of elasticity is multiplied by 1.2 to account for dynamic loading, thus:

$$E_c = (1.2)(33)w_c^{1.5}\sqrt{f'_c} \qquad (4.37)$$

where w_c = unit weight of the concrete, lb/ft³
f'_c = 28-day compressive strength of the concrete, psi

The base equation for the modulus of elasticity is given in ACI 318 [7].

The number of effective floor joists N_{eff} in the system is a function of the longitudinal and transverse stiffness of the floor panel. The presence of transverse blocking or bridging will increase the number of effective joists by transferring a concentrated load to adjacent joists, which reduces the point-load deflection of the floor system. However, the amount of stiffness that the blocking or bridging contributes to the system is very small when a concrete topping is used.

For wood-framed floors, the number of effective joists is determined as follows:

$$N_{eff} = \frac{1}{DF_b - DF_v} \qquad (4.38)$$

TABLE 4.15 Slip Modulus S_{flr} (lb/in./in.) for Wood Deck Fastening

Type of Deck	Nailed	Glued and Nailed
OSB or plywood	600	50,000
Tongue and Groove	5,800	100,000

where $DF_b = 0.0294 + 0.536K_1^{1/4} + 0.516K_1^{1/2} - 0.31K_1^{3/4}$ (4.39)

$DF_v = -0.00253 - 0.0854K_1^{1/4} + 0.079K_2^{1/2} - 0.00327K_2$ (4.40)

$K_1 = \dfrac{K_j}{K_j + \Sigma K_{bi}}$ (4.41)

$K_2 = \dfrac{\Sigma K_{vi}}{\Sigma K_{bi}}$ (4.42)

K_j = longitudinal stiffness = EI_{eff}/L^3 (4.43)

K_{bi} = stiffness of the transverse flexural components, lb/in.

= $0.585 EI_{bi} L/s^3$ for panel or deck components (4.44)

= $(2a/L)^{1.71} EI_x/s^3$ for strongbacks and straps (4.45)

K_{vi} = stiffness of the transverse shear components, lb/in.

= $(2a/L)^{1.71} E_v A/s$ for cross-bridging or solid blocking (4.46)

a = distance of the element to the closest end of the joist, in.

L = joist span, in.

s = joist spacing, in.

E_v = effective shear modulus, psi

= 1000 psi for 2 × 2 cross-bridging without strapping

= 2000 psi for 2 × 2 cross-bridging with strapping

= 2000 psi for 2× blocking without strapping

= 3000 psi for 2× blocking with strapping

A = effective shear area, in^2 = joist depth, in. × 1.5 in.

EI_x = effective flexural stiffness of strongbacks or straps

E = 1.35 × 10^6 psi for 2× strongbacks or strapping (with or without strapping)

EI_{bi} = flexural stiffness of floor panel or deck per unit width

= $EI_{w,\text{perp}}$ (Table 4.14) for wood deck panels

= $EI_{w,\text{perp}} + EI_c + \dfrac{EA_{w,\text{perp}} EA_c h_{cw}^3}{EA_{w,\text{perp}} + EA_c}$ for concrete on wood deck (4.47)

Remedial Measures for Controlling Floor Vibrations in Wood Framed Floors

Walking vibrations was found not to be critical for the two example problems considered in the previous section. However, there might be situations in practice where the vibration threshold as measured by the deflection limit) is exceeded and remedial measures have to be taken to mitigate the effects of excessive floor vibrations. The primary ways for controlling walking vibrations include, increasing the natural frequency of the floor system for floors in the low frequency range; adding mass to the floor to increase damping; and adding full height partitions to the floor also to increase damping. Some available options that could be used to mitigate walking vibration problems include [3, 4, 25]:

1. Reducing the span of the floor joist by adding a new column, if it is architecturally feasible. This increases the natural frequency of the floor system.
2. Using full-height partitions which increases the amount of damping in the floor.
3. Providing additional bridging between joists. This increases the stiffness of the floor in the direction perpendicular to the joist, thus increasing the number of joists participating in resisting the vibration effects.
4. Using a thicker flooring system, preferably one that has some structural strength such as hardwood flooring. This increases the floor stiffness and hence the natural frequency of the floor system.
5. Reinforcing the existing floor joists by using flitch plates on both sides of the joist. In the case of I-joist, plywood panels could be attached to both sides of the joist to increase the floor stiffness.

EXAMPLE 4.9

Floor Vibrations on Floor Framing with Sawn Lumber

Determine if a floor framed with 2 × 10 sawn lumber (DF-L, Select Structural) spaced at 16 in. o.c. and a 14 ft 0 in. simple span with $\frac{3}{4}$-in. plywood nailed to the framing is adequate for walking vibration. Assume a residential occupancy. Neglect the contribution from transverse bridging.

Solution:

Floor stiffness:
From the NDS Supplement, the properties of the 2 × 10 sawn lumber are

$$E = 1.9 \times 10^6 \text{ psi}$$

$$A_m = 13.88 \text{ in}^2$$

$$I_m = 98.9 \text{ in}^4$$

$$EI_{\text{top}} = EI_{w,\text{par}} = 0.016 \times 10^6 \text{ psi/in. (Table 4.14)}$$

$$(EI_{w,\text{par}})(S) = (0.016 \times 10^6)(16 \text{ in.}) = 0.256 \times 10^6 \text{ psi}$$

$$EA_{\text{flr}} = EA_{w,\text{par}} = 0.46 \times 10^6 \text{ lb/in. (Table 4.14)}$$

$$(EA_{w,\text{par}})(S) = (0.46 \times 10^6)(16 \text{ in.}) = 7.36 \times 10^6 \text{ lb}$$

$$S_{\text{flr}} = 600 \text{ lb/in./in. (Table 4.15)}$$

$$C_{\text{fn}} = 1.0 \text{ (simple span, see equation 4.28)}$$

$$L_{\text{flr}} = 48 \text{ in. (not a concrete-topped floor)}$$

From equation 4.34,

$$h_{\text{top}} = \frac{9.25 \text{ in.}}{2} + \frac{\frac{3}{4} \text{ in.}}{2} = 5 \text{ in.}$$

From equation 4.33,

$$EA_{\text{top}} = \frac{EA_{\text{flr}}}{1 + 10EA_{\text{flr}}/S_{\text{flr}}L_{\text{flr}}^2} = \frac{7.36 \times 10^6}{1 + (10)(7.36 \times 10^6)/(600)(48)^2} = 0.135 \times 10^6 \text{ lb}$$

From equation 4.36,

$$y = \frac{EA_{\text{top}} h_{\text{top}}}{EA_m + EA_{\text{top}}} = \frac{(0.135 \times 10^6)(5 \text{ in.})}{(1.9 \times 10^6)(13.88 \text{ in}^2) + (0.135 \times 10^6)} = 0.025 \text{ in.}$$

The relatively small value for y indicates that the plywood decking does not offer much stiffness.

From equation 4.31,

$$EI = EI_m + EI_{\text{top}} + EA_m y^2 + EA_{\text{top}}(h_{\text{top}} - y)^2$$

$$= (1.9 \times 10^6)(98.9) + (0.256 \times 10^6) + (1.9 \times 10^6)(13.88)(0.025)^2 + (0.135 \times 10^6)(5 - 0.025)^2$$

$$= 191 \times 10^6 \text{ lb-in}^2$$

From equation 4.29,

$$\gamma = \frac{14.4}{(L/r)^2} \frac{E}{G} = \frac{14.4}{(14 \times 12)^2 (13.88/98.9)} \left(\frac{1.9 \times 10^6}{0.1 \times 10^6}\right) = 0.069$$

where the radius of gyration is

$$r = \sqrt{\frac{I_m}{A_m}}, \text{ and therefore}$$

$$\frac{1}{r^2} = \frac{A_m}{I_m} = \frac{13.88 \text{ in}^2}{98.9 \text{ in}^4}$$

From equation 4.28, the effective flexural stiffness

$$EI_{\text{eff}} = \frac{EI}{1 + \gamma EI/C_{fn}EI_m} = \frac{191 \times 10^6}{1 + (0.069)(191 \times 10^6)/(1.0)(1.9 \times 10^6)(98.9)} = 178 \times 10^6 \text{ lb-in}^2$$

Number of effective joists:

From equation 4.43, the longitudinal stiffness is

$$K_j = \frac{EI_{\text{eff}}}{L^3} = \frac{178 \times 10^6}{(14 \times 12)^3} = 37.46$$

From equation 4.44, the stiffness of the transverse flexural deck components is

$$K_{bi} = \frac{0.585 EI_{bi} L}{s^3} = \frac{(0.585)(0.040 \times 10^6)(14 \times 12)}{16^3} = 960$$

($EI_{bi} = EI_{w,\text{perp}}$, Table 4.14)

From equation 4.41,

$$K_1 = \frac{37.46}{37.46 + 960} = 0.0376$$

$$K_2 = 0 \text{ (blocking/bridging contribution neglected)}$$

From equations 4.39 and 4.40, we obtain the following parameters:

$$DF_b = 0.0294 + 0.536 K_1^{1/4} + 0.516 K_1^{1/2} - 0.31 K_1^{3/4}$$

$$= 0.0294 + (0.536)(0.0376)^{1/4} + (0.516)(0.0376)^{1/2} - (0.31)(0.0376)^{3/4}$$

$$= \mathbf{0.339}$$

$$DF_v = -0.00253 - 0.0854 K_1^{1/4} + 0.079 K_2^{1/2} - 0.00327 K_2$$

$$= -0.00253 - (0.0854)(0.0376)^{1/4} + 0 - 0$$

$$= \mathbf{-0.0401}$$

From equation 4.38, the number of effective joists is

$$N_{\text{eff}} = \frac{1}{DF_b - DF_v} = \frac{1}{0.339 - (-0.0401)} = 2.64$$

From equation 4.27, the floor deflection under a 225 lb concentrated load is

$$\Delta_p = \frac{C_{\text{pd}}}{N_{\text{eff}}} \frac{PL^3}{48 EI_{\text{eff}}} = \left(\frac{1.0}{2.64}\right) \frac{(225)(14 \times 12)^3}{(48)(178 \times 10^6)} = 0.047 \text{ in.}$$

To control walking vibrations, the floor deflection calculated above must satisfy the following criteria from equation 4.26:

$$\Delta_p \leq 0.024 + 0.1e^{-0.18(L-6.4)} \leq 0.08 \text{ in.}$$

$$L = \text{span of joist, ft} = 14 \text{ ft}$$

$$\Delta_p \leq (0.024) + (0.1)e^{-0.18(14-6.4)}$$

$$= \mathbf{0.049 \text{ in.}} > 0.047 \text{ in.} \quad \text{OK}$$

$$< 0.08 \text{ in.}$$

Using Figure 4.32, the same result is obtained. Therefore, the calculated deflection (0.047 in.) is less than the maximum allowable deflection to control walking vibration (0.049 in.) and the floor is deemed adequate for walking vibrations.

EXAMPLE 4.10

Floor Vibrations on Floor Framing with Preengineered Joists

With reference to the floor section shown in Figure 4.35 and the following design criteria, determine if the floor is adequate for floor vibrations:

- 14-in. I-joists spaced 16 in. o.c.
- $L = 25$ ft 0 in. simple span
- $\frac{3}{4}$-in. plywood glued and nailed to the framing
- 1.5-in. concrete topping, $f'_c = 3000$ psi; $w_c = 110$ lb/ft^3
- 2×14 blocking is provided at midspan

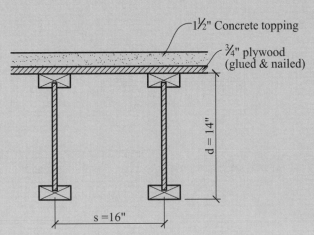

FIGURE 4.35 Floor section for preengineered joists.

Solution:

Floor stiffness:

The flexural and axial stiffness of the 14-in. I-joists are

$$EI_m = 600 \times 10^6 \text{ in}^2\text{-lb}$$

$$EA_m = 20 \times 10^6 \text{ lb}$$

(*Note:* The values above should be obtained from the joist manufacturer.)

$$EI_{w,\text{par}} = (0.016 \times 10^6 \text{ psi/in.})(16 \text{ in.}) = 0.256 \times 10^6 \text{ psi (Table 4.14)}$$

$$EI_{w,\text{perp}} = 0.040 \times 10^6 \text{ psi/in. (Table 4.14)}$$

$$EA_{w,\text{par}} = (0.46 \times 10^6 \text{ lb/in.})(16 \text{ in.}) = 7.36 \times 10^6 \text{ lb (Table 4.14)}$$

$$EA_{w,\text{perp}} = 0.65 \times 10^6 \text{ lb (Table 4.14)}$$

$$S_{\text{flr}} = 50{,}000 \text{ lb/in./in. (Table 4.15)}$$

$$C_{fn} = 1.0 \text{ (simple span, see equation 4.28)}$$

From equation 4.37, the dynamic modulus of elasticity and stiffness of the concrete deck are

$$E_c = (1.2)(33)\, w_c^{1.5} \sqrt{f_c'} = (1.2)(33)(110)^{1.5}\sqrt{3000} = 2.5 \times 10^6 \text{ psi}$$

$$EI_c = \frac{(2.5 \times 10^6)(16 \text{ in.})(1.5)^3}{12} = 11.25 \times 10^6 \text{ lb-in}^2$$

$$EA_c = (2.5 \times 10^6)(16 \text{ in.})(1.5) = 60 \times 10^6 \text{ lb}$$

The concrete deck and plywood sheathing contribute to the axial stiffness of the floor

$$EA_{\text{flr}} = EA_c + EA_{w,\text{par}} = (60 \times 10^6) + (7.36 \times 10^6) = 67.36 \times 10^6 \text{ lb}$$

From equation 4.35, the distance from the top of the concrete to the centroid of the floor deck is

$$y_{\text{deck}} = \frac{EA_{w,\text{par}}\,[(t_d/2) + t_c] + EA_c\,(t_c/2)}{EA_{w,\text{par}} + EA_c} = \frac{(7.36 \times 10^6)[(0.75/2) + 1.5] + (60 \times 10^6)(1.5/2)}{(7.36 \times 10^6) + (60 \times 10^6)} = 0.873 \text{ in.}$$

From equation 4.34, the distance from the centroid of the floor deck to the centroid of the joist is

$$h_{\text{top}} = \frac{d}{2} + t_d + t_c - y_{\text{deck}} = \frac{14}{2} + 0.75 + 1.5 - 0.873 = 8.38 \text{ in.}$$

The effecitve axial stiffness of the floor deck parallel to the joist is obtained from equation 4.33 as

$$EA_{\text{top}} = \frac{EA_{\text{flr}}}{1 + 10 EA_{\text{flr}}/S_{\text{flr}}L_{\text{flr}}^2} = \frac{67.36 \times 10^6}{1 + (10)(67.36 \times 10^6)/(50{,}000)(25 \times 12)^2} = 58.6 \times 10^6 \text{ lb}$$

The flexural stiffness of the floor deck parallel to the joist from equation 4.32 is

$$EI_{\text{top}} = EI_{w,\text{par}} + EI_c + \frac{EA_{w,\text{par}} EA_c h_{cw}^3}{EA_{w,\text{par}} + EA_c}$$

$$= (0.256 \times 10^6) + (11.25 \times 10^6) + \frac{(7.36 \times 10^6)(60 \times 10^6)[(0.75 + 1.5)/2]^3}{(7.36 \times 10^6) + (60 \times 10^6)}$$

$$= 20.9 \times 10^6 \text{ lb-in}^2$$

The distance from the centroid of the system to the centroid of the joist from equation 4.36 is

$$y = \frac{EA_{top}h_{top}}{EA_m + EA_{top}} = \frac{(58.6 \times 10^6)(8.38 \text{ in.})}{(20 \times 10^6) + (58.6 \times 10^6)} = 6.25 \text{ in.}$$

From equation 4.31, the flexural stiffness of the floor panel is

$$EI = EI_m + EI_{top} + EA_m y^2 + EA_{top}(h_{top} - y)^2$$
$$= (600 \times 10^6) + (20.9 \times 10^6) + (20 \times 10^6)(6.25)^2 + (58.6 \times 10^6)(8.38 - 6.25)^2$$
$$= 1668 \times 10^6 \text{ in}^2\text{-lb}$$

From equation 4.30,

$$\gamma = \frac{96 EI_m}{K_s L^2} = \frac{(96)(600 \times 10^6)}{(0.4 \times 10^6)(14 \text{ in.})(25 \times 12)^2} = 0.114$$

Note that $K_s = (0.4 \times 10^6)d$ for preengineered I-joists where d is the depth of the joist

$$EI_{eff} = \frac{EI}{1 + \gamma EI/C_{fn}EI_m} = \frac{1668 \times 10^6}{1 + (0.114)(1668 \times 10^6)/(1.0)(600 \times 10^6)} = 1267 \times 10^6 \text{ in}^2\text{-lb}$$

Number of effective joists:
From equation 4.43,

$$K_j = \frac{EI_{eff}}{L^3} = \frac{1267 \times 10^6}{(25 \times 12)^3} = 46.9$$

The flexural stiffness of the floor panel per unit width from equation 4.47 is

$$EI_{bi} = EI_{w, \text{perp}} + EI_c + \frac{EA_{w, \text{perp}} EA_c h_{cw}^3}{EA_{w, \text{perp}} + EA_c}$$

$$= (0.040 \times 10^6) + (0.704 \times 10^6) + \frac{(0.65 \times 10^6)(3.75 \times 10^6)[(0.75 + 1.5)/2]^3}{(0.65 \times 10^6) + (3.75 \times 10^6)}$$

$$= 1.53 \times 10^6 \text{ lb-in}^2/\text{in.}$$

From equation 4.44, the stiffness of the transverse flexural deck components is

$$K_{bi} = \frac{0.585 EI_{bi} L}{s^3} = \frac{(0.585)(1.53 \times 10^6)(25 \times 12)}{16^3} = 65,674 \text{ lb/in.}$$

From equation 4.46,

$$K_{vi} = \frac{(2a/L)^{1.71} E_v A}{s}$$

$$a = \frac{25 \text{ ft} \times 12}{2} = 150 \text{ in. (blocking at midspan)}$$

$$A = \text{joist depth, inches} \times 1.5 \text{ in.}$$

The effective shear modulus, E_v is 2000 psi for 2× blocking without strapping

$$K_{vi} = \frac{[(2)(150)/(25 \times 12)]^{1.71}(2000)(14)(1.5)}{16} = 2625 \text{ lb/in.}$$

From equation 4.41,

$$K_1 = \frac{K_j}{K_j + \Sigma K_{bi}} = \frac{46.9}{46.9 + 65{,}674} = 0.000714$$

From equation 4.42,

$$K_2 = \frac{\Sigma K_{vi}}{\Sigma K_{bi}} = \frac{2625}{65{,}674} = 0.03997$$

From equations 4.39 and 4.40, we calculate the following parameters:

$$\begin{aligned}
\mathrm{DF}_b &= 0.0294 + 0.536 K_1^{1/4} + 0.516 K_1^{1/2} - 0.31 K_1^{3/4} \\
&= 0.0294 + (0.536)(0.000714)^{1/4} + (0.516)(0.000714)^{1/2} - (0.31)(0.000714)^{3/4} \\
&= \mathbf{0.129} \\
\mathrm{DF}_v &= -0.00253 - 0.0854 K_1^{1/4} + 0.079 K_2^{1/2} - 0.00327 K_2 \\
&= -0.00253 - (0.0854)(0.000714)^{1/4} + (0.079)(0.03997)^{1/2} - (0.00327)(0.03997) \\
&= \mathbf{-0.00083}
\end{aligned}$$

The number of effective joists determined from equation 4.38 is

$$N_{\mathrm{eff}} = \frac{1}{\mathrm{DF}_b - \mathrm{DF}_v} = \frac{1}{(0.129) - (-0.00083)} = 7.68$$

From equation 4.27, the floor deflection under a 225 lb concentrated load is

$$\Delta_p = \frac{C_{pd}}{N_{\mathrm{eff}}} \frac{PL^3}{48 EI_{\mathrm{eff}}} = \left(\frac{1.0}{7.68}\right) \frac{(225)(25 \times 12)^3}{(48)(1267 \times 10^6)} = 0.013 \text{ in.}$$

To control walking vibrations, the floor deflection calculated above must satisfy the following criteria from equation 4.26.

$$\begin{aligned}
\Delta_p &\leq 0.024 + 0.1 e^{-0.18(L-6.4)} \leq 0.08 \text{ in.} \\
&\leq (0.024) + (0.1) e^{-0.18(25-6.4)} \\
&= \mathbf{0.028 \text{ in.}} > 0.013 \text{ in.} \quad \text{OK} \\
&< 0.08 \text{ in.}
\end{aligned}$$

Using Figure 4.32, the same result is obtained. Therefore, the calculated deflection (0.013 in.) is less than the maximum allowable deflection to control walking vibration (0.028 in.) and the floor is deemed adequate for walking vibrations.

REFERENCES

1. ICC (2006), *International Building Code*, International Code Council, Washington, DC.
2. ANSI/AF&PA American Forest & Paper Association, (2005), *National Design Specification for Wood Construction*, Washington, DC.
3. AISC (1997), *Floor Vibrations Due to Human Activity*, AISC/CISC Steel Design Guide Series 11, American Institute of Steel Construction, Chicago.
4. ATC (1997), *Minimizing Floor Vibration*, ATC Design Guide 1, Applied Technology Council, Redwood City, CA.
5. Allen, D. E., and Pernica, G. (1998), *Control of Floor Vibrations*, Construction Technology Update 22, National Research Council of Canada, Ottawa, Ontario, Canada.

6. Onysko, D. M. (1988), Performance criteria for residential floors based on consumer responses, *Proceedings of the 1988 International Conference on Timber Engineering,* Vol. 1, pp. 736–745.
7. ACI (2005), *Building Code Requirements for Structural Concrete and Commentary,* ACI 318-05, American Concrete Institute, Farmington Hills, MI.
8. Willenbrock, Jack H., Manbeck, Harvey B., and Suchar, Michael G. (1998), *Residential Building Design and Construction,* Prentice Hall, Upper Saddle River, NJ.
9. French, Samuel E. (1996), *Determinate Structures: Statics, Strength, Analysis and Design,* Delmar Publishers, Albany, NY.
10. Ambrose, James (1994), *Simplified Design of Wood Structures,* 5th ed., Wiley, New York.
11. Faherty, Keith F., and Williamson, Thomas G. (1995), *Wood Engineering and Construction,* McGraw-Hill, New York.
12. Halperin, Don A. and Bible, G. Thomas (1994), *Principles of Timber Design for Architects and Builders,* Wiley, New York.
13. Stalnaker, Judith J., and Harris, Earnest C. (1997), *Structural Design in Wood,* Chapman & Hall, London.
14. Breyer, Donald et al. (2003), *Design of Wood Structures—ASD,* 5th ed., McGraw-Hill, New York.
15. NAHB (2000), *Residential Structural Design Guide—2000,* National Association of Home Builders Research Center, Upper Marlboro, MD.
16. Cohen, Albert H. (2002), *Introduction to Structural Design: A Beginner's Guide to Gravity Loads and Residential Wood Structural Design,* AHC, Edmonds, WA.
17. Newman, Morton (1995), *Design and Construction of Wood-Framed Buildings,* McGraw-Hill, New York.
18. Kang, Kaffee (1998), *Graphic Guide to Frame Construction—Student Edition,* Prentice Hall, Upper Saddle River, NJ.
19. Kim, Robert H., and Kim, Jai B. (1997), *Timber Design for the Civil and Structural Professional Engineering Exams,* Professional Publications, Belmont, CA.
20. Hoyle, Robert J., Jr. (1978), *Wood Technology in the Design of Structures,* 4th ed., Mountain Press, Missoula, MT.
21. Kermany, Abdy (1999), *Structural Timber Design,* Blackwell Science, London.
22. ICC (2003), *International Residential Code,* International Code Council, Washington, DC.
23. Metwood Building Solutions, Reinforcer technologies by Metwood, http://www.metwood.com.
24. APA (1999), *Glulam Beam Camber,* Technical Note EWS S550E, April.
25. APA (2004), *Minimizing Floor Vibration by Design and Retrofit,* Technical Note E710, September.

PROBLEMS

4.1 Determine the maximum uniformly distributed floor live load that can be supported by a 10 × 24 girder with a simply supported span of 16 ft. Assume that the wood species and stress grade is Douglas fir-larch Select Structural, normal duration loading, normal temperature conditions, with continuous lateral support provided to the compression flange. The girder has a tributary width of 14 ft and a floor dead load of 35 psf in addition to the girder self-weight.

4.2 For the girder in Problem 4.1, determine the minimum bearing length required for a live load of 100 psf assuming that the girder is simply supported on columns at both ends.

4.3 Determine the joist size required to support a dead load of 12 psf (includes the self-weight of the joist) and a floor live load of 40 psf assuming a joist spacing of 24 in. and a joist span of 15 ft. Assume southern pine wood species, normal temperature, and dry service conditions.

4.4 Determine the joist size required to support a dead load of 10 psf (includes the self-weight of the joist) and a floor live load of 50 psf assuming a joist spacing of 16 in. and a joist span of 15 ft. In addition, the joists support an additional 10-ft-high partition wall weighing 10 psf running perpendicular to the floor joists at the midspan of the joists.

Assume spruce-pine-fir Select Structural wood species, normal temperature, and dry service conditions.

4.5 Design a DF-L tongue-and-grooved roof decking to span 14 ft between roof trusses. The decking is laid out in a pattern such that each deck sits on two supports. Assume a dead load of 15 psf, including the self-weight of the deck, and a snow load of 40 psf on a horizontal projected area. Assume that dry service and normal temperature conditions apply.

4.6 A 4 × 14 hem-fir Select Structural lumber joist that is part of an office floor framing is notched 2.5 in. on the tension face at the end supports because of headroom limitations. Determine the maximum allowable end support reaction.

4.7 A simply supported 6 × 12 girder of hem-fir No. 1 species supports a uniformly distributed load (dead plus floor live load) of 600 lb/ft over a span of 24 ft. Assuming normal moisture and temperature conditions and a fully braced beam, determine if the beam is structurally adequate for bending, shear, and deflection.

4.8 A $5\frac{1}{2}$ × 30 in. 24F–1.8E glulam beam spans 32 ft and supports a concentrated moving load of 5000 lb that can be located anywhere on the beam, in addition to its self-weight. Normal temperature and dry service conditions apply, and the beam is laterally braced at the supports and at midspan. Assuming a wood density of 36 lb/ft^3, a load duration factor C_D of 1.0, and the available bearing length l_b of 4 in., is the beam adequate for bending, shear, deflection, and bearing perpendicular to the grain?

4.9 The sloped 2 × 10 rafter shown in Figure 4.36 is supported on a ridge beam at the ridge line and on the exterior stud wall. The reaction from the dead load plus the snow load on the rafter at the exterior stud wall is 2400 lb. Assuming No. 1 hem-fir, normal temperature, and dry service conditions, determine the following:

  (a) The allowable bearing stresses at an angle to the grain F'_θ and the allowable bearing stress or allowable stress perpendicular to the grain $F'_{c\perp}$

  (b) The bearing stress at an angle to the grain, f_θ in the rafter

  (c) The bearing stress perpendicular to the grain $f_{c\perp}$ in the top plate

4.10 Determine if a floor framed with 2 × 12 sawn lumber (hem-fir, No. 1) spaced at 24 in. o.c. and a 15 ft 0 in. simple span with $\frac{3}{4}$-in. plywood glued and nailed to the framing is adequate for walking vibration. Assume a residential occupancy. Assume 2 × solid blocking at $\frac{1}{3}$ points.

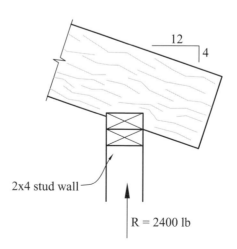

FIGURE 4.36 Sloped rafter detail.

chapter *five*

WOOD MEMBERS UNDER AXIAL AND BENDING LOADS

5.1 INTRODUCTION

Axially loaded members occur in several forms in a timber structure. Some of these include columns, wall studs, truss members, shear wall chords, diaphragm chords, and drag struts or collectors. Columns and wall studs are vertical compression members that carry gravity loads down to the foundation. Additionally, they may be subjected to lateral wind loads if located on the exterior face of the building. In a typical wood-framed building, axially compressed members such as columns and wall studs occur more frequently than members subjected to axial tension. Axial tension members occur most frequently as truss members.

The strength of axially compressed wood members is a function of the unbraced length of the member and the end support conditions, and these members fail either by buckling due to lateral instability caused by the slenderness of the member or by crushing of the wood fibers as they reach their material strength. A more critical load effect that must also be considered for axially loaded members is the case where bending occurs simultaneously with an axial load. In general, the four basic design cases for axially loaded members and the corresponding loading conditions considered in this chapter are as follows:

1. *Tension* (Figure 5.1a): truss members, diaphragm chords, drag struts, shear wall chords
2. *Tension plus bending* (Figure 5.1b): truss bottom chord members
3. *Compression* (Figure 5.1c): columns, wall studs, truss members, diaphragm chords, drag struts, shear wall chords
4. *Compression plus bending* (Figure 5.1d): exterior columns and wall studs, truss top chord members

Each of these load cases is examined in greater detail in several design examples throughout the chapter. In these examples we distinguish between the analysis and design of axially loaded wood members. In the analysis of member strength, the cross-sectional dimensions of the member, the wood species, and the stress grade are usually known, whereas the member strength is unknown and has to be determined. In design, the applied loads are usually known, but the member size, wood species, and stress grade are unknown and have to be determined.

5.2 PURE AXIAL TENSION: CASE 1

For this design case, axial tension parallel to the grain is considered. Examples of this load case may occur in truss web members and truss bottom chord members, diaphragm chords, drag struts, and shear wall chords. The NDS code [1] does not permit tension stress perpendicular to the grain (see NDS Section 3.8.2) because of the negligible tensile strength of wood in that direction. The basic design equation for axial tension stress parallel to grain is

$$f_t = \frac{T}{A_n} \leq F'_t \tag{5.1}$$

where T = applied tension force
A_n = net area at the critical section = $A_g - \Sigma A_{\text{holes}} = A_g - \Sigma(d_{\text{bolt}} + \frac{1}{8}$ in.$)$(thickness, b)
A_g = gross cross-sectional area of the member
ΣA_{holes} = sum of the area of the bolt holes perpendicular to the load
= (no. of bolts perpendicular to load)$(d_{\text{bolt}} + \frac{1}{8}$ in.$)$(member thickness b)
F'_t = allowable tension stress

Using the NDS code applicability table presented in Tables 3.1 and 3.2, the allowable tension stresses are calculated as

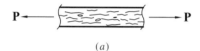

(a)

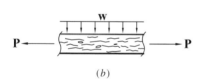

(b)

$$F'_t = \begin{cases} F_t C_D C_M C_t C_F C_i & \text{for sawn lumber} \\ F_t C_D C_M C_t & \text{for glulams} \end{cases} \tag{5.2} \tag{5.3}$$

where F_t = NDS-S tabulated design tension stress
C = adjustment factors, as discussed in Chapter 3

For bolted connections, the NDS code requires that the holes be at least $\frac{1}{32}$ to $\frac{1}{16}$ in. larger than the diameter of the bolt used (see NDS Section 11.1.2.2). However, it is practical to use $\frac{1}{8}$ in. for ease of installation, and this dimension is used in the examples in this book.

Design of Tension Members

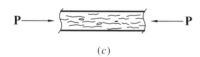

(c)

The design procedure for axially loaded tension members is given below and illustrated in Example 5.3.

1. Perform structural analysis to determine the member force.
2. Assume a trial member size initially.
3. Analyze the trial member to determine the tension stress applied and the tension stress (or tension load), that is allowable, as in Examples 5.1 and 5.2.
4. If the tension stress applied in the trial member is less than the allowable tension stress, the trial member size is deemed adequate. If the tension stress applied is *much* less than the allowable stress, the member is adequate but uneconomical, and in that case, the trial member size should be reduced and the analysis repeated until the tension stress applied is *just* less than the allowable tension stress.
5. If the tension stress applied is greater than the allowable tension stress, the trial size member size must be increased and the analysis repeated until the stress applied is just less than the allowable stress.

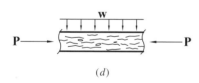

(d)

FIGURE 5.1 Basic design load cases: (a) tension; (b) tension plus bending; (c) compression; (d) compression plus bending.

EXAMPLE 5.1

Analysis of a Sawn-Lumber Tension Member

A 2 × 8 wood member (Figure 5.2) is subjected to an axial tension load of 6400 lb caused by dead load plus snow load. Determine the applied tension stress f_t and the allowable tension stress F'_t and check the adequacy of this member. The lumber is hem-fir No. 2, normal temperature conditions apply, and the moisture content is less than 19%. Assume that the connections are made with one row of $\frac{7}{8}$-in.-diameter bolts.

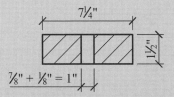

FIGURE 5.2 Cross section of a tension member at a bolt hole.

Solution: The gross area A_g is 10.88 in.² The area of the bolt holes

$$\Sigma A_{\text{holes}} = (1 \text{ hole})(\tfrac{7}{8} \text{ in.} + \tfrac{1}{8} \text{ in.})(1.5 \text{ in.}) = 1.5 \text{ in}^2$$

The net area of the member at the critical section is

$$A_n = A_g - \Sigma A_{\text{holes}} = 10.88 - 1.5 = 9.38 \text{ in}^2$$

The axial tension force applied, $T = 6400$ lb. Therefore, the tension stress applied is

$$f_t = \frac{T}{A_n} = \frac{6400 \text{ lb}}{9.38 \text{ in}^2} = \mathbf{682.3 \text{ psi}}$$

Since a 2 × 8 is dimension lumber, the applicable table is NDS-S Table 4A. The tabulated design tension stress and the adjustment factors from the table are

$F_t = 525$ psi
$C_D = 1.15$ (the C_D value for the shorter-duration load in the load combination is used, i.e. the snow load; see Section 3.1)
$C_M = 1.0$ (dry service conditions)
$C_t = 1.0$ (normal temperature conditions)
$C_F = 1.2$ (for tension stress)
$C_i = 1.0$ (assumed since no incision is prescribed)

Using the NDS applicability table (Table 3.1), the allowable tension stress is given as

$$F'_t = F_t C_D C_M C_t C_F C_i$$

$$= (525)(1.15)(1.0)(1.0)(1.2)(1.0) = \mathbf{725 \text{ psi}}$$

$$f_t = 682.3 \text{ psi} < F'_t = 725 \text{ psi} \quad \mathbf{OK}$$

Since the applied tension stress f_t is less than the allowable tension stress F'_t, the 2 × 8 hem-fir No. 2 is adequate.

EXAMPLE 5.2

Analysis of a Glulam Tension Member

For a $2\frac{1}{2} \times 6$ (four laminations) 5DF glulam axial combination member (Figure 5.3) subject to a pure tension load of 15,500 lb caused by dead load plus floor live load, calculate the applied tension stress f_t and the allowable tension stress F'_t, and check the structural adequacy of the member. Assume that normal temperature conditions apply, the moisture content is greater than 16%, and that the connection is made with one row of $\frac{3}{4}$-in.-diameter bolts.

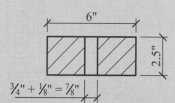

FIGURE 5.3 Cross section of a tension member at a bolt hole.

Solution: The gross area A_g for a $2\frac{1}{2} \times 6$ in. glulam is 15 in.² (*Note:* The glulam is specified using the actual size.) The area of the bolt holes

$$\Sigma A_{\text{holes}} = (1 \text{ hole})(\tfrac{3}{4} \text{ in.} + \tfrac{1}{8} \text{ in.})(2\tfrac{1}{2} \text{ in.}) = 2.19 \text{ in}^2$$

The net area at the critical section is

$$A_n = A_g - \Sigma A_{\text{holes}} = 15 - 2.19 = 12.81 \text{ in}^2$$

The tension stress applied f_t is

$$f_t = \frac{T}{A_n} = \frac{15{,}500 \text{ lb}}{12.81 \text{ in}^2} = \mathbf{1210 \text{ psi}}$$

For a 5DF glulam axial combination, use NDS-S Table 5B. The tabulated design tension stress and the applicable adjustment factors from the table are obtained as follows:

$F_t = 1600$ psi for 5DF
$C_{M(Ft)} = 0.80$ (wet service since MC > 16%)
$C_D = 1.0$ (the C_D value for the shortest-duration load in the load combination is used, i.e., floor live load; see Chapter 3)
$C_t = 1.0$ (normal temperature conditions)

Using the NDS applicability table (Table 3.2), the allowable tension stress is given as

$$F'_t = F_t C_D C_M C_t$$
$$= (1600)(1.0)(0.8)(1.0) = \mathbf{1280 \text{ psi}} > f_t \quad \mathbf{OK}$$

A $2\frac{1}{2} \times 6$ 5DF glulam tension member is adequate.

EXAMPLE 5.3

Design of a Wood Tension Member

Design member AF of the typical interior truss shown in Figure 5.4 assuming the following design parameters:

The trusses are spaced 4 ft 0 in. o.c.

The dead load is 22.5 psf and the snow load is 40 psf on a horizontal plan area.

The ends of the truss members are connected with one row of $\frac{1}{2}$-in.-diameter bolts.

The wood species/stress grade is No. 2 southern pine.

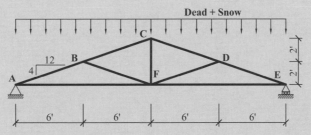

FIGURE 5.4 Roof truss profile.

Solution:

1. Analyze the truss to obtain the member forces (Figure 5.5). The tributary width of the truss is 4 ft.

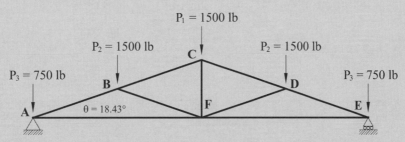

FIGURE 5.5 Free-body diagram of the roof truss.

$$\text{Dead load } D = 22.5 \text{ psf} \times 4 \text{ ft (tributary width of truss)} = 90 \text{ lb/ft}$$
$$\text{Snow load } S = 40 \text{ psf} \times 4\text{ft} = 160 \text{ lb/ft}$$
$$\text{Total load } w_{TL} = \mathbf{250 \text{ lb/ft}}$$

Calculate the joint loads:

$$P_1 = (6 \text{ ft}/2 + 6 \text{ ft}/2)(250 \text{ lb/ft}) = 1500 \text{ lb}$$
$$P_2 = (6 \text{ ft}/2 + 6 \text{ ft}/2)(250 \text{ lb/ft}) = 1500 \text{ lb}$$
$$P_3 = (6 \text{ ft}/2)(250 \text{ lb/ft}) = 750 \text{ lb}$$

Analyzing the truss using the method of joints or a structural analysis software program, the force in the truss bottom chord (member AF) is obtained as

$$T_{AF} = 6750 \text{ lb (tension)}$$

2. Assume the member size. Try a 2 × 8 (dimension lumber). Therefore, use NDS-S Table 4B.

3. From NDS-S Table 4B (southern pine dimension lumber), we obtain the tabulated design tension stress and the adjustment factors as follows. The design tension stress F_t is 650 psi. The adjustment or C factors are

$C_D = 1.15$ (the C_D value for the shortest-duration load in the load combination is used, i.e., the snow load; see Section 3.1)
$C_t = 1.0$ (normal temperature conditions)
$C_M = 1.0$ (dry service, since the truss members are protected from weather)
$C_F = 1.0$ (for Southern pine dimension lumber, for $d \leq 12$ in., $C_F = 1.0$) For $d > 12$ in., see the adjustment factors section of NDS-S Table 4B
$C_i = 1.0$ (assumed)

Using the NDS applicability table (Table 3.1), the allowable tension stress is given as

$$F_t' = F_t C_D C_M C_t C_F C_i$$

$$= (650)(1.15)(1.0)(1.0)(1.0)(1.0) = \mathbf{747.5 \text{ psi}}$$

From NDS-S Table 1B, the gross area A_g for a 2 × 8 is 10.88 in.² The available net area at the critical section is

$$A_n = A_g - \text{(no. of bolt holes perpendicular to force)} \Sigma (d_{\text{bolt}} + \tfrac{1}{8} \text{ in.})(\text{thickness } b)$$

$$= 10.88 - (1)(\tfrac{1}{2} \text{ in.} + \tfrac{1}{8} \text{ in.})(1.5 \text{ in.}) = 9.94 \text{ in}^2$$

4. Tension stress applied

$$f_t = \frac{T}{A_n} = \frac{6750 \text{ lb}}{9.94 \text{ in}^2} = \mathbf{679 \text{ psi}} < F_t' \qquad \mathbf{OK}$$

2 × 8 No. 2 southern pine is adequate for the truss bottom chord.

EXAMPLE 5.4

Design of a Wood Tension Member

Design member AF of the truss in Example 5.3 assuming that the wood species and stress grade is Douglas fir-larch Select Structural and the total load (caused by dead load plus snow load) is 500 lb/ft.

Solution:

1. Analyze the truss to obtain the member forces. The given total load (dead plus snow) w_{TL} is 500 lb/ft. Using the truss in Figure 5.5, we calculate the joint loads as follows:

$$P_1 = (6 \text{ ft}/2 + 6 \text{ ft}/2)(500 \text{ lb/ft}) = 3000 \text{ lb}$$

$$P_2 = (6 \text{ ft}/2 + 6 \text{ ft}/2)(500 \text{ lb/ft}) = 3000 \text{ lb}$$

$$P_3 = (6 \text{ ft}/2)(500 \text{ lb/ft}) = 1500 \text{ lb}$$

Analyzing the truss using the method of joints or a structural analysis software program, the force in the truss bottom chord (member AF) is obtained as

$$T_{AF} = 13{,}500 \text{ lb (tension)}$$

2. Try a 2 × 8, which is dimension lumber. Therefore, use NDS-S Table 4A.
3. From NDS-S Table 4A (DF-L Select Structural) we obtain the tabulated design tension stress and the adjustment factors as follows. The design tension stress F_t is 1000 psi. The adjustment or C factors are

C_D = 1.15 (the C_D value for the shortest-duration load in the load combination is used, i.e., the snow load)
C_t = 1.0 (normal temperature conditions)
C_M = 1.0 (dry service, since the truss members are protected from the weather)
$C_F(F_t)$ = 1.2
C_i = 1.0 (assumed)

Using the NDS applicability table (Table 3.1), the allowable tension stress is given as

$$F'_t = F_t C_D C_M C_t C_F C_i$$

$$= (1000)(1.15)(1.0)(1.0)(1.2)(1.0) = \mathbf{1380 \text{ psi}}$$

From NDS-S Table 1B, the gross area A_g for a 2 × 8 = 10.88 in.² The available net area at the critical section is

$$A_n = A_g - \text{(no. of bolt holes perpendicular to force)} \times \Sigma \, (d_{\text{bolt}} + \tfrac{1}{8} \text{ in.})(\text{thickness } b)$$

$$= 10.88 - (1)(\tfrac{1}{2} \text{ in.} + \tfrac{1}{8} \text{ in.}) \times (1.5 \text{ in.}) \simeq 9.94 \text{ in}^2$$

4. Applied tension stress

$$f_t = \frac{T}{A_n} = \frac{13{,}500 \text{ lb}}{9.94 \text{ in}^2} = \mathbf{1358 \text{ psi}} < F'_t \quad \mathbf{OK}$$

A 2 × 8 DF-L Select Structural is adequate for the truss bottom chord.

5.3 AXIAL TENSION PLUS BENDING: CASE 2

For members subjected to combined axial tension and bending loads, such as truss bottom chords, five design checks need to be investigated. It should be noted that for each design check, the controlling load duration factor based on the pertinent load combination for each design check should be used (see Chapter 3). The required design checks are as follows.

Design Check 1: Tension on the Net Area

Tension on the net area (Figure 5.6a) occurs at the ends of the member where the gross area of the member may be reduced due to the presence of bolt holes at the end connections (e.g., at the ends of truss members). The tension stress applied on the net area

$$f_t = \frac{T}{A_n} \leq F'_t \tag{5.4}$$

where T = tension force applied
allowable tension stress $F'_t = F_t C_D C_M C_t C_F C_i$
A_n = net area at the critical section = $A_g - \Sigma A_{\text{holes}}$
A_g = gross cross-sectional area of the tension member
ΣA_{holes} = sum of the area of the bolt holes perpendicular to the load
= (no. of bolts perpendicular to load)$(d_{\text{bolt}} + \tfrac{1}{8} \text{ in.})$(member thickness b)
F'_t = allowable tension stress

Using the NDS code applicability table (Tables 3.1 and 3.2), the allowable tension stresses are calculated as

$$F'_t = \begin{cases} F_t C_D C_M C_t C_F C_i & \text{for sawn lumber} \\ F_t C_D C_M C_t & \text{for glulams} \end{cases} \tag{5.2}$$
$$\tag{5.3}$$

where F_t = NDS-S tabulated design tension stress
$C_D, C_M, \ldots$ = adjustment factors, as discussed in Chapter 3

Design Check 2: Tension on the Gross Area

Tension on the gross area (Figure 5.6b) occurs at the midspan of the member where there are no bolt holes, hence the gross area is applicable (e.g., at the midspan of truss members and wall studs). The tension stress applied on the gross area

$$f_t = \frac{T}{A_g} \leq F'_t \tag{5.5}$$

The parameters in equation (5.5) are as defined in design check 1.

Design Check 3: Bending

The maximum bending moment and bending stress occur at the midspan of the member (Figure 5.6c) and the required design equation is

$$f_{bt} = \frac{M}{S_x} \leq F'_b \tag{5.6}$$

where f_{bt} = bending stress in the tension fiber
M = maximum bending moment in the member (typically, occurs at midspan)
S_x = section modulus about the axis of bending
F'_b = allowable bending stress

Using the applicability table (Tables 3.1 and 3.2), the allowable bending stress is obtained as

$$F'_b = \begin{cases} F_b C_D C_M C_t C_L C_F C_i C_r C_{fu} & \text{(for sawn lumber)} \\ F_b C_D C_M C_t C_L C_v C_{fu} C_c & \text{(for glulam)} \end{cases}$$

where F_b is the NDS-S tabulated design bending stress.

Design Check 4: Bending plus Tension

Bending plus tension (Figure 5.6d) occurs in the bottom or tension fibers at the midspan of the member where the moment and tension force are at their maximum values, creating the maximum tension stress in the tension or bottom fiber of the member. The interaction equation is given as

$$\frac{f_t}{F'_t} + \frac{f_{bt}}{F^*_b} \leq 1.0 \quad \text{(NDS code equation 3.9-1)} \tag{5.7}$$

where f_t = tension stress applied due to the axial tension force
f_{bt} = tension stress applied due to bending of the member
F'_t = allowable tension stress (see design check 1)

The allowable bending stress is

$$F^*_b = \begin{cases} F_b C_D C_M C_t C_F C_i C_r C_{fu} & \text{(for sawn lumber)} \tag{5.8} \\ F_b C_D C_M C_t C_V C_{fu} C_c & \text{(for glulam)} \tag{5.9} \end{cases}$$

It should be noted that the allowable bending stress above does not include the beam stability factor C_L because the fibers in the tension zone of a member subjected to combined bending plus axial tension are not susceptible to buckling instability since this can only be caused by compressive stresses. Therefore, the allowable bending stress for design is denoted as F^*_b.

Design Check 5: Bending minus Tension

Bending plus tension (Figure 5.6e) occurs in the top or compression fibers of the member subjected to combined axial tension plus bending, with the compression stresses caused by the bending of the member. The compression stress f_{bc} caused by the bending moment applied is opposed or counteracted by the tension stress resulting from the axial tension force P on the member. For this design check, the beam stability factor C_L has to be considered since there is a possibility that a net compression stress could exist in the top fibers

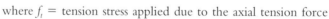
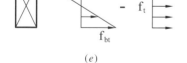

FIGURE 5.6 Stress diagrams for load case 2: (a) design check 1; (b) design check 2; (c) design check 3; (d) design check 4; (e) design check 5.

of the member. Where the axial tension stress f_t exceeds the compression stress due to bending f_{bc}, this design check will not be critical and can be ignored since this condition will produce a net tension stress in the top fibers which is less than the tension stress from design check 4.

The interaction equation for this design check is given as

$$\frac{f_{bc} - f_t}{F_b^*} \leq 1.0 \qquad \text{(NDS code equation 3.9-2)} \tag{5.10}$$

where the bending stress in the compression fibers $f_{bc} = M/S_x$. Note that f_{bc} is equal to f_{bt} from design check 3 because the wood member has a rectangular cross section. The allowable bending stress for sawn lumber is given as

$$F_b^{**} = F_b C_D C_M C_t C_F C_i C_r C_L C_{fu} \tag{5.11}$$

The allowable bending stress for glulam for this case is:

$$F_b^{**} = F_b C_D C_M C_t C_L C_{fu} C_c \tag{5.12}$$

It is important to note that it may not be necessary to perform all of the foregoing design checks for a given problem since some of the design checks could be eliminated by inspection.

Euler Critical Buckling Stress

The Euler critical buckling load for a column is the minimum load at which the column first bows out, and this load is given as

$$P_{cr} = \frac{\pi^2 EI}{(K_e l)^2} \tag{5.13}$$

where E = modulus of elasticity
I = moment of inertia about the axis of bending
$K_e l$ = effective length of the column = l_e
K_e = effective length factor (= 1.0 for column with pinned end supports)
l = unbraced length or height of column

For a rectangular column section with width b and depth d, the Euler critical buckling load becomes

$$P_{cr} = \frac{\pi^2 E(bd^3/12)}{(K_e l)^2} \tag{5.14}$$

Dividing both sides of the equation above by the cross-sectional area A (= bd) yields the Euler critical buckling stress

$$F_{cE} = \frac{\pi^2 E(bd^3/12)}{(K_e l)^2 bd} = \frac{(\pi^2/12)Ed^2}{(K_e l)^2} \tag{5.15}$$

This equation is simplified further to yield the NDS code equation for the Euler critical buckling stress, which is given as

$$F_{cE} = \frac{0.822 E}{(K_e l/d)^2} \tag{5.16}$$

In the NDS code, the modulus of elasticity E in equation (5.16) is replaced by the allowable buckling modulus of elasticity E'_{min}; hence, the equation becomes

$$F_{cE} = \frac{0.822 E'_{min}}{(K_e l/d)^2} \tag{5.17}$$

5.4 PURE AXIAL COMPRESSION: CASE 3

There are two categories of columns: solid columns and built-up columns. Examples of *solid columns* include interior building columns, struts (such as trench shores), studs (such as wall studs),

EXAMPLE 5.5

Axial Tension plus Bending—Truss Bottom Chord

The roof section of a building shown in Figure 5.7 has 40-ft-span trusses spaced at 2 ft 0 in. o.c. The roof dead load is 15 psf, the snow load is 35 psf, and the ceiling dead load is 15 psf of horizontal plan area. Design a typical bottom chord for the truss assuming the following design parameters:

- No. 1 & Better hem-fir.
- Moisture content ≤ 19%, and normal temperature conditions apply.
- The members are connected with a single row of $\frac{3}{4}$-in.-diameter bolts.

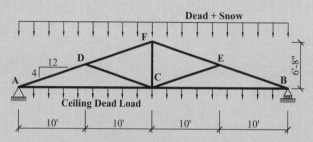

FIGURE 5.7 Roof truss elevation.

Solution: Calculate the joint loads. The given loads in psf of horizontal plan area are as follows:

$$\text{Roof dead load } D \text{ (roof)} = 15 \text{ psf}$$

$$\text{Ceiling dead load } D \text{ (ceiling)} = 15 \text{ psf}$$

$$\text{Snow load } S = 35 \text{ psf}$$

Since the tributary area for a typical roof truss is 80 ft² (i.e., 2-ft tributary width × 40-ft span), the roof live load L_r from equation (2.4) will be 20 psf, which is less than the snow load S. Therefore, the controlling load combination from Chapter 2 will be dead load plus snow load (i.e., $D + S$).

$$\text{Total load on roof } w_{TL} = (15 + 35 \text{ psf})(2\text{-ft tributary width}) = 100 \text{ lb/ft}$$

$$\text{Total ceiling load} = (15 \text{ psf})(2 \text{ ft}) = 30 \text{ lb/ft}$$

The concentrated gravity loads at the truss joints are calculated as follows:

Roof dead load + snow load:

$$P_A(\text{top}) = (10 \text{ ft}/2)(100 \text{ lb/ft}) = 500 \text{ lb}$$

$$P_D = (10 \text{ ft}/2 + 10 \text{ ft}/2)(100 \text{ lb/ft}) = 1000 \text{ lb}$$

$$P_F = (10 \text{ ft}/2 + 10 \text{ ft}/2)(100 \text{ lb/ft}) = 1000 \text{ lb}$$

$$P_E = (10 \text{ ft}/2 + 10 \text{ ft}/2)(100 \text{ lb/ft}) = 1000 \text{ lb}$$

$$P_B(\text{top}) = P_A(\text{top}) = 500 \text{ lb}$$

Ceiling dead loads:

$$P_A(\text{bottom}) = (20 \text{ ft}/2)(30 \text{ lb/ft}) = 300 \text{ lb}$$

$$P_B(\text{bottom}) = (20 \text{ ft}/2)(30 \text{ lb/ft}) = 300 \text{ lb}$$

$$P_C = (20 \text{ ft}/2 + 20 \text{ ft}/2)(30 \text{ lb/ft}) = 600 \text{ lb}$$

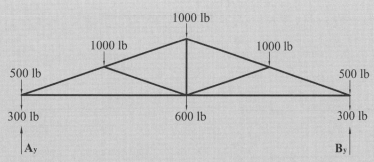

FIGURE 5.8 Free-body diagram of the roof truss.

Analyzing the truss in Figure 5.8 using the method of joints *or* computer analysis software, we obtain the tension force in the bottom chord member (member *AC* or *CB*) as $T_{AC} = 5400$ lb. In addition to the tension force on member *AC*, there is also a uniform ceiling load of 30 lb/ft acting on the truss bottom chord. The free-body diagram of member *AC* is shown in Figure 5.9.

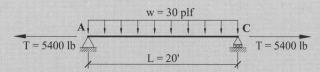

FIGURE 5.9 Free-body diagram of member *AC*.

Design

Assume 2 × 10 sawn lumber. The gross cross-sectional area and the section modulus can be obtained from NDS-S Table 1B as follows:

$$A_g = 13.88 \text{ in}^2$$

$$S_{xx} = 21.4 \text{ in}^3$$

Since the trial member is dimension lumber, NDS-S Table 4A is applicable. From the table we obtain the tabulated design stresses and stress adjustment factors.

$F_b = 1100$ psi (tabulated bending stress)
$F_t = 725$ psi (tabulated tension stress)
$C_F(F_b) = 1.1$ (size adjustment factor for bending stress)
$C_F(F_t) = 1.1$ (size adjustment factor for tension stress)
$C_D = 1.15$ (for design check with snow load)
$ = 0.9$ (for design check with dead load only)
$C_L = 1.0$ (assuming lateral buckling is prevented by the ceiling and bridging)
$C_r = 1.15$ (all three repetitive member requirements are met; see Section 3.1.9)

Design Check 1:

The tension force applied *T* is 5400 lb (caused by dead load + snow load). The net area

$$A_n = A_g - \Sigma A_{\text{holes}} = 13.88 - (1 \text{ hole})(\tfrac{3}{4} \text{ in.} + \tfrac{1}{8} \text{ in.})(1.5 \text{ in.}) = 12.57 \text{ in}^2$$

From equation (5.4) the tension stress applied at the supports

$$f_t = \frac{T}{A_n} = \frac{5400 \text{ lb}}{12.57 \text{ in}^2} = 430 \text{ psi}$$

Using the NDS applicability table (Table 3.1), the allowable tension stress is given as

$$F'_t = F_t C_D C_M C_t C_F C_i$$

$$= (725)(1.15)(1.0)(1.0)(1.1)(1.0) = \mathbf{917 \text{ psi}} > f_t \quad \mathbf{OK}$$

Design Check 2:
From equation (5.5), the tension stress applied at the midspan

$$f_t = \frac{T}{A_g} = \frac{5400 \text{ lb}}{13.88 \text{ in}^2} = \mathbf{389 \text{ psi}} < F'_t = 917 \text{ psi} \quad \mathbf{OK}$$

Design Check 3:
The maximum moment which for this member occurs at the midspan is given as

$$M_{max} = \frac{(30 \text{ lb/ft})(20 \text{ ft})^2}{8} = 1500 \text{ ft-lb} = 18,000 \text{ in.-lb}$$

This moment is caused by the ceiling *dead* load only (since no ceiling live load is specified), therefore, the controlling load duration factor C_D is 0.9. From equation (5.6) the bending stress applied (i.e., tension stress due to bending) is

$$f_{bt} = \frac{M_{max}}{S_x} = \frac{18,000 \text{ in.-lb}}{21.4 \text{ in}^3} = 842 \text{ psi}$$

Using the NDS applicability table (Table 3.1), the allowable bending stress is given as

$$F'_b = F_b C_D C_M C_t C_L C_F C_i C_r C_{fu}$$

$$= (1100)(0.9)(1.0)(1.0)(1.0)(1.1)(1.0)(1.15)(1.0)$$

$$= \mathbf{1252 \text{ psi}} > f_{bt} \quad \mathbf{OK}$$

Design Check 4:
For this case, the applicable load is the dead load plus the snow load; therefore, the controlling load duration factor C_D is 1.15 (see Chapter 3). Using the NDS applicability table (Table 3.1), the allowable axial tension stress and the allowable bending stress are calculated as follows:

$$F'_t = F_t C_D C_M C_t C_F C_i$$

$$= (725)(1.15)(1.0)(1.0)(1.1)(1.0) = 917 \text{ psi}$$

$$F^*_b = F_b C_D C_M C_t C_F C_i C_r C_{fu}$$

$$= (1100)(1.15)(1.0)(1.0)(1.1)(1.0)(1.15)(1.0) = 1600 \text{ psi}$$

The axial tension stress applied f_t at the midspan is 389 psi, as calculated for design check 2, while the tension stress applied due to bending f_{bt} is 842 psi, as calculated in design check 3. From equation (5.7) the combined axial tension plus bending interaction equation for the stresses in the tension fiber of the member is given as

$$\frac{f_t}{F'_t} + \frac{f_{bt}}{F^*_b} \leq 1.0$$

Substitution into the interaction equation yields

$$\frac{389 \text{ psi}}{917 \text{ psi}} + \frac{842 \text{ psi}}{1600 \text{ psi}} = \mathbf{0.95} < 1.0 \quad \mathbf{OK}$$

Design Check 5:
The compression stress applied due to bending is

$$f_{bc} = \frac{M_{max}}{S_x} = \frac{18{,}000 \text{ in.-lb}}{21.4 \text{ in}^3} = 842 \text{ psi} \ (= f_{bt} \text{ since this is a rectangular cross section})$$

It should be noted that f_{bc} and f_{bt} are equal for this problem because the wood member is rectangular in cross section. The load duration factor for this case C_D is 1.15 (combined dead + snow loads). Using the NDS applicability table (Table 3.1), the allowable compression stress due to bending is

$$F_b^{**} = F_b C_D C_M C_t C_L C_F C_i C_r C_{fu}$$

$$= (1100)(1.15)(1.0)(1.0)(1.0)(1.1)(1.0)(1.15)(1.0) = 1600 \text{ psi}$$

From equation (5.10) the interaction equation for this design check is given as

$$\frac{f_{bc} - f_t}{F_b^{**}} \leq 1.0$$

Substituting in the interaction equation yields

$$\frac{842 \text{ psi} - 389 \text{ psi}}{1600 \text{ psi}} = \mathbf{0.283 \ <<< \ 1.0} \qquad \mathbf{OK}$$

From all of the steps above, we find that all five design checks are satisfied; therefore, use 2 × 10 hem-fir No. 1 & Better for the bottom chord. Note that if any one of the design checks above had not been satisfied, we would have had to increase the member size and/or stress grade until all five design checks are satisfied.

---

and truss members. *Built-up columns* are made up of individual laminations joined mechanically with no spaces between laminations. The mechanical connection between the laminations is made with through-bolts or nails. Section 15.3 of the NDS code gives the mechanical connection requirements for built-up columns. Built-up columns are commonly used within interior or exterior walls of wood-framed buildings because the component members that make up the built-up column are already available on site and can readily be fastened together on site.

The strength of a column is affected by buckling if the column length becomes excessive. There are two types of columns with respect to the column length: short columns and long columns. A *short column* is one in which the axial strength depends only on the crushing or material capacity of the wood. A *long column* is one in which buckling occurs before the column reaches the crushing or material capacity $F_c' A$ of the wood member. The strength of a long column is a function of the slenderness ratio $K_e l/r$ of the column, which is defined as the ratio of the effective length $K_e l$ of the column to the least radius of gyration r of the member. The radius of gyration is given as

$$r = \sqrt{\frac{I}{A}} \qquad (5.18)$$

where I = moment of inertia of the column member about the axis of buckling
A = cross-sectional area of the column

The required allowable stress design (ASD) equation for this load case is

$$f_c = \frac{P}{A_g} \leq F_c' \qquad (5.19)$$

where f_c = axial compression stress
P = axial compression force applied in the wood column
A_g = gross cross-sectional area of the tension member
F_c' = allowable compression stress

Using the NDS adjustment factors applicability tables (Tables 3.1 and 3.2), the allowable compressive stress is obtained as

$$F'_c = \begin{cases} F_c C_D C_M C_t C_F C_i C_P & \text{(sawn lumber)} \\ F_c C_D C_M C_t C_P & \text{(glulam)} \end{cases} \quad \begin{array}{l}(5.20a)\\(5.20b)\end{array}$$

The adjustment factors C_D, C_M, C_t, C_i, and C_F are as discussed in Chapter 3, and the column stability factor C_P depends on the unbraced length of the column. The higher the column slenderness ratio ($K_e l/r$), the smaller the column stability factor C_P. For columns that are laterally braced over their entire length, the column stability factor C_P is 1.0. For all other cases, calculate C_P using

$$C_P = \frac{1 + F_{cE}/F_c^*}{2c} - \sqrt{\left(\frac{1 + F_{cE}/F_c^*}{2c}\right)^2 - \frac{F_{cE}/F_c^*}{c}} \quad (5.21)$$

where F_{cE} = Euler critical buckling stress = $\dfrac{0.822 E'_{min}}{(l_e/d)^2}$ [see equation (5.17)]

$$E'_{min} = E_{min} C_M C_t C_i \quad (5.22)$$

(E_{min} is the tabulated minimum elastic modulus for buckling calculations and is a function of the pure bending modulus of elasticity E with a factor of safety of 1.66 applied, in addition to a 5% lower exclusion value on the pure bending modulus of elasticity)

l_e = effective length of column = $K_e l$

l = unbraced column length (about the axis of buckling)

l_x = unbraced length for buckling about the x–x axis (see Figure 5.10)

l_{y1} or l_{y2} = unbraced length for buckling about the y–y axis (see Figure 5.10)

(the larger of l_{y1} or l_{y2} is used in the buckling calculations because the larger the unbraced length, the lower the allowable compressive stress)

K_e = buckling length coefficient (see Table 5.1)

(for most columns in wood buildings, pinned conditions are usually assumed at both ends of the column, and consequently, the effective length factor K_e is 1.0)

d = dimension of the column section *perpendicular* to the axis of buckling. For x–x axis buckling, use $d = d_x$; for y–y axis buckling, use $d = d_y$; (the NDS code limits the slenderness ratio l_e/d to 50 to minimize the impact of slenderness on wood columns; when the l_e/d ratio is greater than 50, the designer would need to increase the column size and/or reduce the unbraced length of the column by providing lateral bracing, if possible, to bring the slenderness ratio to within the limit of 50;

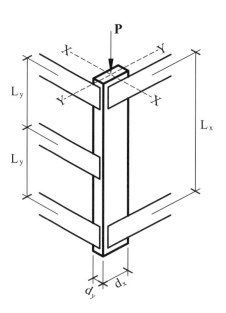

FIGURE 5.10 Column with unequal unbraced lengths.

TABLE 5.1 Buckling Length Coefficients, K_e

Buckling modes						
Theoretical K_e value	0.50	0.70	1.0	1.0	2.0	2.0
Recommended design K_e when ideal conditions are approximated	0.65	0.80	1.2	1.0	2.1	2.4
End condition legend	Rotation fixed, translation fixed Rotation free, translation fixed Rotation fixed, translation free Rotation free, translation free					

Source: Adapted from Ref. 1.

it should be noted that the NDS code allows slenderness ratios of up to 75, but only during construction)

F_c^* = crushing strength of column at zero slenderness ratio (i.e., strength when $l_e/d = 0$)

= F_c (All C-factors except C_p) (5.23)

c = interaction factor = 0.8 (for sawn lumber)
= 0.9 (for glulam)
= 0.85 (for round timber poles and piles)

Built-up Columns

Built-up columns (Figure 5.11) are commonly used at the end of shear walls as tension or compression chord members. They can also be used in lieu of a solid post within a stud wall to support heavy concentrated loads. Section 15.3 of the NDS code [1] lists the minimum basic provisions that apply to built-up columns, and these are listed here for reference:

- Each lamination is to be rectangular in cross section.
- Each lamination is to be at least 1.5 in. thick.
- All laminations are to have the same depth.
- Faces of the adjacent laminations are to be in contact.
- All laminations are full column length.
- The nailed or bolted connection requirements of Sections 15.3.3 and 15.3.4 of the NDS code are to be met.

The allowable compressive stress in a built-up column is calculated as

$$F_c' = F_c C_D C_M C_t C_F C_i C_P$$

where $C_P = K_f \left[\dfrac{1 + F_{cE}/F_c^*}{2c} - \sqrt{\left(\dfrac{1 + F_{cE}/F_c^*}{2c}\right)^2 - \dfrac{F_{cE}/F_c^*}{c}} \right]$ (5.24)

$F_c^* = F_c C_D C_M C_t C_F C_i$

EXAMPLE 5.6

Column Axial Load Capacity in Pure Compression

Design a column for an axial compression dead load plus snow load of 17 kips given the following parameters:

No. 1 Douglas fir-larch wood species.

Column unbraced lengths $l_x = l_y = 12$ ft.

Dry service and normal temperature conditions apply.

Solution: Several design aids for selecting an appropriate column size for a given load are given in Appendix B. Entering Figure B.10 with an applied load of 17 kips and an unbraced length of 12 ft, a trial size of 6 × 6 is indicated. It is important to note that the load duration factor C_D is 1.0 for this chart. Since a load duration factor C_D of 1.15 is given for this problem, it can be seen by inspection that the actual capacity of the 6 × 6 would be slightly higher than 17 kips for this example.

A 6 × 6 member is a post and timber (P&T); therefore, NDS-S Table 4D is applicable. From the table the stress adjustment or C factors are

$$C_M = 1.0 \text{ (normal moisture conditions)}$$

$$C_t = 1.0 \text{ (normal temperature conditions)}$$

$$C_D = 1.15 \text{ (snow load controls for the given load combination of } D + S)$$

$$C_i = 1.0 \text{ (assumed)}$$

From the table the tabulated design stresses are obtained as follows:

$$F_c = 1000 \text{ psi (compression stress parallel to grain)}$$

$$E = 1.6 \times 10^6 \text{ psi (reference modulus of elasticity)}$$

$$E_{min} = 0.58 \times 10^6 \text{ psi (buckling modulus of elasticity)}$$

$$C_F = (12/d)^{1/9} \leq 1.0$$

$$= 1.0 \text{ for } d < 12 \text{ in.; therefore, } C_F = 1.0$$

The column dimensions are $d_x = 5.5$ in. and $d_y = 5.5$ in. The cross-sectional area $A = 30.25$ in^2. The effective length factor K_e is assumed to be 1.0, as discussed earlier in the chapter; therefore, the slenderness ratios about the x–x and y–y axes are given as

$$\frac{l_e}{d_x} = \frac{(1.0)(12 \text{ ft}) \times (12 \text{ in./ft})}{5.5 \text{ in.}} = 26.2 < 50 \quad \text{OK}$$

$$\frac{l_e}{d_y} = \frac{(1.0)(12 \text{ ft}) \times (12 \text{ in./ft})}{5.5 \text{ in.}} = 26.2 < 50 \quad \text{OK}$$

Using the applicability table (Table 3.1), the allowable modulus of elasticity for buckling calculations is given as

$$E'_{min} = E_{min} C_M C_t C_i$$

$$= (0.58 \times 10^6)(1.0)(1.0)(1.0) = 0.58 \times 10^6 \text{ psi}$$

$$c = 0.8 \text{ (sawn lumber)}$$

The Euler critical buckling stress is calculated as

$$F_{cE} = \frac{0.822 E'_{min}}{(l_e/d)^2} = \frac{(0.822)(0.58 \times 10^6)}{(26.2)^2} = 695 \text{ psi}$$

$$F_c^* = F_c C_D C_M C_t C_F C_i$$
$$= (1000)(1.15)(1.0)(1.0)(1.0)(1.0) = 1150 \text{ psi}$$

$$\frac{F_{cE}}{F_c^*} = \frac{695 \text{ psi}}{1150 \text{ psi}} = 0.604$$

From equation (5.21) the column stability factor is calculated as

$$C_P = \frac{1 + 0.604}{(2)(0.8)} - \sqrt{\left[\frac{1 + 0.604}{(2)(0.8)}\right]^2 - \frac{0.604}{0.8}} = 0.502$$

The allowable compression stress

$$F'_c = F_c^* C_P = (1150)(0.502) = 577.3 \text{ psi}$$

The allowable compression load capacity of the column is

$$P_{allowable} = F'_c A_g = (577.3 \text{ psi})(30.25 \text{ in}^2)$$
$$= 17{,}463 \text{ lb} = 17.5 \text{ kips} > 17 \text{ kips} \quad \text{OK}$$

Use a 6 × 6 No. 1 DF-L column.

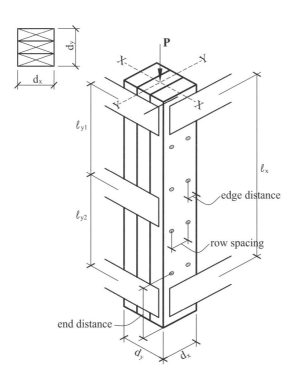

d_x = column dimension parallel to the face of the laminations in contact with each other

d_y = column dimension perpendicular to the face of the laminations

ℓ_x = effective column length for buckling about the x–x axis = $K\ell_x$

ℓ_y = effective column length for buckling about the y–y axis = $K\ell_y$

FIGURE 5.11 Built-up column elevation and cross section.

K_f = 0.60 (for nailed columns when calculating F_{cE} using L_{ey}/d_y)
0.75 (for bolted columns when calculating F_{cE} using L_{ey}/d_y)
1.0 (for nailed or bolted columns when calculating F_{cE} using L_{ex}/d_x)

The first two K_f values above are used when the weak axis of the column controls (i.e., when l_{ey}/d_y is greater than l_{ex}/d_x). Where buckling about the strong axes controls or when l_{ex}/d_x is greater than l_{ey}/d_y, K_f is 1.0.

It should be noted that where the built-up column is embedded within a sheathed wall such that it is continuously braced for buckling about the weak axis, the column stability factor C_{pyy} will be 1.0. The K_f factor in equation (5.24) applies only to weak-axis buckling and accounts for the reduction in axial load capacity of the column due to interface shear between the laminations of the built-up column. For the minimum nailing or bolting requirements for built-up columns, the reader is referred to Sections 15.3.3 and 15.3.4 of the NDS code [1].

P–Delta Effects in Members under Combined Axial and Bending Loads

The bending moments in a member supported at both ends and subjected to combined axial compression plus bending are amplified by the presence of that axial load. The magnitude of the moment amplification is a function of the slenderness ratio of the member. The amplification of the bending moment in the presence of axial compression load is known as the *P–δ effect*. This concept is illustrated by the free-body diagram in Figure 5.13.

Consider a pin-ended column subjected to combined axial load P and a lateral bending load W at the midheight as shown in Figure 5.13. The lateral load causes lateral deflection of the column, which is further amplified by the presence of the axial compression load P until equilibrium is achieved at a deflection δ; hence, it is called the P–δ effect.

Summing moments about point B in the free-body diagram of Figure 5.13b, we obtain

$$-P\delta - \frac{W}{2}\frac{L}{2} + M = 0$$

therefore,

$$M = \frac{WL}{4} + P\delta \quad (5.25)$$

where M = maximum second-order or final moment
$WL/4$ = maximum first-order or maximum moment in the absence of the axial load P
δ = maximum lateral deflection

The maximum second-order moment M includes the moment magnification or the *P–δ moment*. Thus, the presence of the axial compression load leads to an increase in the column moment, but the magnitude of the increase depends on the slenderness ratio of the column. The moment magnification effect is accounted for in the NDS code by multiplying the bending stress term f_b/F_b' by the magnifier $1/(1 - f_c/F_{cE})$, (see equation 5.31) where f_c is the applied axial compression stress in the column and F_{cE} is the Euler critical buckling stress about the axis of buckling as calculated in equation (5.17).

5.5 AXIAL COMPRESSION PLUS BENDING: CASE 4

For members subjected to combined axial compression and bending loads, there are four design checks that need to be investigated. It is important to note that for each design check, the controlling load duration factor based on the pertinent load combination for each design check should be used (see Chapter 3). The four design checks required are as follows.

Design Check 1: Compression on the Net Area

Compression on the net area (Figure 5.14a) occurs at the ends of the member (i.e., at the connections) and the column stability factor C_p is 1.0 for this case since there can be no buckling at the ends of the member. The design equation required is

EXAMPLE 5.7

Axial Load Capacity of a Built-up Column

Calculate the axial load capacity of a 12-ft-long nailed built-up column consisting of three 2 × 6's (Figure 5.12), assuming Douglas fir-larch No. 1 lumber, normal temperature and moisture conditions, and a load duration factor C_D of 1.0.

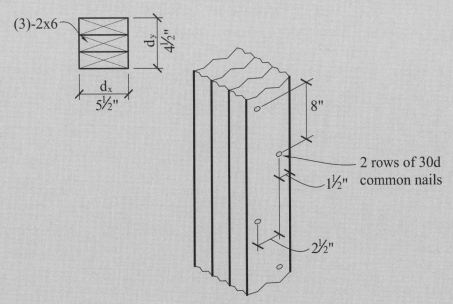

FIGURE 5.12 Built-up column cross section.

Solution: For three 2 × 6's, $d_x = 5.5$ in., $d_y = (3)(1.5\text{ in.}) = 4.5$ in. (see NDS-S Table 1B). Since a 2 × 6 is dimension lumber, NDS-S Table 4A is applicable. From the table the tabulated design stresses are obtained as

$$F_c = 1500 \text{ psi}$$

$$E_{\min} = 0.62 \times 10^6 \text{ psi}$$

From NDS-S Table 4A, the size factor for axial compression stress parallel to grain is

$$C_{F(F_c)} = 1.1 \text{ (size factor for axial compression stress parallel to the grain)}$$

The following stress adjustment or C factors were specified in the problem:

$$C_M = 1.0 \text{ (normal moisture conditions)}$$

$$C_t = 1.0 \text{ (normal temperature conditions)}$$

$$C_D = 1.0$$

$$C_i = 1.0 \text{ (assumed)}$$

The unbraced length of the column $l_x = l_y = 12$ ft. $K_e = 1.0$ (building columns are typically assumed to be pinned at both ends).

$$\frac{l_{ex}}{d_x} = \frac{K_e l_x}{d_x} = \frac{(1.0)(12 \text{ ft})(12 \text{ in./ft})}{5.5} = 19.9 < 50 \quad \textbf{OK}$$

$$\frac{l_{ey}}{d_y} = \frac{K_e l_y}{d_y} = \frac{(1.0)(12 \text{ ft})(12 \text{ in./ft})}{4.5} = \textbf{32} < 50 \quad \textbf{OK}$$

Since $l_{ey}/d_y > l_{ex}/d_x$, the weak-axis slenderness ratio (l_{ey}/d_y) governs; therefore, the controlling column stability factor is about the weak axis (i.e., C_{Pyy})

$$K_f = 0.60 \text{ (nailed built-up column)}$$

$$c = 0.8 \text{ (visually graded lumber)}$$

$$E'_{min} = E_{min}C_M C_t C_i$$

$$= (0.62 \times 10^6)(1.0)(1.0)(1.0) = 0.62 \times 10^6 \text{ psi}$$

$$F_{cEyy} = \frac{0.822 E'_{min}}{(l_{ey}/d_y)^2} = \frac{(0.822)(0.62 \times 10^6)}{(32)^2} = 498 \text{ psi}$$

$$F_c^* = F_c C_D C_M C_t C_F C_i$$

$$= (1500)(1.0)(1.0)(1.0)(1.1)(1.0) = \mathbf{1650 \text{ psi}}$$

$$\frac{F_{cE}}{F_c^*} = \frac{498 \text{ psi}}{1650 \text{ psi}} = 0.302$$

From equation (5.24) the controlling column stability factor (i.e., C_{Pyy}) is calculated as

$$C_P = 0.60\left[\frac{1 + 0.302}{(2)(0.8)} - \sqrt{\left[\frac{1 + 0.302}{(2)(0.8)}\right]^2 - \frac{0.302}{0.8}}\right] = 0.168$$

The allowable compression stress is given as

$$F'_c = F_c^* C_P = (1650)(0.168) = 277.0 \text{ psi}$$

The axial load capacity of the built-up column is calculated as

$$P_{\text{allowable}} = F'_c A_g = (277.0)(4.5 \text{ in.})(5.5 \text{ in.})$$

$$= \mathbf{6858 \text{ lb}}$$

From Section 15.3.3 of the NDS code [1], the minimum nailing requirement is 30d nails spaced vertically at 8 in. o.c. and staggered $2\frac{1}{2}$ in. with an edge distance of $1\frac{1}{2}$ in. and an end distance of $3\frac{1}{2}$ in. (see Figure 5.12).

$$f_c = \frac{P}{A_n} \leq F'_c \qquad (5.26)$$

where f_c = axial compression stress applied
P = tension force applied
A_n = net area at the critical section = $A_g - \Sigma A_{\text{holes}}$
A_g = gross cross-sectional area of the tension member
ΣA_{holes} = sum of the area of the bolt holes perpendicular to the load
= (no. of bolts perpendicular to load)($d_{\text{bolt}} + \frac{1}{8}$ in.)(member thickness b)
F'_c = allowable compression stress parallel to the grain

Using the NDS code applicability table (Tables 3.1 and 3.2), the allowable compression stress parallel to the grain is as calculated in equations 5.20a and 5.20b.

In calculating the column stability factor C_P, the larger slenderness ratio is used since this represents the most critical condition with respect to buckling.

Design Check 2: Compression on the Gross Area

Compression on the gross area (Figure 5.14b) occurs at the midspan of the member (i.e., where no connections exist) and the gross cross-sectional area of the member is applicable. For this location, however, column stability is a factor, and thus the column stability factor C_P may be less than 1.0 and has to be calculated. The required design equation is

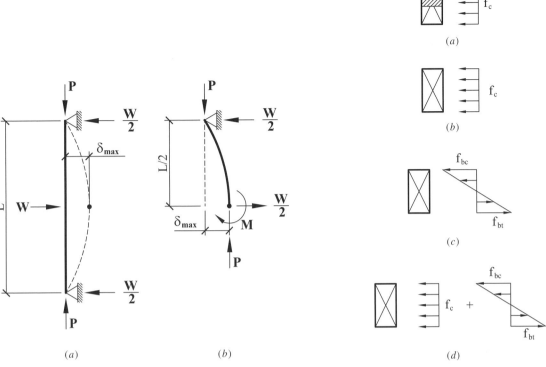

FIGURE 5.13 P–δ effect in axially loaded compression members.

FIGURE 5.14 Stress diagrams for load case 4: (*a*) design check 1; (*b*) design check 2; (*c*) design check 3; (*d*) design check 4.

$$f_c = \frac{P}{A_g} \leq F'_c \tag{5.27}$$

The parameters in equation (5.27) are as defined in design check 1.

Design Check 3: Bending Only

Bending only (Figure 5.14*c*) occurs at the point of maximum moment (i.e., at the midspan of the member) and the required design equation is

$$f_{bc} = \frac{M}{S_x} \leq F'_b \tag{5.28}$$

where f_{bc} = bending stress in the compression fiber
 M = maximum bending moment in the member (typically, occurs at midspan)
 S_x = section modulus about the axis of bending
 F'_b = allowable bending stress

Using the adjustment factors applicability table (Tables 3.1 and 3.2), the allowable bending stress is obtained as

$$F'_b = \begin{cases} F_b C_D C_M C_t C_L C_F C_i C_r C_{fu} & \text{(for sawn lumber)} \\ F_b C_D C_M C_t C_L C_v C_{fu} C_c & \text{(for glulam)} \end{cases}$$

where F_b is the NDS-S tabulated design bending stress. Note that for dimension-sawn lumber built-up columns with three or more laminations embedded within a sheathed wall and subjected to bending about its strong axis, the repetitive member factor C_r is 1.15.

Design Check 4: Bending plus Compression

Bending plus compression (Figure 5.14*d*) occurs at the point of maximum moment (i.e., at the midspan) where the moment magnification (or P-δ) factor is at its maximum value. The general design interaction equation for combined *concentric* axial compression load plus *biaxial* bending is given in the NDS code as

$$\left(\frac{f_c}{F'_c}\right)^2 + \frac{f_{bx}/F'_{bx}}{1 - f_c/F_{cEx}} + \frac{f_{by}/F'_{by}}{1 - f_c/F_{cEy} - (f_{bx}/F_{bEx})^2} \leq 1.0 \qquad (5.29)$$

where f_c = axial compression stress
F'_c = allowable compression stress parallel to the grain
f_{bx} = bending stress due to *x–x* axis bending
F'_{bx} = allowable bending stress for bending about the *x–x* axis
F_{cEx} = Euler critical stress for buckling about the *x–x* axis
F_{bEx} = Euler critical stress for buckling about the *x–x* axis
f_{by} = bending stress due to *y–y* axis bending
F'_{by} = allowable bending stress for bending about the *y–y* axis
F_{cEy} = Euler critical stress for buckling about the *y–y* axis

$f_c < F_{cEx} = \dfrac{0.822E'_{min}}{(l_{ex}/d_x)^2}$ (Euler critical column buckling stress about the *x–x* axis)

$f_c < F_{cEy} = \dfrac{0.822E'_{min}}{(l_{ey}/d_y)^2}$ (Euler critical column buckling stress about the *y–y* axis)

$f_{by} < F_{bE} = \dfrac{1.20E'_{min}}{(R_B)^2}$ (Euler critical lateral-torsional buckling stress for *y–y* axis bending)

$R_B = \sqrt{\dfrac{l_e d_x}{b^2}} \leq 50$

l_e = effective length for bending members from Table 3.9
d_x = column dimension perpendicular to the *x–x* axis (see Figure 5.15)
b = column dimension perpendicular to the *y–y* axis

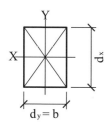

FIGURE 5.15
Column cross section.

For combined *concentric* axial load plus *uniaxial* bending about the strong (*x–x*) axis of the member, the interaction equation above reduces to

$$\left(\frac{f_c}{F'_c}\right)^2 + \frac{f_{bx}/F'_{bx}}{1 - f_c/F_{cEx}} \leq 1.0 \qquad (5.30)$$

Using the adjustment factors applicability table (Tables 3.1 and 3.2), the allowable bending stress for each axis of bending is obtained from the equations

$$F'_b = \begin{cases} F_b C_D C_M C_t C_L C_F C_i C_r C_{fu} & \text{(for sawn lumber)} \\ F_b C_D C_M C_t C_L C_v C_{fu} C_c & \text{(for glulam)} \end{cases}$$

where F_b is the NDS-S tabulated design bending stress.

Using the adjustment factors applicability table (Tables 3.1 and 3.2), the allowable compressive stress is calculated from equations 5.20a and 5.20b.

In the calculation of the column stability factor C_P and the allowable compression stress parallel to grain F'_c, the larger of the slenderness ratios about both orthogonal axes (l_{ex}/d_x or l_{ey}/d_y) should be used since the column will buckle about the axes with the larger slenderness ratio. The higher the column slenderness ratio $K_e l/r$, the smaller the column stability factor C_P. For columns that are laterally braced over their entire length, the column stability factor C_P is 1.0. For all other cases, calculate C_P using equation 5.21.

$$C_P = \frac{1 + F_{cE}/F^*_c}{2c} - \sqrt{\left(\frac{1 + F_{cE}/F^*_c}{2c}\right)^2 - \frac{F_{cE}/F^*_c}{c}} \qquad (5.21)$$

where F_{cE} = Euler critical buckling stress = $\dfrac{0.822E'_{min}}{(l_e/d)^2}$ [see equation (5.17)]

$$E'_{min} = E_{min} C_M C_t C_i \qquad (5.22)$$

EXAMPLE 5.8

Combined Compression plus Bending—Truss Top Chord

Design the top chord member AD of the roof truss in Example 5.5 (Figure 5.16).

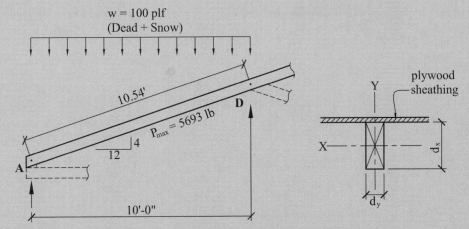

FIGURE 5.16 Truss top chord load diagram and cross section.

Solution: The maximum axial compression force in member AD is

P_{max} (from computer analysis, method of joints, or method of sections) = 5693 lb

The maximum moment in member AD is

$$M_{max} = \frac{(100 \text{ lb/ft})(10 \text{ ft})^2}{8} = 1250 \text{ ft-lb} = 15{,}000 \text{ in.-lb}$$

Select a Trial Member Size

Try 2 × 10 No. 1 & Better hem-fir. A 2 × 10 is dimension lumber; therefore, use NDS-S Table 4A. From NDS-S Table 1B, we obtain the section properties for the trial member size:

gross cross-sectional area $A_g = 13.88 \text{ in}^2$

section modulus $S_x = 21.4 \text{ in}^3$

$d_x = 9.25$ in.

$d_y = 1.5$ in.

From NDS-S Table 4A, we obtain the tabulated design stress values and the stress adjustment or C factors as follows:

$$F_c = 1350 \text{ psi}$$

$$F_b = 1100 \text{ psi}$$

$$E_{\min} = 0.55 \times 10^6 \text{ psi}$$

$$C_F(F_c) = 1.0$$

$$C_F(F_b) = 1.1$$

$$C_M = 1.0 \text{ (normal moisture conditions)}$$

$$C_t = 1.0 \text{ (normal temperature conditions)}$$

$$C_{fu} = 1.0$$

$$C_r = 1.15 \text{ (the condition for repetitiveness is discussed in Chapter 3)}$$

$$C_D = 1.15 \text{ (snow load controls for the load combination } D + S\text{)}$$

$$C_i = 1.0 \text{ (assumed)}$$

It should be noted that for each design check, the controlling load duration factor value will correspond to the C_D value of the shortest-duration load in the load combination for that design case. Thus, the C_D value for the various design cases may vary.

Design Check 1: Compression on Net Area
This condition occurs at the ends (i.e., supports) of the member. For this load case, the axial load P is caused by dead load plus snow load ($D + S$); therefore, the load duration factor C_D from Chapter 3 is 1.15. The compression stress applied

$$f_c = \frac{P}{A_n} \leq F'_c$$

$$\text{net area } A_n = A_g - \Sigma A_{\text{holes}}$$

$$= 13.88 - (1 \text{ hole})(\tfrac{3}{4} \text{ in.} + \tfrac{1}{8} \text{ in.})(1.5 \text{ in.})$$

Therefore,

$$A_n = 12.57 \text{ in}^2$$

The compressive axial stress applied

$$f_c = \frac{P}{A_n} = \frac{5693 \text{ lb}}{12.57 \text{ in}^2} = 453 \text{ psi}$$

At support locations, there can be no buckling; therefore, $C_P = 1.0$.
The allowable compression stress parallel to the grain is

$$F'_c = F_c C_D C_M C_t C_F C_i C_P$$

$$= (1350)(1.15)(1.0)(1.0)(1.0)(1.0)(1.0)$$

Therefore,

$$F'_c = 1553 \text{ psi} > f_c = 453 \text{ psi} \quad \textbf{OK}$$

Design Check 2: Compression on Gross Area
This condition occurs at the *midspan* of the member. The axial load P for this load case is also caused by dead load plus the snow load ($D + S$); therefore, the load duration factor C_D from Chapter 3 is 1.15. The applied axial compressive stress

$$f_c = \frac{P}{A_g} = \frac{5693 \text{ lb}}{13.88 \text{ in}^2} = 411 \text{ psi}$$

Unbraced length. For x–x axis buckling, the unbraced length $l_{ux} = 10.54$ ft. For y–y axis buckling, the unbraced length $l_{uy} = 0$ ft (plywood sheathing braces top chord).

Effective length. For building columns that are supported at both ends, it is usual practice to assume an effective length factor K_e of 1.0. For x–x axis buckling, effective length

$$l_{ex} = K_e l_{ux} = (1.0)(10.54 \text{ ft}) = 10.54 \text{ ft}$$

For y–y axis buckling, effective length

$$l_{ey} = K_e l_{uy} = (1.0)(0) = 0 \text{ ft}$$

Slenderness ratio. For x–x axis buckling, the effective length is

$$\frac{l_{ex}}{d_x} = \frac{(10.54 \text{ ft})(12)}{9.25 \text{ in.}} = \mathbf{13.7} < 50 \text{ (larger slenderness ratio controls)}$$

For y–y axis buckling, the effective length is

$$\frac{l_{ey}}{d_y} = \frac{0}{1.5 \text{ in.}} = 0 < 50$$

Therefore,

$$(l_e/d)_{max} = 13.7.$$

This slenderness ratio will be used in the calculation of the column stability factor C_P and the allowable compression stress parallel to the grain F_c'. The buckling modulus of elasticity

$$E'_{min} = E_{min} C_M C_t C_i$$
$$= (0.55 \times 10^6)(1.0)(1.0)(1.0) = 0.55 \times 10^6 \text{ psi}$$
$$c = 0.8 \text{ (visually graded lumber)}$$

The Euler critical buckling stress about the weaker axis (i.e., the axis with the higher slenderness ratio), which for this problem happens to be the x–x axis, is

$$F_{cE} = \frac{0.822 E'_{min}}{(l_e/d)_{max}^2} = \frac{(0.822)(0.55 \times 10^6)}{(13.7)^2} = 2409 \text{ psi}$$

$$F_c^* = F_c C_D C_M C_t C_F C_i$$
$$= (1350)(1.15)(1.0)(1.0)(1.0)(1.0) = \mathbf{1553 \text{ psi}}$$

$$\frac{F_{cE}}{F_c^*} = \frac{2409 \text{ psi}}{1553 \text{ psi}} = 1.551$$

From equation (5.21) the column stability factor is calculated as

$$C_P = \frac{1 + 1.551}{(2)(0.8)} - \sqrt{\left[\frac{1 + 1.551}{(2)(0.8)}\right]^2 - \frac{(1.551)}{(0.8)}} = 0.818$$

The allowable compression stress parallel to the grain is

$$F_c' = F_c^* C_P = (1553)(0.818) = \mathbf{1270 \text{ psi}} > f_c = 411 \text{ psi} \quad \mathbf{OK}$$

Design Check 3: Bending Only
This condition occurs at the point of maximum moment (i.e., at the *midspan* of the member), and the moment is caused by the uniformly distributed gravity load on the top chord (member AD), and since this load is the dead load plus the snow load, the controlling load duration factor (see Chapter 3) is 1.15. The maximum moment in member AD

$$M_{max} = \frac{w_{D+S}l^2}{8}$$

$$= \frac{(100 \text{ lb/ft})(10 \text{ ft})^2}{8} = 1250 \text{ ft-lb} = 15{,}000 \text{ in.-lb.}$$

The compression stress applied in member AD due to bending

$$f_{bc} = \frac{M}{S_x} = \frac{15{,}000 \text{ in.-lb}}{21.4 \text{ in}^3} = 701 \text{ psi}$$

The allowable bending stress

$$F'_b = F_b C_D C_M C_t C_L C_F C_i C_r C_{fu}$$

$$= (1100)(1.15)(1.0)(1.0)(1.0)(1.1)(1.0)(1.15)(1.0)$$

$$= \mathbf{1600 \text{ psi}} > f_b = 701 \text{ psi} \quad \mathbf{OK}$$

Design Check 4: Bending plus Axial Compression Force
This condition occurs at the *midspan* of the member. For this load case, the loads causing the combined stresses are the dead load plus the snow load ($D + S$); therefore, the controlling load duration factor C_D is 1.15. It should be noted that because the load duration factor C_D for this combined load case is the same as for load cases 2 and 3, the parameters to be used in the combined stress interaction equation can be obtained from design checks 2 and 3. The reader is cautioned to be aware that the C_D value for all four design checks are not necessarily always equal.

The interaction equation for combined concentric axial load plus uniaxial bending is obtained from equation (5.30) as

$$\left(\frac{f_c}{F'_c}\right)^2 + \frac{f_{bx}/F'_{bx}}{1 - f_c/F_{cEx}} \leq 1.0$$

where $f_c = 411$ psi (occurs at the midspan; see design check 2)
$F'_c = 1267$ psi (occurs at the midspan; see design check 2)
$f_{bx} = 701$ psi (occurs at the midspan; see design check 3)
$F'_{bx} = 1600$ psi (occurs at the midspan; see design check 3)

The reader should note that in the interaction equation above, it is the Euler critical buckling stress about the x–x axis, F_{cEx}, that is required since the bending of the truss top chord member is about that axis.

The effective length for x–x axis buckling is

$$\frac{l_{ex}}{d_x} = \frac{(10.54 \text{ ft})(12)}{9.25 \text{ in.}} = \mathbf{13.7} < 50$$

The Euler critical buckling stress about the x–x axis is

$$F_{cEx} = \frac{0.822 E'_{min}}{(l_{ex}/d_x)^2} = \frac{(0.822)(0.55 \times 10^6)}{(13.7)^2} = 2409 \text{ psi}$$

Substituting the parameters above into the interaction equation yields

$$\left(\frac{411}{1267}\right)^2 + \left(\frac{701/1600}{1 - 411/2409}\right) = \mathbf{0.63} < 1.0 \quad \mathbf{OK}$$

2×10 No. 1 & Better hem-fir is adequate for the truss top chord.

EXAMPLE 5.9

Combined Axial Compression plus Bending—Exterior Stud Wall

Design the ground floor exterior stud wall for the two-story building with the roof and floor plans shown in Figure 5.17. The floor-to-floor height is 10 ft and the wall studs are spaced at 2 ft o.c. Assume Douglas fir-larch (DF-L) wood species and the design loads shown.

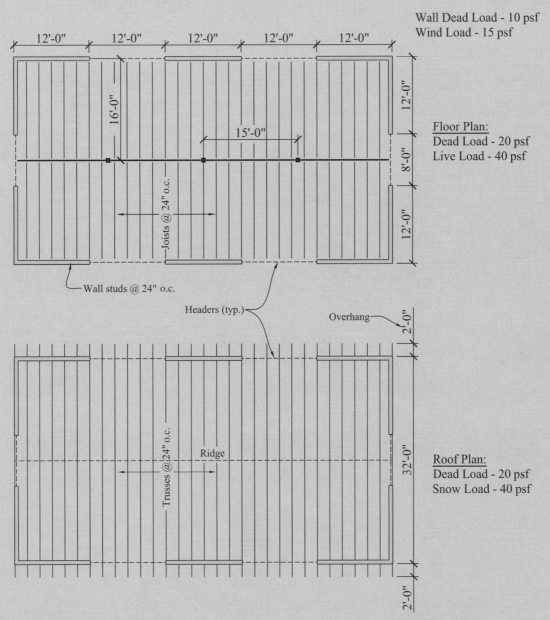

Wall Dead Load - 10 psf
Wind Load - 15 psf

Floor Plan:
Dead Load - 20 psf
Live Load - 40 psf

Roof Plan:
Dead Load - 20 psf
Snow Load - 40 psf

FIGURE 5.17 Roof and floor plans.

Solution: First, we calculate the gravity and lateral loads acting on a typical wall stud (Figure 5.18).

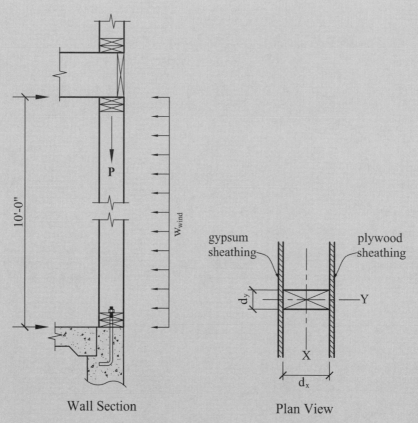

FIGURE 5.18 Stud wall section and plan view.

Tributary width of wall stud = 2 ft

Roof dead load D = 20 psf

Roof snow load S = 40 psf

Roof live load L_r = 20 psf (see Chapter 2)

Since $L_r < S$, the snow load controls. The lateral wind load W = 15 psf (acts horizontally). The net vertical wind load is assumed to be zero for this building.

Second-floor dead load D = 20 psf

Second-floor live load L = 40 psf

Second-floor wall self-weight = 10 psf

Ground-floor wall self-weight = 10 psf

Gravity loads:

Tributary area per stud at the *roof* level = $\left(\dfrac{32 \text{ ft}}{2} + 2 \text{ ft}\right)(2 \text{ ft})$ = **36 ft^2**

Tributary area per stud at the *second-floor* level = $\dfrac{16 \text{ ft}}{2} (2 \text{ ft})$ = **16 ft^2**

The total *dead load* on the ground-floor studs is

$$P_D = P_{D(\text{roof})} + P_{D(\text{floor})} + P_{D(\text{wall})}$$

$$= (20 \text{ psf})(36 \text{ ft}^2) + (20 \text{ psf})(16 \text{ ft}^2) + (10 \text{ psf})(2 \text{ ft stud spacing})(10 \text{ ft wall height})$$

$$= \mathbf{1440 \text{ lb}}$$

The total *snow load* on the ground-floor studs is

$$P_S = (40 \text{ psf})(36 \text{ ft}^2) = \mathbf{1440 \text{ lb}}$$

The total *floor live load* on the ground-floor studs is

$$P_L = (40 \text{ psf})(16 \text{ ft}^2) = \mathbf{640 \text{ lb}}$$

Note that the roof live load L_r has been neglected in this example since it will not govern because its maximum value of 20 psf is less than the specified snow load ($S = 40$ psf) for this building.

Lateral loads. The wind load acts perpendicular to the face of the stud wall, causing bending of the stud about the x–x (strong) axis (see Figure 5.18). The lateral wind load

$$w_{\text{wind}} = (15 \text{ psf})(2 \text{ ft tributary width}) = 30 \text{ lb/ft}$$

The maximum moment due to wind load is calculated as

$$M_w = \frac{wL^2}{8}$$

$$= \frac{(30)(10^2)}{8} = \mathbf{375 \text{ ft-lb}} = \mathbf{4500 \text{ in.-lb}}$$

The most critical axial load combination and the most critical combined axial and bending load combination are now determined following the normalized load procedure outlined in Chapter 3. The axial load P in the wall stud will be caused by the gravity loads D, L, and S, while the bending load and moment will be caused by the lateral wind load W. The load values to be used in the load combinations are as follows:

$$D = 1440 \text{ lb}$$
$$S = 1440 \text{ lb}$$
$$L = 640 \text{ lb}$$
$$W = 4500 \text{ in.-lb}$$

All other loads are assumed to be zero and are therefore neglected in the load combinations. The applicable load combinations with all the zero loads neglected are shown in Table 5.2.

TABLE 5.2 Applicable and Governing Load Combinations

Load Combination	Axial Load, P (lb)	Moment, M (in.-lb)	C_D	Normalized Load and Moment P/C_D	M/C_D
D	1440	0	0.9	1600	0
$D + L$	$1440 + 640 = 2080$	0	1.0	2080	0
$D + S$	$1440 + 1440 = 2880$	0	1.15	2504	0
$D + 0.75L + 0.75S$	$1440 + (0.75)(640 + 1440) = \mathbf{3000}$	0	1.15	2609	0
$D + 0.75W + 0.75L + 0.75S$	$1440 + (0.75)(640 + 1440) = \mathbf{3000}$	$0.75(4500) = \mathbf{3375}$	1.6	1875	2110
$0.6D + W$	$(0.6)(1440) = \mathbf{864}$	4500	1.6	540	2813

It will be recalled from Chapter 3 that the necessary condition to use the normalized load method is that all the loads be similar and of the same type. We can separate the load cases in Table 5.2 into two types of loads: pure axial load cases and combined load cases. The most critical load combinations are the load cases with the highest normalized load or moment. Since not all the loads on this wall stud are pure axial loads only or bending loads only, the normalized load method discussed in Chapter 3 can then be used only to determine the most critical pure axial load case, not the most critical combined load case. The most critical combined axial load plus bending load case would have to be determined by carrying out the design (or analysis) for all the combined load cases, with some of the load cases eliminated by inspection.

Pure axial load case. For the pure axial load cases, the load combination with the highest normalized load P/C_D) from Table 5.2 is

$$D + 0.75(L + S) \qquad P = 3000 \text{ lb with } C_D = 1.15$$

Combined load case. For the combined load cases, the load combinations with the highest normalized load P/C_D and normalized moment M/C_D from Table 5.2 are

$$D + 0.75(L + S + W) \qquad P = 3000 \text{ lb} \qquad M = 3375 \text{ in.-lb} \qquad C_D = 1.6$$

or

$$0.6D + W \qquad P = 864 \text{ lb} \qquad M = 4500 \text{ in.-lb} \qquad C_D = 1.6$$

By inspection it would appear as if the load combination $D + (L + S + W)$ will control for the combined load case, but this should be verified through analysis, and so both of these combined load cases will be investigated.

Select a Trial Member Size

Several design aids are given in Appendix B to help the reader with preliminary sizing of wall studs. Using Figure B.23, a 2 × 6 DF-L No. 2 appears to be adequate to resist the pure axial load case and the combined load cases above for the given floor-to-floor height of 10 ft. Try a 2 × 6 DF-L No. 2. A 2 × 6 is dimension lumber; therefore, use NDS-S Table 4A. From NDS-S Table 1B we obtain the section properties for the trial member size:

$$\text{gross cross-sectional area } A_g = 8.25 \text{ in}^2$$

$$\text{section modulus } S_x = 7.56 \text{ in}^3$$

$$d_x = 5.5 \text{ in.}$$

$$d_y = 1.5 \text{ in. (fully braced by sheathing)}$$

From NDS-S Table 4A, we obtain the tabulated design stress values and the stress adjustment or C factors as follows:

$$F_c = 1350 \text{ psi}$$

$$F_b = 900 \text{ psi}$$

$$F_{c\perp} = 625 \text{ psi}$$

$$E_{\min} = 0.58 \times 10^6 \text{ psi}$$

$$C_F(F_c) = 1.1$$

$$C_F(F_b) = 1.3$$

$$C_M = 1.0 \text{ (normal moisture conditions)}$$

$$C_t = 1.0 \text{ (normal temperature conditions)}$$

$$C_r = 1.15 \text{ (the condition for repetitiveness is discussed in Chapter 3)}$$

$$C_{fu} = 1.0$$

$$C_i = 1.0$$

$$C_L = 1.0$$

$$C_D = \text{to be determined for each design check}$$

The beam stability factor C_L will be 1.0 for bending of the wall stud about the x–x (strong) axis because the wall stud is braced against lateral torsional buckling by the plywood sheathing. The load duration factor C_D will be determined for each design check, and for each design check the controlling load duration factor value will correspond to the C_D value of the shortest-duration load in the load combination for that design case. Thus, the C_D value for the various design cases may vary.

Design Check 1: Compression on Net Area

This condition occurs at the ends (i.e., supports) of the member. Note that for this design check, which is a pure axial load case, the load combination with the highest normalized axial load P/C_D will be most critical. This controlling load combination from Table 5.2 is $D + 0.75(L + S)$, for which **$P = 3000$ lb** and **$C_D = 1.15$**. Since the ends of the wall studs are usually connected to the sill plates and top plates with nails rather than bolts, the net area will be equal to the gross area since there are no bolt holes to consider.

$$\text{Net area } A_n = A_g - \Sigma A_{\text{holes}} = A_g = 8.25 \text{ in}^2$$

$$\text{Compressive axial stress applied } f_c = \frac{P}{A_n} = \frac{3000 \text{ lb}}{8.25 \text{ in}^2} = 364 \text{ psi}$$

At support locations, there can be no buckling; therefore, $C_P = 1.0$.

The allowable compression stress parallel to the grain is

$$F'_c = F_c C_D C_M C_t C_F C_i C_P$$

$$= (1350)(1.15)(1.0)(1.0)(1.1)(1.0)(1.0)$$

$$= \textbf{1708 psi} > f_c = 364 \text{ psi} \quad \textbf{OK}$$

It should be noted that design check 1 could have been eliminated by inspection when compared to design check 2, since the column stability factor C_P for design check 1 is 1.0, which will be greater than that for design check 2, and thus the allowable stresses in design check 2 will be smaller.

Design Check 2: Compression on Gross Area

This is also a pure axial load case that occurs at the midheight of the wall stud, and the most critical load combination will be the load combination with the highest normalized axial load P/C_D. This controlling load combination from Table 5.2 is $D + 0.75(L + S)$, for which **$P = 3000$ lb** and **$C_D = 1.15$**. The axial compression stress

$$f_c = \frac{P}{A_g} = \frac{3000 \text{ lb}}{8.25 \text{ in}^2} = 364 \text{ psi}$$

The compressive stress perpendicular to the grain in the sill plate is

$$f_{c\perp} = \frac{P}{A_g} = \frac{3000 \text{ lb}}{8.25 \text{ in}^2} = 364 \text{ psi}$$

Unbraced length. For x–x axis buckling, the unbraced length l_{ux} is 10 ft. For y–y axis buckling, the unbraced length l_{uy} is 0 ft (plywood sheathing braces y–y axis buckling).

Effective length. Wall studs are usually supported at both ends, therefore, it is usual practice to assume an effective length factor K_e of 1.0.

For x–x axis buckling, effective length $l_{ex} = K_e l_{ux} = (1.0)(10 \text{ ft}) = 10$ ft.

For y–y axis buckling, effective length $l_{ey} = K_e l_{uy} = (1.0)(0) = 0$ ft.

Slenderness ratio. For x–x axis buckling, the effective length is

$$\frac{l_{ex}}{d_x} = \frac{(10 \text{ ft})(12)}{5.5 \text{ in.}} = \textbf{21.8} < 50 \text{ (larger slenderness ratio controls)}$$

For y–y axis buckling, the effective length is

$$\frac{l_{ey}}{d_y} = \frac{0}{1.5 \text{ in.}} = 0 < 50$$

Therefore,

$$(l_e/d)_{\max} = 21.8.$$

This slenderness ratio will be used in the calculation of the column stability factor C_P and the allowable compression stress parallel to the grain F_c':

$$\text{Buckling modulus of elasticity } E_{\min}' = E_{\min} C_M C_t C_i$$

$$= (0.58 \times 10^6)(1.0)(1.0)(1.0) = 0.58 \times 10^6 \text{ psi}$$

$$c = 0.8 \text{ (visually graded lumber)}$$

The Euler critical buckling stress about the "weaker" axis (i.e., the axis with the higher slenderness ratio), which for this problem happens to be the x–x axis, is

$$F_{cE} = \frac{0.822 E_{\min}'}{(l_e/d)_{\max}^2} = \frac{(0.822)(0.58 \times 10^6)}{(21.8)^2} = 1003 \text{ psi}$$

$$F_c^* = F_c C_D C_M C_t C_F C_i$$

$$= (1350)(1.15)(1.0)(1.0)(1.1)(1.0) = \mathbf{1708 \text{ psi}}$$

$$\frac{F_{cE}}{F_c^*} = \frac{1003 \text{ psi}}{1708 \text{ psi}} = 0.587$$

From Equation (5.21) the column stability factor is calculated as

$$C_P = \frac{1 + 0.587}{(2)(0.8)} - \sqrt{\left[\frac{1 + 0.587}{(2)(0.8)}\right]^2 - \frac{0.587}{0.8}} = 0.492$$

The allowable compression stress *parallel* to the grain is

$$F_c' = F_c^* C_P = (1708)(0.492) = \mathbf{840 \text{ psi}} > f_c = 364 \text{ psi} \quad \mathbf{OK}$$

The allowable compression stress *perpendicular* to the grain in the sill plate is

$$F_{c\perp}' = F_{c\perp} C_M C_t C_i = (625)(1.0)(1.0)(1.0)$$

$$= 625 \text{ psi} > f_{c\perp} = 364 \text{ psi} \quad \mathbf{OK}$$

Design Check 3: Bending Only
This condition occurs at the *midheight* of the wall stud, and the bending moment is caused by the lateral wind load. Only the two controlling combined load cases in Table 5.2 need be considered for this design check. The load combination with the maximum normalized moment M/C_D is $0.6D + W$, for which the maximum moment $\mathbf{M = 4500 \text{ in.-lb}, P = 864 \text{ lb, and } C_D = 1.6.}$

The compression stress applied in the wall stud due to bending

$$f_{bc} = \frac{M}{S_x} = \frac{4500 \text{ in.-lb}}{7.56 \text{ in}^3} = 595 \text{ psi}$$

The allowable bending stress

$$F_b' = F_b C_D C_M C_t C_L C_F C_i C_r C_{fu}$$

$$= (900)(1.6)(1.0)(1.0)(1.0)(1.3)(1.0)(1.15)(1.0)$$

$$= 2153 \text{ psi} > f_b = 595 \text{ psi} \quad \mathbf{OK}$$

Design Check 4: Bending plus Axial Compression Force
This condition occurs at the *midheight* of the wall stud, and only the two controlling combined load cases in Table 5.2 need be considered for this design check, and these are

$$D + 0.75(L + S + W) \quad P = 3000 \text{ lb} \quad M = 3375 \text{ in.-lb} \quad C_D = 1.6$$

and

$$0.6D + W \quad P = 864 \text{ lb} \quad M = 4500 \text{ in.-lb} \quad C_D = 1.6$$

We investigate these load combinations separately to determine the most critical.

Load combination $D + 0.75(L + S + W)$:

$$P_{\max} = 3000 \text{ lb} \quad M_{\max} = 3375 \text{ in.-lb} \quad C_D = 1.6$$

The bending stress applied is

$$f_{bx} = \frac{M_{\max}}{S_{xx}} = \frac{3375 \text{ in.-lb}}{7.56 \text{ in}^3} = 446 \text{ psi (lateral wind load causes bending about the } x\text{--}x \text{ axis)}$$

The axial compression stress applied at the midheight is

$$f_c = 364 \text{ psi (from design check 2)}$$

The allowable bending stress

$$F_b' = F_b C_D C_M C_t C_L C_F C_i C_r C_{fu}$$

$$= (900)(1.6)(1.0)(1.0)(1.0)(1.3)(1.0)(1.15)(1.0)$$

$$= 2153 \text{ psi} > f_b = 446 \text{ psi} \quad \textbf{OK}$$

Note that the beam stability factor C_L is 1.0 because the compression edge of the stud is braced laterally by the wall sheathing for bending due to loads acting perpendicular to the face of the wall. We now proceed to calculate the column stability factor C_P.

$$F_c^* = F_c C_D C_M C_t C_F C_i$$

$$= (1350)(1.6)(1.0)(1.0)(1.1)(1.0) = \textbf{2376 psi}$$

The Euler critical buckling stress about the "weaker" axis (i.e., the axis with the higher slenderness ratio), which for this problem happens to be the x--x axis, is

$$F_{cE(\max)} = \frac{0.822 E_{\min}'}{(l_e/d)_{\max}^2} = \frac{(0.822)(0.58 \times 10^6)}{(21.8)^2} = 1003 \text{ psi}$$

$$\frac{F_{cE}}{F_c^*} = \frac{1003 \text{ psi}}{2376 \text{ psi}} = 0.422$$

From equation (5.21) the column stability factor is calculated as

$$C_P = \frac{1 + 0.422}{(2)(0.8)} - \sqrt{\left[\frac{1 + 0.422}{(2)(0.8)}\right]^2 - \frac{0.422}{0.8}} = 0.377$$

The allowable compression stress *parallel* to the grain is

$$F_c' = F_c^* C_P = (2376)(0.377) = \textbf{896 psi} > f_c = 364 \text{ psi} \quad \textbf{OK}$$

The interaction equation for combined concentric axial load plus uniaxial bending is obtained from equation (5.30) as

$$\left(\frac{f_c}{F_c'}\right)^2 + \frac{f_{bx}/F_{bx}'}{1 - f_c/F_{cEx}} \leq 1.0$$

The Euler critical buckling stress about the x–x axis (i.e., the axis of bending of the wall stud due to lateral wind loads) is

$$F_{cEx} = \frac{0.822 E'_{min}}{(l_{ex}/d_x)^2} = \frac{(0.822)(0.58 \times 10^6)}{(21.8)^2} = 1003 \text{ psi}$$

Substituting the parameters above into the interaction equation yields

$$\left(\frac{364}{896}\right)^2 + \frac{446/2153}{1 - 364/1003} = 0.49 < 1.0 \qquad \text{OK}$$

Load combination $0.6D + W$:

$$P_{max} = 864 \text{ lb} \qquad M_{max} = 4500 \text{ in.-lb} \qquad C_D = 1.6$$

The bending stress applied is

$$f_{bx} = \frac{M_{max}}{S_{xx}} = \frac{4500 \text{ in.-lb}}{7.56 \text{ in}^3} = 595 \text{ psi (lateral wind load causes bending about the } x\text{–}x \text{ axis)}$$

The axial compression stress

$$f_c = \frac{P}{A_g} = \frac{864 \text{ lb}}{8.25 \text{ in}^2} = 105 \text{ psi}$$

Except for the applied bending and axial stresses, all the other parameters used in the interaction equation for the preceding combined load case would also apply to this load case. Therefore, the interaction equation for this load case will be

$$\left(\frac{105}{896}\right)^2 + \frac{595/2153}{1 - 105/1003} = 0.32 < 1.0 \qquad \text{OK}$$

It should be noted that because the interaction values obtained are much less than 1.0, a more efficient design can be obtained by using a lower stress grade for the 2×6 wall stud, such as No. 3 grade.

Eccentrically Loaded Columns

Wood columns with axial compression loads applied eccentrically are not common in regular wood-framed building structures and therefore are not covered in this book. The interaction equation for this special loading condition, which is more cumbersome than the interaction equation for combined concentric axial load plus bending load presented earlier, can be found in Section 15.4 of the NDS code.

5.6 PRACTICAL CONSIDERATIONS FOR ROOF TRUSS DESIGN

The application of the four load cases discussed in this chapter to typical roof trusses such as those shown in Figure 5.19 is discussed next. The slope or pitch of a truss is usually noted as a ratio of the vertical to the horizontal dimension, with the horizontal number equal to 12 (e.g.,

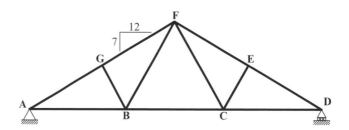

FIGURE 5.19 Typical roof truss.

4:12 or 7:12). The slope or pitch of a roof truss is usually determined by the owner or architect, not by the engineer.

The spacing of the roof truss is usually selected by the engineer and is typically 16 or 24 in. o.c. The truss spacing should be a multiple of the stud spacing in the bearing walls to ensure that the truss bears directly on a wall stud, otherwise the top plates have to be designed to resist the truss reactions resulting in bending of the plates about their weak axis. The analysis of a roof truss is usually carried out with structural analysis software, and in many cases the design of the truss members is also performed with computer software. When performing the design of the typical roof truss in Figure 5.19 for gravity loads, the following *four* load cases have to be considered, and the compression members of the truss must be sized such that the maximum slenderness ratio l_e/d does not exceed 50.

1. Pure tension (web members *BF* and *CF*)
2. Combined tension plus bending (bottom chord members *AB*, *BC*, and *CD*)
3. Pure compression (web members *BG* and *CE*)
4. Combined compression plus bending (top chord members *AG*, *GF*, *FE*, and *ED*)

In practice, roof trusses are preengineered and built by truss manufacturers such as TrusJoist. However, the designer or engineer of record has to specify the truss spacing and profile, the support locations, and the gravity and wind loads applied, including those causing uplift of the truss. The engineer will, however, have to check and approve the shop drawings submitted by the truss manufacturer.

Wind Uplift on Roof Trusses If there is net wind uplift on a roof truss, stress reversals may occur in the bottom chord members, (e.g. members *AB*, *BC*, and *CD*). These members which are normally in tension will be subjected to compression forces because of the net wind uplift. In addition, the roof truss supports will have to be tied down to the wall studs using hurricane clips.

Types of Roof Trusses

There are two main categories of roof trusses: lightweight and heavy timber roof trusses. Lightweight roof trusses are made from dimension-sawn lumber, and heavy timber roof trusses are made from larger sawn-lumber sizes (e.g., 4 × 6 to 8 × 12). Lightweight roof trusses can be used to span up to 60 to 70 ft and are typically spaced at 16 or 24 in. o.c. Heavy timber trusses were used in the past to span longer distances, up to 120 ft, and are typically spaced much farther apart than are lightweight roof trusses. A maximum span-to-depth ratio of 8 to 10 is typically specified for flat roof heavy timber trusses and 6 to 8 for bowstring trusses. Heavy timber trusses are no longer used frequently in wood buildings.

Several common profiles of lightweight roof trusses were shown in Figure 1.7. The pitch or rise per foot of roof trusses varies from 3 in 12 to 12 in 12, depending on the span of the roof truss. Roof pitches shallower than 4 in 12 are adequate for shedding rainwater, but pitches greater than 4 in 12 are required for proper shedding of snow and ice loads in cold climates. For a given span, the higher the roof pitch, the smaller the axial forces in the compression and tension chords of the truss. The members of lightweight roof trusses are usually connected with toothed light-gage plate connectors or nailed plate connectors designed in accordance with the Truss Plate Institute's *National Design Standard for Metal Plated Connected Wood Trusses* ANSI/TPI 1-2002 [17]. This specification is used by preengineered truss manufacturers for the computer-aided design of factory-built roof trusses.

Bracing and Bridging of Roof Trusses

Lateral bracing and bridging are required for lightweight roof trusses during and after erection to ensure lateral stability. During erection of roof trusses, lateral bracing or bridging perpendicular to the truss span are required to tie the top and bottom chords and web members of adjacent trusses together for lateral stability and to prevent the trusses from tipping or rolling over on their sides before attachment of the roof diaphragms. Lateral braces and bridging should be provided per BCSI 1-06 [18]. Typically, the top chord members are temporarily braced to maintain stability until the roof sheathing is installed while the web and bottom chord members are permanently braced. The temporary top chord bracing consists of horizontal 2× continuous wood members placed in the same plane as the top chord and running perpendicular to the truss span. They are usually spaced 10 ft on centers for truss spans up to 30 ft; 8 ft on centers for truss spans between 30 and 45 ft; and 6 ft on centers for truss spans between 45 and 60 ft. Permanent braces for the truss bottom chord members consists of continuous horizontal 2× wood members placed perpendicular to the truss span and in the plane of the bottom chord, and spaced at 8 to 10 ft on centers, and preferably located at a truss web member. These continuous members are located on top of the bottom chord to avoid interference and should be lapped over at least two trusses. In addition, 2× diagonal braces should be provided in the plane of the bottom chord between all the horizontal braces and these should occur every 10 truss spaces or at 20 ft on centers maximum. The permanent bracing for the truss web members include diagonal braces in the vertical plane (perpendicular to the truss span) that is connected to the webs of several trusses, and the location of these diagonal braces should coincide with the location of the bottom chord horizontal braces. The diagonal braces should occur at every 10 truss spaces or at 20 ft on centers maximum. Several collapses of roof trusses during erection have been reported, so it is imperative that roof trusses be braced adequately to prevent collapse.

In some cases, roof trusses may be erected in modules of three or more adjacent trusses laterally braced together on the ground for stability with or without the roof sheathing attached. Without adequate lateral bracing or bridging, it is highly possible for the compression chord members to buckle out of the plane of the truss even under the self-weight of the truss and buckling failures are usually catastrophic and sudden [18]. The force in the lateral bracing for compression is usually assumed to be at least 2% of the compression force in the braced member.

During construction, temporary braces should be provided to brace the first truss to the ground. In addition, diagonal bridging in the vertical plane has to be provided to tie the compression chord to the tension chord of adjacent trusses. The temporary continuous horizontal bridging is located at the bottom of the top chord to avoid interference with installation of the roof sheathing. The slenderness ratio of the 2× bracing or bridging elements must not be greater than 50. A minimum of two 16d nails are used to connect the bridging to the truss member. Figure 5.20 shows typical bridging for a roof truss bottom chord. For more information, the reader should refer to the Truss Plate Institute (TPI) publication, *Guide to Good Practice for Handling, Installing and Bracing Metal Connected Wood Trusses* (BCSI 1-03) [18].

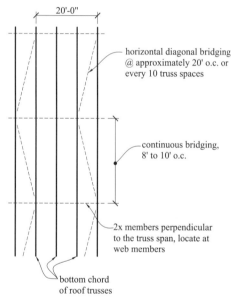

FIGURE 5.20 Typical truss bottom chord bridging.

REFERENCES

1. ANSI/AF&PA. (2005), *National Design Specification for Wood Construction,* American Forest & Paper Association, Washington, DC.
2. Faherty, Keith F., and Williamson, Thomas G. (1995), *Wood Engineering and Construction,* McGraw-Hill, New York.
3. AITC (American Institute for Timber Construction) (1994), *Timber Construction Manual,* 4th ed., Wiley, New York.
4. Willenbrock, Jack H., Manbeck, Harvey B., and Suchar, Michael G. (1998), *Residential Building Design and Construction,* Prentice Hall, Upper Saddle River, NJ.
5. French, Samuel E. (1996), *Determinate Structures: Statics, Strength, Analysis and Design,* Delmar Publishers, Albany, NY.
6. Ambrose, James (1994), *Simplified Design of Wood Structures,* 5th ed., Wiley, NY.
7. Halperin, Don A., and Bible, G. Thomas (1994), *Principles of Timber Design for Architects and Builders,* Wiley, NY.

8. Stalnaker, Judith J., and Harris, Earnest C. (1997), *Structural Design in Wood,* Chapman & Hall, London.
9. Breyer, Donald et al. (2003), *Design of Wood Structures—ASD,* 5th ed., McGraw-Hill, New York.
10. NAHB (2000), *Residential Structural Design Guide—2000,* National Association of Home Builders Research Center, Upper Marlboro, MD.
11. Cohen, Albert H. (2002), *Introduction to Structural Design: A Beginner's Guide to Gravity Loads and Residential Wood Structural Design,* AHC, Edmonds, WA.
12. Newman, Morton (1995), *Design and Construction of Wood-Framed Buildings,* McGraw-Hill, New York.
13. Kang, Kaffee (1998), *Graphic Guide to Frame Construction—Student Edition,* Prentice Hall, Upper Saddle River, NJ.
14. Kim, Robert H., and Kim, Jai B. (1997), *Timber Design for the Civil and Structural Professional Engineering Exams,* Professional Publications, Belmont, CA.
15. Hoyle, Robert J., Jr. (1978), *Wood Technology in the Design of Structures,* 4th ed., Mountain Press, Missoula, MT.
16. Kermany, Abdy (1999), *Structural Timber Design,* Blackwell Science, London.
17. ANSI/TPI 1-2002, *National Design Standard for Metal Plated Connected Wood Truss Construction,* Truss Plate Institute, Alexandria, VA.
18. BCSI 1-06, *Guide to Good Practice for Handling, Installing, and Bracing Metal Connected Wood Trusses,* Truss Plate Institute, Alexandria, VA.

PROBLEMS

5.1 A 4 × 8 wood member is subjected to an axial tension load of 8000 lb caused by dead load plus the snow load plus the wind load. Determine the applied tension stress f_t and the allowable tension stress F'_t, and check the adequacy of this member. The lumber is Douglas fir-larch No. 1, normal temperature conditions apply, and the moisture content is greater than 19%. Assume that the end connections are made with a single row of $\frac{3}{4}$-in.-diameter bolts.

5.2 A 2 × 10 Hem-fir No. 3 wood member is subjected to an axial tension load of 6000 lb caused by the dead load plus the snow load. Determine the applied tension stress f_t and the allowable tension stress F'_t, and check the adequacy of this member. Normal temperature and dry service conditions apply, and the end connections are made with a single row of $\frac{3}{4}$-in.-diameter bolts.

5.3 A $2\frac{1}{2}$ × 9 (six laminations) 5DF glulam axial combination member is subjected to an axial tension load of 20,000 lb caused by the dead load plus the snow load. Determine the tension stress applied f_t and the allowable tension stress F'_t, and check the adequacy of this member. Normal temperature and dry service conditions apply, and the end connections are made with a single row of $\frac{7}{8}$-in.-diameter bolts.

5.4 For the roof truss elevation shown in Figure 5.21, determine if a 2 × 10 spruce-pine-fir No. 1 member is adequate for the bottom chord of the typical truss, assuming that the roof dead load is 20 psf, the snow load is 40 psf, and the ceiling dead load is 15 psf of the horizontal plan area. Assume that normal temperature and dry service conditions apply and that the members are connected with a single row of $\frac{3}{4}$-in.-diameter bolts. The trusses span 36 ft and are spaced at 2 ft o.c.

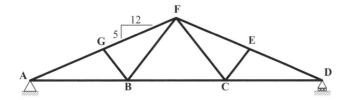

FIGURE 5.21 Roof truss elevation.

5.5 For the roof truss in Problem 5.4, determine if a 2 × 10 spruce-pine-fir No. 1 member is adequate for the top chord.

5.6 Determine if 10-ft-high 2 × 6 hem-fir No. 1 interior wall studs spaced at 16 in. o.c. is adequate to support a dead load of 300 lb/ft and a snow load of 600 lb/ft. Assume that the wall is sheathed on both sides and that normal temperature and dry service conditions apply. Neglect interior wind loads.

5.7 Determine the axial load capacity of a nailed built-up column of four 2 × 6's with a 12-ft unbraced height. Assume hem-fir No. 2, normal temperature and dry service conditions, and a load duration factor C_D of 1.0.

5.8 Determine if 2 × 6 Douglas fir-larch Select Structural studs spaced at 2 ft o.c. are adequate at the ground-floor level to support the following loads acting on the exterior stud wall of a three-story building.

Roof: dead load = 300 lb/ft; snow load = 800 lb/ft; roof live load = 400 lb/ft.

Third floor: dead load = 400 lb/ft; live load = 500 lb/ft.

Second floor: dead load = 400 lb/ft; live load = 500 lb/ft.

Lateral wind load acting perpendicular to the face of the wall = 15 psf. The floor-to-floor height is 10 ft, the wall studs are spaced at 2 ft o.c., and assume that the studs are sheathed on both sides.

chapter *six*

Roof and Floor Sheathing under Vertical and Lateral Loads (Horizontal Diaphrams)

6.1 INTRODUCTION

There are numerous uses for plywood in wood structures as well as in other structure types. In a wood structure, plywood is used primarily as roof and floor sheathing and wall sheathing. The floor and roof sheathing supports gravity loads and also acts as a horizontal diaphragm to support lateral wind and seismic loads. Horizontal diaphragms behave like deep horizontal beams spanning between the vertical lateral force resisting systems, (i.e., shear walls). The wall sheathing supports transverse wind loads and acts as a vertical diaphragm to support lateral wind and seismic loads. Plywood is also used in composite structural components, such as preengineered I-joists or as sheathing in a preengineered insulated roof panel. *Plywood* is manufactured by peeling a log of wood into thin plies or veneers that are approximately $\frac{1}{16}$ to $\frac{5}{16}$ in. thick. Each ply is bonded together with adhesive, heat, and pressure, and is typically oriented 90° to adjacent plies. The thickness of the final product typically varies between $\frac{1}{4}$ and $1\frac{1}{8}$ in.

The use of *oriented strand board* (OSB) has also gained wide acceptance. OSB is composed of thin strands of wood that are pressed and bonded into mats oriented 90° to adjacent mats. The strength characteristics of OSB are similar to those of plywood and the two are often used interchangeably.

The standard size of plywood or OSB is in 4 × 8 ft sheets. Most plywood conforms to Product Standard PS 1-95, *Construction and Industrial Plywood,* published by the American Plywood Association (APA) [3]. OSB typically conforms to Product Standard PS 2-92, *Performance Standard for Wood-Based Structural-Use Panels,* also published by the APA [5]. PS-1 and PS-2 are voluntary product standards that contain information pertaining to the use and design properties of plywood and OSB panels.

Plywood Grain Orientation

The inner plies of plywood are typically oriented such that the grain of each adjacent ply is at an angle 90° with adjacent plies. The grain in the outer plies is oriented parallel to the 8-ft edge, which is the strongest configuration when the sheet of plywood is oriented perpendicular to the supporting structure (see Figure 6.1). In three-ply construction, there is one inner band oriented 90° with the outer plies or face grain. In four-ply construction, the two inner plies are oriented in the same direction. In five-ply construction the bands typically alternate in grain orientation (see Figure 6.2).

Plywood Species and Grades

Plywood can be manufactured from a variety of wood species. PS-1 defines the various species and classifies them into five groups, with group 1 representing the strongest and group 5 rep-

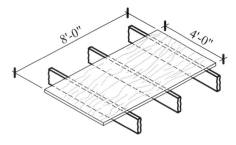

FIGURE 6.1 Plywood orientation on framing members.

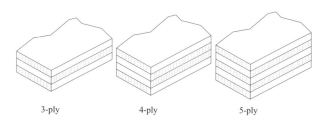

FIGURE 6.2 Plywood grain orientation in cross section.

resenting the weakest. The term *Structural I* refers to plywood made from a group 1 species and these panels are quite frequently specified by designers because of the certainty of the quality of wood species from which the component plies of the plywood are derived. The grade of the plywood is a function of the quality of the veneer. In PS-1 there are several veneer grades, which are summarized in Table 6.1 in order of decreasing quality. Grades C and D are the most commonly used for structural applications. Grade C is the minimum grade required for exterior use, and grade D is not permitted for exterior use. A plywood panel could have different veneer grades on each face. For example, the most common structural types of plywood are C–C (C grade on both outer plies) and C–D (C grade on the face, D grade on the back).

There are four main exposure categories for plywood: Exterior, Exposure 1, Exposure 2 (Intermediate Glue), and Interior. This classification is based on the ability of the bond in the glue to resist moisture. The plies of Exterior plywood are glued with fully waterproof glue and are C grade or better. Exterior plywood should be used when the moisture content in service exceeds 18%, or when the panel is exposed permanently to the weather. The plies of Exposure 1 plywood are glued with fully waterproof glue, but the veneer may be D grade. Exposure 1 plywood is recommended for use when the moisture content in service is relatively high or when the panels experience prolonged exposure to the weather during construction. The plies of Exposure 2 or Intermediate Glue (IMG) plywood are glued with an adhesive that has intermediate resistance to moisture. Exposure 2 plywood is recommended for protected applications where the panel is not exposed continuously to high humidity. Interior plywood is recommended for permanently protected interior applications. The glue used for Interior plywood has a moderate resistance to moisture and can sustain only short periods of high humidity in service.

Section 2304.7.2 of the IBC [12] requires that roof sheathing be constructed with exterior glue. For wall panels, exterior glue is to be used when the panel is exposed to the weather (see IBC Section 2304.6.1). When the wall panels are on an exterior wall but not exposed, the glue can be Exposure 1. Interior structural panels use Exposure 2 glue. Table 6.2 shows examples of various designations that could be used to identify plywood. Each sheet of plywood is typically stamped for identification. Sample grade stamps are shown in Figure 6.3.

TABLE 6.1 Plywood Veneer Grades

Grade	Description
N	Intended for natural finish; free from knots and other defects; synthetic fillers permitted for small defects
A	Suitable for painting; free from knots and other defects; synthetic fillers permitted for small defects
B	Slightly rough grain permitted; minor sanding and patching for up to 5% of panel area; small open defects permitted
C-plugged	Improved C grade; knotholes limited to $\frac{1}{4} \times \frac{1}{2}$ in.
C	Knotholes less than 1 in. in diameter permitted; knotholes of 1.5 in. permitted in limited cases
D	Knotholes less than 2 in. in diameter permitted; knotholes of 2.5 in. permitted in limited cases

TABLE 6.2 Plywood Sheathing Grades

C–C EXT	Exterior plywood with C-grade veneer on the face, back, and cross-band plies; uses *Exterior* glue	Used for exterior exposed applications (exposed permanently to the weather)
C–C EXT STRUCT I	Same as above but uses group 1 wood species	Same as above
C–D INT	Interior plywood with C-grade veneer on the face plies; D-grade veneer on the back and cross-band plies; uses *Interior* glue	Interior applications only
C–DX	Same as C–D INT but uses *Exterior* glue	Used for interior applications and exterior protected applications (e.g., roof of covered building)
C–D INT STRUCT I	Same as C–D INT but uses only group 1 wood species	Interior applications only

Span Rating

The span rating for floor and roof sheathing appears as two numbers separated by a sloped line (see Figure 6.3b and d). The left-hand number or numerator indicates the maximum recommended spacing of supports in inches when the panel is used for *roof sheathing* spanning over three or more supports with the grains perpendicular to the supports. The right-hand number represents the maximum recommended spacing of supports in inches when the panel is used as *floor sheathing* (subfloor).

The span rating for APA-rated STURD-I-FLOOR and APA-rated siding appears as a single number (see Figure 6.3a and c). APA-rated STURD-I-FLOOR is intended for single-floor applications (such as carpet and pad) and are typically manufactured with span ratings of 16, 20, 24, 32, and 48 in. The span ratings for APA-rated STURD-I-FLOOR are valid when the panel spans over three or more supports with the grains perpendicular to the supports. APA-rated siding is typically manufactured with span ratings of 16 and 24 in. The span rating for APA-rated siding is valid when the panel is oriented parallel or perpendicular to the supports, provided that blocking is used on the unsupported edges.

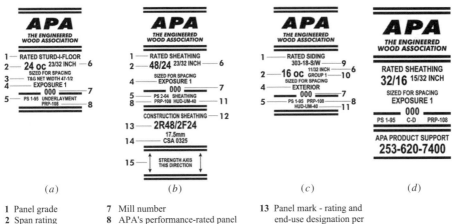

1 Panel grade
2 Span rating
3 Tongue-and-groove
4 Bond classification
5 Product standard
6 Thickness
7 Mill number
8 APA's performance-rated panel standard
9 Siding face grade
10 Species group number
11 HUD recognition
12 Panel grade, Canadian standard
13 Panel mark - rating and end-use designation per the Canadian standard
14 Canadian performance-rated panel standard
15 Panel face orientation indicator

FIGURE 6.3 Plywood grade stamps. (Courtesy of APA–The Engineered Wood Association, Tacoma, WA.)

6.2 ROOF SHEATHING: ANALYSIS AND DESIGN

To design roof sheathing for gravity loads, the designer has two basic options: NDS or IBC Tables. In some limited applications, the section and material properties are used directly to determine the stresses and deflections, similar to the design of beams. The *Wood Structural Panels Supplement* [16] (separate supporting document for the NDS) gives the section properties for various plywood span ratings and panel thickness (see Table 3.1 of the *Wood Structural Panels Supplement*). There are also uniform load tables (Tables 7.1 to 7.3 of the *Wood Structural Panels Supplement*) for various span ratings when the panels are either parallel or perpendicular to the supports. These tables assume that the panel edges perpendicular to the supports are supported. The edges of the panels can be supported by one of the following: tongue-and-grooved edges, panel edge clips, or wood blocking (see Figure 6.4). When panel edge clips are used, they are to be placed halfway between the supports, except that two panel clips are required at third points when the supports are 48 in. apart [see IBC Table 2304.7(3)]. If the panels are without edge support, Table 6.2 of the *Wood Structural Panels Supplement* should be used.

As an alternative to using the *Wood Structural Panels Supplement*, the table in Section 2304 of the IBC could be used. This is the approach taken in this book. For roof sheathing continuous over two or more spans perpendicular to the supports, Table 2304.7(3) should be used. This table correlates the span rating and panel thickness with uniform total and live loads, depending on the edge support condition (see Example 6.1). The load capacity shown is based on deflections of $L/180$ for total loads and $L/240$ for live loads.

For roof sheathing continuous over two or more spans parallel to the supports, Table 2304.7(5) should be used. The deflection criteria are based on $L/180$ for total loads and $L/240$ for live loads, and the design must include edge supports. This table is also based on five-ply panels, with the exception that four-ply panels can be used with a 15% reduction in some cases, as noted in the table.

6.3 FLOOR SHEATHING: ANALYSIS AND DESIGN

Two types of systems are used for floor sheathing. The first is a single-layer system, which is also referred to as a *single-floor* or *combination subfloor–underlayment*. In this system the panel serves as the structural support and the underlayment for the floor finish. The span rating for panels intended for single-floor use are designated by a single number (see Figure 6.3a). The APA-rated STURD-I-FLOOR panels are intended for single-floor use. Single-floor systems are typically used when floor finishes have some structural strength and can therefore bridge between adjacent plywood panels (e.g., hardwood floor, concrete, or gypcrete topping).

The second type is a *two-layer system* in which the bottom layer, or *subfloor*, acts as the primary structural support. The upper layer, referred to as the *underlayment*, provides a surface for the floor finish and is typically $\frac{1}{4}$ to $\frac{1}{2}$ in. thick. *Two-layer systems* are typically used where floor finishes have little structural strength (e.g., terrazzo).

As discussed previously, the *Wood Structural Panels Supplement* [16] could be used to design floor sheathing for gravity loads when the panel is oriented parallel or perpendicular to the support framing. Alternatively, IBC Tables 2304.7(3) and 2304.7(4) could be used directly. These tables use the single-floor and two-layer span rating designations and are based on panels oriented perpendicular to the supports continuous over two or more spans. The total uniform load capacity is 100 psf, which is typically adequate for any wood floor structure, and the deflection is based on $L/360$. There is a unique case when the support spacing is 48 in., where the total

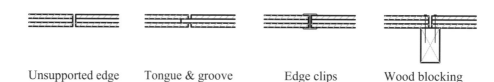

FIGURE 6.4 Plywood edge supports.

Unsupported edge   Tongue & groove   Edge clips   Wood blocking uniform load capacity is reduced to 65 psf. When using these tables, the panels are required to have edge support (see Figure 6.4) unless one of the following is provided: $\frac{1}{4}$-in.-thick underlayment with the panel edges staggered over the subfloor, $1\frac{1}{2}$ in. of lightweight concrete topping over the subfloor, or if the floor finish is $\frac{3}{4}$-in. wood strips.

EXAMPLE 6.1

Roof and Floor Sheathing Under Gravity Loads

For the framing plan shown in Figure 6.5 design (a) the roof and (b) the floor sheathing for gravity loads. The vertical loads are as follows:

roof dead load = 20 psf (includes mechanical/electrical and framing)

roof snow load = 40 psf (on a horizontal plane)

floor dead load = 20 psf (includes mechanical/electrical and framing)

floor live load = 40 psf

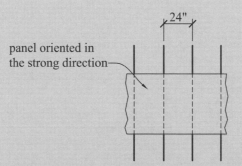

FIGURE 6.5 Roof and floor framing plans.

Solution:

(a) *Roof sheathing*. The total dead load on the sheathing would exclude the weight of the mechanical and electrical fixtures and the framing. For this problem, that weight will be assumed to be 10 psf. Therefore, the dead load on the plywood sheathing is

$$DL_{sheathing} = 20 \text{ psf} - 10 \text{ psf} = \textbf{10 psf}$$

$$\text{Snow load} \qquad\qquad = \textbf{40 psf}$$

$$\text{Total load} \qquad\qquad = \textbf{50 psf}$$

The span of the plywood sheathing is 24 in. (spacing of framing).

Enter IBC 2000 Table 2304.7(3) under the "Roof" section and look for plywood with a maximum span greater than or equal to the span required (i.e., 24 in.). From this table, the following is selected:

Span rating = 24/16

$\frac{7}{16}$-in.-thick C–DX plywood (recall that roof panels have exterior glue)

Edges without support

$$\text{Allowable total load} = \textbf{50 psf} = 50 \text{ psf applied} \qquad \textbf{OK}$$

$$\text{Allowable live load} = \textbf{40 psf} = 40 \text{ psf applied} \qquad \textbf{OK}$$

Use $\frac{7}{16}$-in.-thick C–DX plywood with a span rating of 24/16 and unsupported edges. Note that a $\frac{1}{2}$-in.-thick panel could also have been selected, but it is more economical to select the smaller panel for the same span rating. Since the span rating is the same with or without edge support, it is more economical to select a system without edge supports. Note also that the selection of a 24/0-rated panel would not be adequate since the total and live load capacities are 40 and 30 psf, respectively, which is less than the loads applied.

(b) *Floor sheathing*. The floor joists are spaced at 24 in. o.c. Similar to the roof sheathing solution, the total dead load on the sheathing would exclude the weight of the mechanical and electrical fixtures and the framing. Assuming that weight to be 10 psf, the dead load on the plywood sheathing is

$$DL_{sheathing} = 20 \text{ psf} - 10 \text{ psf} = \mathbf{10 \text{ psf}}$$

$$\underline{\text{Live load} \qquad\qquad\qquad = \mathbf{40 \text{ psf}}}$$

$$\text{Total load} \qquad\qquad\qquad = \mathbf{50 \text{ psf}}$$

Two types of floor systems are available.

1. *Two-layer floor system* [Use IBC Table 2304.7(3)]. To design a two-layer system, enter the last column of IBC 2000 Table 2304.7(3) and look for a row with a plywood span greater than or equal to 24 in. From the table, the following is selected:

Span rating = 48/24.

C–D plywood $\frac{23}{32}$ in. thick.

Allowable total load (see footnote in Table 2304.7(3)) = 100 psf > 50 psf   **OK**

According to footnote (d) in IBC Table 2304.7(3), edge support is required unless one of the following is provided: $\frac{1}{4}$-in.-thick underlayment with the panel edges staggered over the subfloor, $1\frac{1}{2}$ in. of lightweight concrete topping over the subfloor, or a floor finish of $\frac{3}{4}$-in. wood strips.

2. *Single-layer floor system* [either IBC Table 2304.7(3) or 2304.7(4)]. Using IBC Table 2304.7(3) to find a panel with a maximum span greater than or equal to the span required (i.e., 24 in.), the following is selected:

Span rating = 24.

C–D plywood 23 in./32 thick.

Allowable total load (see footnote in Table 2304.7(3)) = 100 psf > 50 psf   **OK**

Alternatively, the following is selected from IBC Table 2304.7(4):

Assuming group 1 species plywood ⇒ $\frac{3}{4}$-in.-thick C–D plywood

Assuming group 2 or 3 species plywood ⇒ $\frac{7}{8}$-in.-thick C–D plywood

Assuming group 4 species plywood ⇒ 1-in.-thick C–D plywood

Allowable total load (see footnote in Table 2304.7(4)) = **100 psf** > 50 psf

Similar to the two-layer system, edge supports are required unless one of the following is provided: $\frac{1}{4}$-in.-thick underlayment with the panel edges staggered over the subfloor, $1\frac{1}{2}$ in. of lightweight concrete topping over the subfloor, or a floor finish of $\frac{3}{4}$-in. wood strips.

Extended Use of the IBC Tables for Gravity Loads on Sheathing

To extend the use of IBC Tables 2304.7(3) and 2304.7(4) beyond the loads and span combinations given in the table, the following allowable total and live load equations could be used assuming moment equivalence:

$$W_{aTL} = \left(\frac{L_{max}}{L_a}\right)^2 W_{TL}$$

$$W_{aLL} = \left(\frac{L_{max}}{L_a}\right)^2 W_{LL}$$

where W_{aTL} = adjusted total load capacity, psf
W_{aLL} = adjusted live load capacity, psf
W_{TL} = total load capacity corresponding to the span rating, psf
W_{LL} = live load capacity corresponding to the span rating, psf
L_{max} = maximum allowable span (span rating), in.
L_a = actual span (joist spacing), in.

6.4 PANEL ATTACHMENT

The *Wood Structural Panels Supplement* [16] gives the minimum nailing requirements for roof and floor sheathing. Note that greater nailing requirements may be required, based on the panel strength as a diaphragm (see Section 6.5). In addition to nailing, the panels can be glued to the supports, which reduces squeaking due to the plywood slipping on the supports, and it also helps in the control of floor vibrations (see Section 4.7). A summary of the nailing requirements is shown in Table 6.3. Nails at the intermediate supports are typically referred to as *field nailing*. Nails at the edges could be along a main framing member and at the blocking, if blocking is provided. Fasteners should be spaced at least $\frac{3}{8}$ in. from the panel edges (see IBC Section 2305.1.2.1).

EXAMPLE 6.2

Extended Use of the IBC Tables for Gravity Loads on Sheathing

(a) Given a $\frac{7}{16}$-in.-thick C–DX plywood (edges unsupported) with a span rating 24/16, determine the allowable total load and allowable live load if it is used as a roof sheathing with a span of 20 in.

(b) Repeat part (a) assuming that the plywood is used as floor sheathing (two-layer system) with a span of 12 in.

Solution:

(a) *Roof sheathing.* From IBC Table 2304.7(3), the following is obtained for 24/16 roof sheathing (edges unsupported):

L_{max} = 24 in. (maximum span of the plywood when used as a roof sheathing)

W_{TL} = 50 psf

W_{LL} = 40 psf

The actual span of the plywood sheathing, L_a = 20 in. The adjusted allowable total and live loads are calculated as follows:

$$W_{aTL} = \left(\frac{L_{max}}{L_a}\right)^2 W_{TL} = \left(\frac{24}{20}\right)^2 (50) = \textbf{72 psf} \text{ (adjusted total load capacity)}$$

$$W_{aLL} = \left(\frac{L_{max}}{L_a}\right)^2 W_{LL} = \left(\frac{24}{20}\right)^2 (40) = \textbf{57 psf} \text{ (adjusted live-load capacity)}$$

(b) *Floor sheathing.* From IBC Table 2304.7(3), the following is obtained for 24/16 floor sheathing:

L_{max} = 16 in. (maximum span of the plywood when used as a floor sheathing)

L_a = 12 in.

W_{TL} = 100 psf (see footnote for tabulated total load capacity)

The actual span of the plywood sheathing, L_a = 16 in. The adjusted allowable total load is calculated as follows:

$$W_{aTL} = \left(\frac{L_{max}}{L_a}\right)^2 W_{TL} = \left(\frac{16}{12}\right)^2 (100) = \textbf{178 psf} \text{ (adjusted total load capacity)}$$

TABLE 6.3 Minimum Nailing Requirements for Roof and Floor Sheathing

Panel Description (Span Rating, Panel Thickness)	Nail Size	Nail Spacing (in.) Panel Edges	Intermediate Supports
Single floor			
16, 20, 24, $\frac{3}{4}$ in. thick or less	6d	6	12
24, $\frac{7}{8}$, or 1 in. thick	8d	6	12
32-, 48-, and 32-in. c-c span or less	8d	6	12
48-in. and 48-in. c-c span or less	8d	6	6
Subfloor			
$\frac{7}{16}$ to $\frac{1}{2}$ in. thick	6d	6	12
$\frac{7}{8}$ in. thick or less	8d	6	12
Greater than $\frac{7}{8}$ in. thick	10d	6	6
Roof			
$\frac{5}{16}$ to 1 in. thick	8d	6	12
Greater than 1 in. thick	8d	6	12
Spacing 48 in. and greater	8d	6	6

Source: Adapted from Ref. 2, Table 6.1.

6.5 HORIZONTAL DIAPHRAGMS

In a general structural sense, a diaphragm is a platelike structural element that is subjected to in-plane loading. In concrete building or a steel-framed building with a concrete slab, the concrete slab acts as the diaphragm. In a wood structure, the diaphragm is mainly the roof and floor sheathing, but drag struts and chords are also critical members in the diaphragm and load path (Figure 6.6). The sequence of the load path can be summarized as follows:

- Lateral wind or seismic loads are imposed on the diaphragm.
  - For wind loads, the wind pressure is applied to the wall sheathing and wall studs, which is then transferred to the diaphragm and the foundation at the base.
  - For seismic loads, the lateral force is assumed to transfer directly to the diaphragm.
- Loads are transferred from the diaphragm sheathing to the floor framing through the diaphragm fasteners.

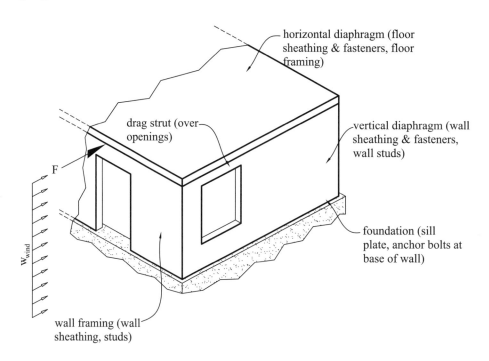

FIGURE 6.6 Load path on a building.

- Loads are transferred from the framing to the drag struts (where drag struts are not present, the loads are transferred directly to the shear walls).
- Loads are transferred from the drag struts to the shear walls (vertical diaphragms).
- Loads are transferred from wall sheathing to the wall framing through the shearwall fasteners.
- Loads are transferred from the wall framing to the sill plate.
- Loads are transferred from the sill plate to the anchor bolts.
- Loads are transferred from the anchor bolts to the foundation.

The design of each of these elements and connections is necessary to maintain a continuous load path to the foundation. Each of these items is covered later in the book.

The diaphragm can be modeled as a beam with simple supports at the shear wall locations. This beam consists of a web, which resists the shear (sheathing), flanges that resist the tension and compression forces resulting from a moment couple (chords), and supports that transfer the load to the shear walls (drag struts or collectors). Figure 6.7 illustrates the beam action of the diaphragm.

In a typical wood-framed structure, the shear force is resolved into a unit shear across the diaphragm. The unit shear can be defined as

$$v_d = \frac{V}{b}$$

where v_d = unit shear in the diaphragm, lb/ft
V = shear force ($=wL/2$ at supports), lb
b = width of the diaphragm parallel to the lateral force (distance between the chords), ft
L = length of the diaphragm perpendicular to the lateral force (distance between the shear walls), ft

The shear force V transferred to the shear walls is distributed uniformly in a similar manner. The unit shear in the shear walls can be defined as

$$v_w = \frac{V}{\Sigma L_w}$$

where v_w = unit shear in the shear walls along the diaphragm support line, lb/ft
V = total shear force along the diaphragm support line, lb
ΣL_w = summation of the shear wall lengths along diaphragm support line, ft

The top plates in the walls act as the chords and drag struts. The chords are oriented perpendicular to the lateral load and carry the bending moment in the diaphragm. The chord force can be defined as follows:

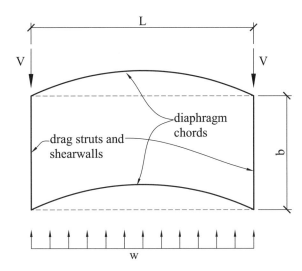

FIGURE 6-7 Beam action of a diaphragm.

$$T = C = \frac{M}{b}$$

where T = maximum tension chord force, lb
 C = maximum compression chord force, lb
 M = maximum diaphragm moment = $wL^2/8$, ft-lb

The drag struts or collectors are structural elements that are parallel to the lateral load and they help to drag or collect the diaphragm shears to the shear walls across door and window openings and at reentrant corners of buildings, and thus prevent stress concentrations at the ends of the shear walls. The total force in the drag strut can be defined as

$$F_{ds} = v_d L_{ds}$$

where F_{ds} = total force in the drag strut, lb
 v_d = unit shear in the diaphragm, lb/ft
 L_{ds} = length of the drag strut, ft

It should be noted that the total force F_{ds} can either be entirely a tension or compression force, or it can be distributed as a tension force at one end of the drag strut and a compression force at the other end. The total force is distributed in this latter manner when the drag strut is located between two shear walls and is due to the combined "pushing" and "pulling" of each shear wall. See Example 6.3 for further explanation of the calculation of drag strut forces.

Horizontal Diaphragm Strength

In the course of a building design, the floor and roof sheathing thicknesses are usually sized for gravity loads prior to calculating lateral loads. For most cases this is typically economical since the panel thickness is not the most critical factor in the strength of the diaphragm. The limiting factor is usually the capacity of the nails or staples that fasten the sheathing to the framing members. In general, there is a direct relationship between the amount of fasteners and the shear strength of the diaphragm.

There are two basic types of diaphragms: blocked and unblocked (see Figure 6.12). A *blocked diaphragm* is one in which the sheathing is fastened to a wood framing member under all four panel edges as well as to the intermediate supports. An *unblocked diaphragm* is fastened only at two panel edges and at the intermediate supports. It is usually more economical to design an unblocked diaphragm because of the added field labor required to provide blocking.

With reference to Table 6.3, the minimum nailing requirements are the same for blocked and unblocked diaphragms. However, decreasing the fastener spacing at the panel edges can increase the diaphragm capacity. IBC Table 2306.3.1 lists the shear capacities for various diaphragm types. Several diaphragm configurations and loading conditions are also identified in this table (load cases 1 through 6). The loading direction and sheathing orientation will define which load case to use (see Figure 6.13).

It should be noted that the diaphragm shear strength values shown in IBC Table 2306.3.1 could be increased by 40% for wind design (IBC Section 2306.3.2). The reason for this increase is due to the historically good performance of wood diaphragms during high-wind events. Seismic loads should be converted to service-level design values before entering IBC Table 2306.3.1.

The diaphragm shear strength values shown in the table are based on framing members of Douglas fir-larch or southern pine. For staples in framing members of other species, multiply the following specific gravity adjustment factors by the tabulated allowable unit shear for *structural* I grade, regardless of the plywood grade specified.

$$\text{SGAF} = \begin{cases} 0.82 & \text{for staples when } G \geq 0.42 \\ 0.65 & \text{for staples when } G < 0.42 \end{cases}$$

For nails in framing members of other species, multiply the following specific gravity adjustment factor by the tabulated unit shear for the *actual* grade of plywood specified.

$$\text{SGAF} = 1 - (0.5 - G) \leq 1.0 \qquad \text{for nails}$$

EXAMPLE 6.3

Horizontal Diaphragm Forces

Based on the floor plan and diaphragm loading due to wind shown in Figure 6.8, calculate (a) the force in the diaphragm chords, (b) the unit shear in the diaphragm, (c) the unit shear in the shear walls, and (d) the drag strut forces.

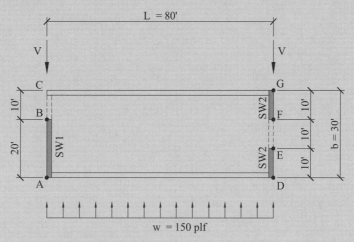

FIGURE 6.8 Example floor plan and diaphragm loading.

Solution: The diaphragm can be modeled as a simple span beam, as shown in Figure 6.9.

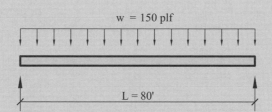

FIGURE 6.9 Example diaphragm loading.

(a) *Force in the diaphragm chords:*

$$M = \frac{wL^2}{8} = \frac{(150)(80)^2}{8} = 120{,}000 \text{ ft-lb}$$

$$T_{CG} = C_{AD} = \frac{M}{b} = \frac{120{,}000}{30} = \mathbf{4000 \text{ lb}}$$

Note: Since the lateral wind load can act in the reverse direction, chords *AD* and *CG* would have to be designed for both tension and compression loads.

(b) *Unit shear in the diaphragm:*

$$V_{\max} = \frac{wL}{2} = \frac{(150)(80)}{2} = 6000 \text{ lb}$$

$$v_d = \frac{V}{b} = \frac{6000}{30} = \mathbf{200\ lb/ft}\ \text{(maximum unit shear)}$$

The variation in unit shear can be shown as in Figure 6.10.

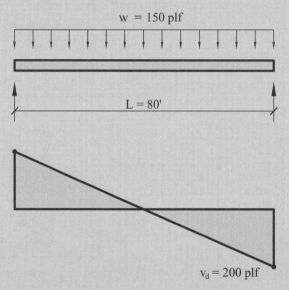

FIGURE 6.10 Variation in unit shear across a diaphragm.

(c) *Unit shear in the shear walls.* Along line *ABC*:

$$v_w = \frac{V}{\Sigma L_w} = \frac{6000}{20} = \mathbf{300\ lb/ft}$$

Along line *DEFG*:

$$v_w = \frac{V}{\Sigma L_w} = \frac{6000}{10 + 10} = \mathbf{300\ lb/ft}$$

(d) *Drag strut forces.* The total force in each drag strut is

$$F_{CB} = F_{EF} = v_d L_{ds} = (200)(10) = \mathbf{2000\ lb}$$

To determine the maximum drag strut force, the variation in diaphragm and shear wall unit shears has to be plotted as shown in Figure 6.11. Therefore, the maximum drag strut forces are

$$F_{CB} = \mathbf{2000\ lb}\ \text{(tension or compression)}$$
$$F_{EF} = \mathbf{1000\ lb}\ \text{(tension or compression)}$$

The maximum drag strut force can also be determined by using the free body diagrams in Figure 6.11. Equilibrium of forces in the free body diagrams yield the same results as was obtained using the unit shear diagram. Note that the total force in drag strut *CB* is 2000 lb and can either be tension or compression.

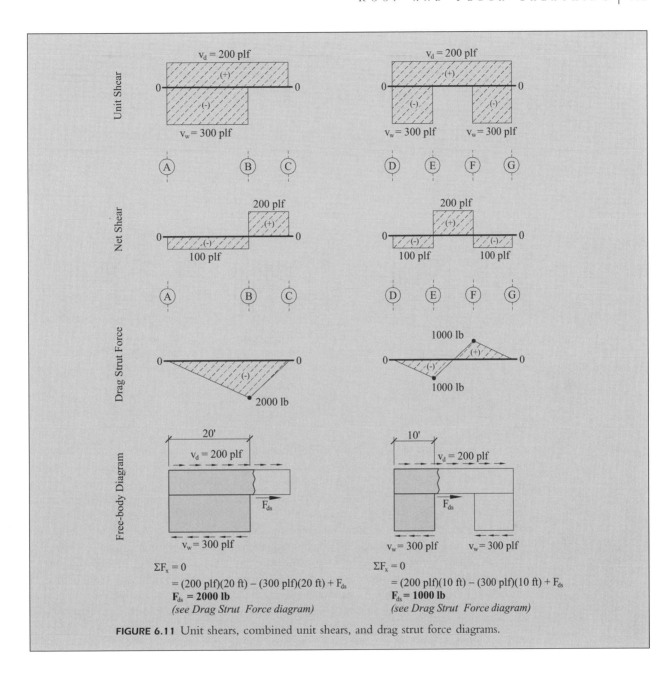

FIGURE 6.11 Unit shears, combined unit shears, and drag strut force diagrams.

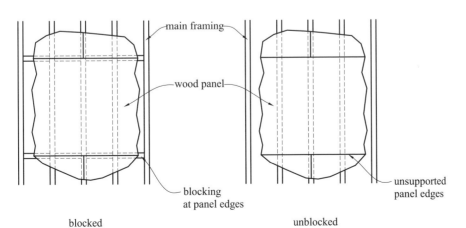

FIGURE 6.12 Blocked and unblocked diaphragms.

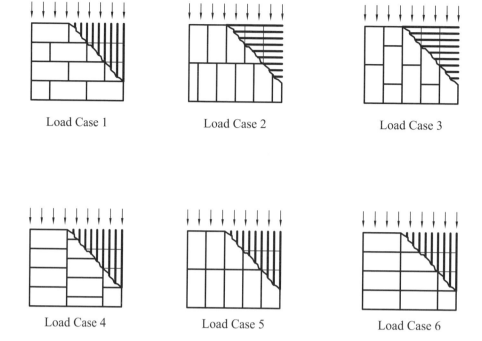

FIGURE 6.13
Diaphragm load cases.

where SGAF = specific gravity adjustment factor
G = specific gravity of wood (NDS Table 11.3.2A)

The code limits the aspect ratio of horizontal diaphragms. When all of the panel edges are supported and fastened, the maximum ratio of the diaphragm length to the width (L/b), (see Figure 6.14) is limited to 4. For unblocked diaphragms, the ratio is limited to 3 (IBC Section 2305.2.3). The distribution of lateral forces to the shear walls in a building is dependent on the rigidity of the horizontal diaphragm, and the diaphragm rigidity can be classified as flexible or rigid based on the ratio of diaphragm lateral deflection to the shear wall lateral deflection. For the design of wood-framed buildings, it is common in practice to assume a flexible diaphragm, and this assumption has been confirmed by other authors [11–14]. IBC 2006 permits wood structural panels to be classified as flexible diaphragms provided they have no more than $1\frac{1}{2}$ inches of non-structural topping. Furthermore, the simplified analysis procedure for seismic loads presented in Chapter 2 assumes a flexible diaphragm (see IBC Section 1613.6 and ASCE 7 Section 12.3). This procedure is permitted for wood-framed structures that are three stories and less in height, which is the category that most wood structures fall into.

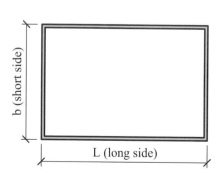

FIGURE 6.14 Diaphragm aspect ratios.

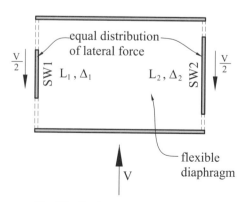

FIGURE 6.15 Flexible diaphragm versus shear wall deflection.

For a flexible diaphragm, the distribution of lateral seismic forces is based on the plan tributary widths or areas of each shear wall. For wind loads, the tributary area of the shear walls associated with the vertical exterior wall surfaces perpendicular to the wind direction are used. With reference to the flexible diaphragm shown in Figure 6.15, the lateral force is distributed equally to each shear wall, even though one wall is longer than the other. Since the lateral force to each shear wall is equal, a shorter wall will have a larger lateral deflection. In this case, the diaphragm is assumed to be flexible enough to absorb this unbalanced shear wall deflection.

For a rigid diaphragm, a torsional moment is developed and must be considered in the design of shear walls. This moment develops when the centroid of the lateral load is eccentric to the center of rigidity of the shear walls. One approximate method for resisting this torsional moment is a couple in the shear walls oriented perpendicular to the lateral load. This is shown in Figure 6.16, where the center of rigidity of the transverse shear walls does not coincide with the centroid of the lateral load. This results in a torsional moment that is resolved by forces in the longitudinal walls as shown.

In some cases a rigid diaphragm is necessary, such as an open-front structure. IBC Section 2305.2.5 limits the aspect ratios of rigid diaphragms in open-front structures. The length of the diaphragm normal to the opening cannot exceed 25 ft, and the length-to-width ratio cannot exceed 1.0 for single-story structures and 0.67 for structures taller than one story (see Figure 6.17).

The approach taken in this book is to assume a flexible diaphragm. However, the reader is encouraged to examine various diaphragm types to confirm this assumption. One approach is to analyze the diaphragm as both flexible and rigid, and use the larger forces for design.

Openings in Horizontal Diaphragms

Floor openings such as for stairs, elevators, or mechanical chases, and roof openings for skylights, may occur in a wood structure and must be considered in the design of the horizontal diaphragm. As with any break in structural continuity, proper attention should be given to the design and detailing of these diaphragms due to the localized stress increases that occur, and it is recommended that drag struts be provided along the edges of the opening. The ASCE 7 load specification classifies a building with area of diaphragm opening greater than 50% of the total diaphragm area as having plan structural irregularity for the purposes of seismic load calculations (ASCE 7 Section 12.3). Most wood buildings have diaphragm openings that are much smaller

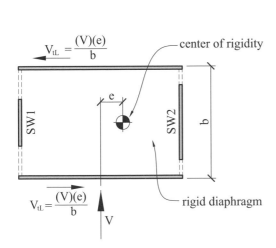

FIGURE 6.16 Rigid diaphragm with eccentric lateral load.

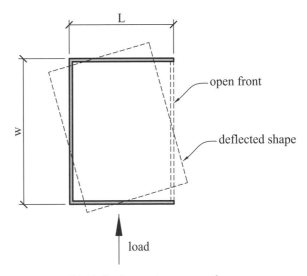

FIGURE 6.17 Rigid diaphragm in an open-front structure.

than 50%, and thus can be classified as regular. Consider the diaphragm loading and shear diagram shown in Figure 6.18.

Where, the shear forces are

$$V_1 = \frac{(150 \text{ plf})(80 \text{ ft})}{2} = 6000 \text{ lb}$$

$$V_2 = \left(\frac{40 \text{ ft} - 10 \text{ ft}}{40 \text{ ft}}\right)(6000 \text{ lb}) = 4500 \text{ lb}$$

$$V_3 = \left(\frac{40 \text{ ft} - 14 \text{ ft}}{40 \text{ ft}}\right)(6000 \text{ lb}) = 3900 \text{ lb}$$

Note that the total shear along each side of the opening must be distributed over a reduced length of diaphragm that excludes the opening. For a flexible diaphragm, the fastener layout may have to be modified to achieve the required shear capacity on each side of the opening. For a rigid diaphragm, the increased forces in the chords due to the diaphragm opening have to be transferred into the diaphragm, and this will lead to an increase in the number of fasteners used to connect the diaphragm to the chords. A unit shear diagram can be constructed as shown in Figure 6.19.

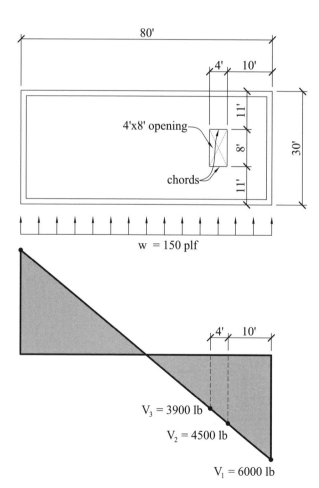

FIGURE 6.18 Opening in a diaphragm.

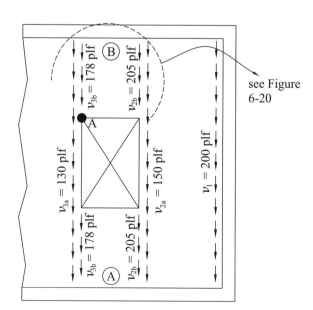

FIGURE 6.19 Unit shear in a flexible diaphragm with an opening.

The unit shears are

$$v_1 = \frac{6000 \text{ lb}}{30 \text{ ft}} = 200 \text{ plf}$$

$$v_{2a} = \frac{4500 \text{ lb}}{30 \text{ ft}} = 150 \text{ plf}$$

$$v_{2b} = \frac{4500 \text{ lb}}{(30 \text{ ft} - 8 \text{ ft opening})} = 205 \text{ plf}$$

$$v_{3a} = \frac{3900 \text{ lb}}{30 \text{ ft}} = 130 \text{ plf}$$

$$v_{3b} = \frac{3900 \text{ lb}}{(30 \text{ ft} - 8 \text{ ft opening})} = 178 \text{ plf}$$

These unit shear values would now be compared to IBC Table 2306.3.1 to select an appropriate fastener layout for the flexible diaphragm case.

For a rigid diaphragm, the unit shear diagram in the transverse direction remains as shown in Figure 6.19. However, there is a torsional moment that is developed in each diaphragm panel on each side of the opening. This torsional moment can be resolved into a couple of forces along the panel edges that are perpendicular to the lateral load (see Figure 6.20). In the figure

$$V_L = \text{total force along left side of panel}$$

$$= \frac{11 \text{ ft}}{11 \text{ ft} + 11 \text{ ft}} (3900 \text{ lb}) = 1950 \text{ lb}$$

$$V_R = \text{total force along right side of panel}$$

$$= \frac{11 \text{ ft}}{11 \text{ ft} + 11 \text{ ft}} (4500 \text{ lb}) = 2250 \text{ lb}$$

The total force on the diaphragm along each side of the diaphragm opening is proportional to the length of the diaphragm beyond the opening and parallel to the lateral load.

$$v_L = \text{unit shear along left side of panel}$$

$$= \frac{1950 \text{ lb}}{11 \text{ ft}} = 178 \text{ plf}$$

$$v_R = \text{unit shear along right side of panel}$$

$$= \frac{2250 \text{ lb}}{11 \text{ ft}} = 205 \text{ plf}$$

Considering the equilibrium of panel B and summing moments about point A yields

$$\Sigma M_A = (75 \text{ plf})(4 \text{ ft})\left(\frac{4 \text{ ft}}{2}\right) + (V_0)(11 \text{ ft}) - (2250 \text{ lb})(4 \text{ ft}) = 0$$

Therefore,

$$V_0 = 764 \text{ lb (total force to be added to the chords)}$$

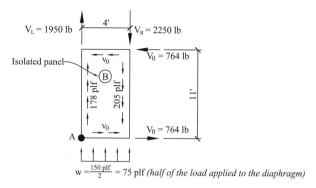

FIGURE 6.20 Unit shear in a rigid diaphragm with an opening.

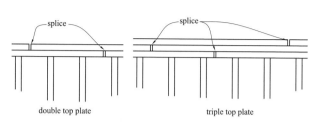

FIGURE 6.21 Double and triple top plates.

The unit shear in the diaphragm due to this additional force is

$$v_0 = \frac{764 \text{ lb}}{4'} = 191 \text{ plf}$$

This value of 191 plf for the unit shear values along the panel sides would now be compared to IBC Table 2306.3.1 to select an appropriate fastener layout.

Chords and Drag Struts

The chords and drag struts are the top plates in horizontal diaphragms. The chords are perpendicular to the direction of the lateral loads, while the drag struts are parallel to the lateral load. The drag struts are used to ensure continuity of the lateral load path in a diaphragm, and the chords act as the tension and compression flanges of the horizontal diaphragm as it bends in-plane due to the lateral loads. These top plates are usually not continuous; that is, they have to be spliced because of length limitations. Staggered splices are normally used (Figure 6.21) and the plates are commonly nailed together in the field.

For nailed or bolted double top plates, only one member is effective to resist the axial tension loads because of the splice. However, for double top plates connected with splice plates, both members will be effective in resisting the axial tension forces. For large chord or drag strut forces, more than two plates may be required. For a triple-top plate system, only two members are effective in resisting the axial tension loads.

The chords and drag struts must be designed for the worst-case tension and compression loads. Typically, the tension load controls because the tension strength is generally less than compression strength. Furthermore, the tension force cannot be transferred across a splice, whereas the compression force can be transferred by end bearing. The ASCE 7 load standard requires that certain structural elements, such as drag struts, be designed for special seismic forces E_m, but light-framed construction, which includes wood buildings, are exempt from this requirement.

EXAMPLE 6.4

Design of Horizontal Diaphragm Elements

Design the roof diaphragm of the building shown in Figure 6.22 for a lateral wind load of 20 psf. The floor-to-floor height of the building is 10 ft, and the height from the roof datum or level to the peak of the roof truss is 6 ft. Assume hem-fir for all wood framing and that normal temperature and moisture conditions apply.

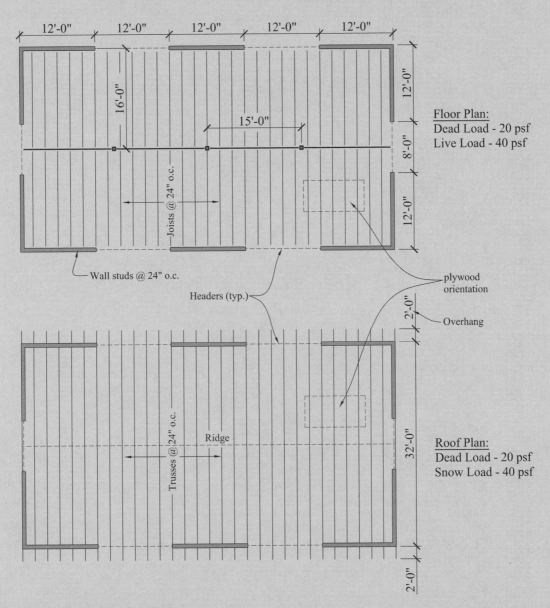

FIGURE 6.22 Roof and floor plan.

Roof dead load = 20 psf

Roof snow load = 40 psf (on a horizontal plane)

Floor dead load = 20 psf

Floor live load = 40 psf

Solution: (*Note:* Only the roof diaphragm has been designed in this example problem; the reader should follow the same procedure to design the floor diaphragm.)

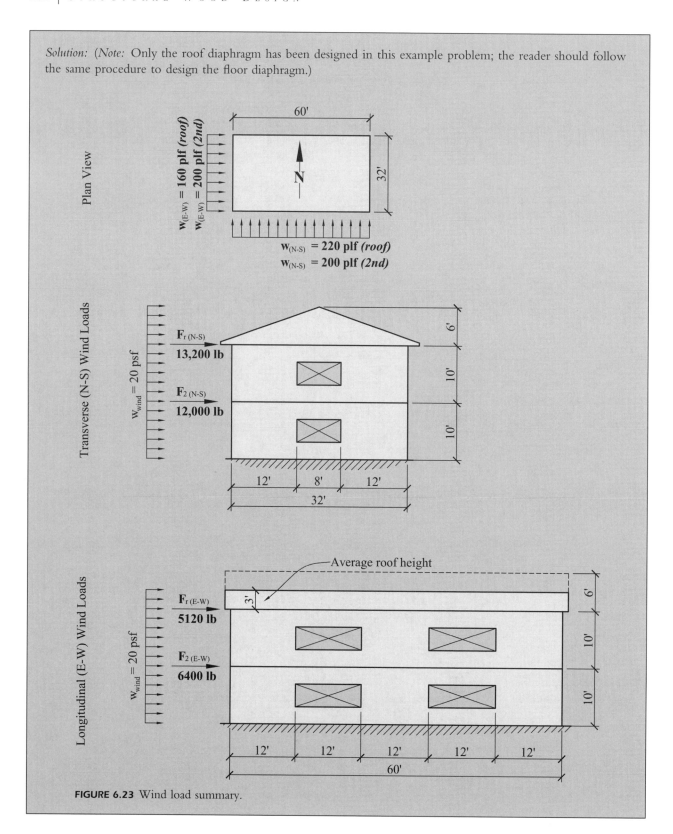

FIGURE 6.23 Wind load summary.

First calculate the wind loads (see Figure 6.23). In the N–S direction the uniformly distributed lateral loads in the horizontal diaphragm are

$$\text{roof lateral load } w_r = (20 \text{ psf})\left(6 \text{ ft} + \frac{10 \text{ ft}}{2}\right) = \mathbf{220 \text{ lb/ft}}$$

$$\text{second-floor lateral load } w_2 = (20 \text{ psf})\left(\frac{10 \text{ ft}}{2} + \frac{10 \text{ ft}}{2}\right) = \mathbf{200 \text{ lb/ft}}$$

$$\text{total lateral force at the roof level } F_{\text{roof}} = (220 \text{ plf})(60 \text{ ft}) = \mathbf{13{,}200 \text{ lb}}$$

$$\text{total lateral force at the second-floor level } F_{\text{2nd}} = (200 \text{ plf})(60 \text{ ft}) = \mathbf{12{,}000 \text{ lb}}$$

In the E–W direction the uniformly distributed lateral loads in the horizontal diaphragm are

$$\text{roof lateral load } w_r = (20 \text{ psf})\left(\frac{6 \text{ ft}}{2} + \frac{10 \text{ ft}}{2}\right) = \mathbf{160 \text{ lb/ft}}$$

$$\text{second-floor lateral load } w_2 = (20 \text{ psf})\left(\frac{10 \text{ ft}}{2} + \frac{10 \text{ ft}}{2}\right) = \mathbf{200 \text{ lb/ft}}$$

$$\text{total lateral force at the roof level } F_{\text{roof}} = (160 \text{ plf})(32 \text{ ft}) = \mathbf{5120 \text{ lb}}$$

$$\text{total lateral force at the second-floor level } F_{\text{2nd}} = (200 \text{ plf})(32 \text{ ft}) = \mathbf{6400 \text{ lb}}$$

Roof Diaphragm Design

In designing the roof diaphragm for lateral loads, we consider the horizontal projection of the roof; thus, the roof slope or pitch is not a factor in designing for lateral loads (Figure 6.24). Check the diaphragm aspect ratio:

$$\frac{60 \text{ ft}}{32 \text{ ft}} = 1.88 < 3 \qquad \mathbf{OK}$$

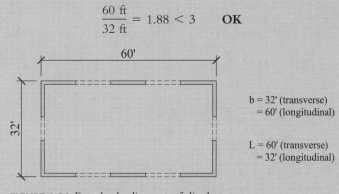

FIGURE 6.24 Free-body diagram of diaphragm:

Note: IBC Table 2305.2.3 requires that the diaphragm aspect be less than 4 if blocking is provided, and 3 if blocking is omitted. Since the aspect ratio is less than 3 for this example, we can proceed with the design with or without blocking.

N–S direction. The lateral load w_r is perpendicular to continuous panel joints and perpendicular to unblocked panel edges; therefore, use load case 1 from Figure 6.13 (or IBC Table 2306.3.1):

$$\text{maximum horizontal shear (reaction) } V_{\text{N-S}} = \frac{w_{\text{N-S}} L}{2} = \frac{(220)(60)}{2} = \mathbf{6600 \text{ lb}}$$

$$\text{maximum moment } M_{\text{N-S}} = \frac{w_{\text{N-S}} L^2}{8} = \frac{(220)(60)^2}{8} = \mathbf{99{,}000 \text{ ft-lb}}$$

The maximum unit shear in the diaphragm is

$$v_d = \frac{V}{b} \text{ (note that the overhangs are excluded)}$$

$$= \frac{6600}{32 \text{ ft}} = \mathbf{206.3 \text{ lb/ft}} \approx 207 \text{ lb/ft}$$

Note: The exact value of 206.3 lb/ft has to be used when calculating the drag strut forces to retain accuracy in the calculations.

For load case 1 with $v_d \approx 207$ lb/ft and assuming an unblocked diaphragm, the following is obtained from IBC 2000 Table 2306.3.1 (Figure 6.25):

$\frac{3}{8}$-in.-thick C–DX plywood, Structural I.

8d common nails.

Nails spaced at 6 in. (supported edges) and 12 in. (intermediate framing; see footnote b).

$v_{\text{allowable}} = 240$ lb/ft.

Allowable shear (lb/ft) for wood structural panel diaphragms with framing of Douglas fir-larch or southern pine for wind or seismic loading

Panel Grade	Common Nail Size or Staple Length and Gage	Minimum Fastener Penetration in Framing (in.)	Minimum Nominal Panel Thickness (in.)	Minimum Nominal Width of Framing Member (in.)	Blocked Diaphragms				Unblocked Diaphragms	
					\multicolumn{4}{c}{Fastener spacing (inches) at diaphragm boundaries (all cases) at continuous panel edges parallel to load Cases 3, 4), and at all panel edges (Cases 5 and 6)}	\multicolumn{2}{c}{Fasteners spaced 6" max. at supported edges}
					6	4	2½	2	Case 1 (No unblocked edges or continuous joints parallel to load)	All other configurations (Cases 2,3,4,5, & 6)
					\multicolumn{4}{c}{Fastener spacing (in.) at other panel edges (Cases 1,2,3,&4)}		
					6	6	4	3		
Structural I Sheathing	6d	1¼	5⁄16	2	185	250	375	420	165	125
				3	210	280	420	475	185	140
	1½" 16 Gage	1		2	155	205	310	350	135	105
				3	175	230	345	390	155	115
	8d	1⅜	3⁄8	2	270	360	530	600	**240**	180
				3	300	400	600	675	265	200
	1½" 16 Gage	1		2	175	235	350	400		
				3	200	265				

FIGURE 6.25 Partial view of IBC Table 2306.3.1 (N–S direction). (Adapted from Ref. 3.)

The values above are valid only if the framing is Douglas fir-larch or southern pine. Since hem-fir wood species is specified, the value obtained from the table must be adjusted as follows:

$$\text{SGAF} = 1 - (0.5 - G) \leq 1.0 \quad \text{for nails} \quad G = 0.43 \text{ (NDS Table 11.3.2A)}$$

$$= 1 - (0.5 - 0.43) = 0.93$$

Adjusted allowable shear:

Note: Section 2306.3.1 of the IBC allows a 40% increase in the allowable shear value, therefore:

$$v_{\text{allowable}} = 1.4(0.93)(240 \text{ lb/ft}) = \mathbf{312 \text{ lb/ft}} > v_d = 207 \text{ lb/ft} \quad \mathbf{OK}$$

However, by inspection of IBC Table 2304.7(3), the minimum panel thickness is at least $\frac{3}{8}$ in. for roof framing members spaced 24 in. Therefore, it would be necessary to select a $\frac{3}{8}$-in. panel for gravity loads.

E–W direction. The lateral load w_r is parallel to continuous panel joints and parallel to unblocked panel edges; therefore, use load case 3 from Figure 6.13 (or IBC Table 2306.3.1):

$$\text{maximum horizontal shear (reaction) } V_{\text{E-W}} = \frac{w_{\text{E-W}}L}{2} = \frac{(160)(32)}{2} = 2560 \text{ lb}$$

$$\text{maximum moment } M_{\text{E-W}} = \frac{w_{\text{E-W}}L^2}{8} = \frac{(160)(32)^2}{8} = 20{,}480 \text{ ft-lb}$$

The maximum unit shear in the diaphragm is

$$v_d = \frac{V}{b}$$

$$= \frac{2560}{60 \text{ ft}} = \mathbf{42.7 \text{ lb/ft}} \approx 43 \text{ lb/ft}$$

Note: The exact value of 42.7 lb/ft has to be used when calculating the drag strut forces to retain accuracy in the calculations.

Allowable shear (lb/ft) for wood structural panel diaphragms with framing of Douglas fir-larch or southern pine for wind or seismic loading

					Blocked Diaphragms				Unblocked Diaphragms	
					Fastener spacing (inches) at diaphragm boundaries (all cases) at continuous panel edges parallel to load Cases 3, 4), and at all panel edges (Cases 5 and 6)				Fasteners spaced 6" max. at supported edges	
Panel Grade	Common Nail Size or Staple Length and Gage	Minimum Fastener Penetration in Framing (in.)	Minimum Nominal Panel Thickness (in.)	Minimum Nominal Width of Framing Member (in.)	6	4	2½	2	Case 1 (No unblocked edges or continuous joints parallel to load)	All other configurations (Cases 2,3,4,5, & 6)
					Fastener spacing (in.) at other panel edges (Cases 1,2,3,&4)					
					6	6	4	3		
Structural I Sheathing	6d	1¼	⁵⁄₁₆	2	185	250	375	420	165	125
				3	210	280	420	475	185	140
	1½" 16 Gage	1		2	155	205	310	350	135	105
				3	175	230	345	390	155	115
	8d	1⅜	⅜	2	270	360	530	600	240	180 ⇐
				3	300	400	600	675	265	200
	1½" 16 Gage	1		2	175	235	350	400		
				3	200	265				

FIGURE 6.26 Partial view of IBC Table 2306.3.1 (E–W direction). (Adapted from Ref. 3.)

For load case 3 with $v_d \approx 43$ lb/ft and assuming an unblocked diaphragm, the following is obtained from IBC 2000 Table 2306.3.1 (Figure 6.26):

$\frac{3}{8}$-in.-thick C–DX plywood, Structural I.

8d common nails.

Nails spaced at 6 in. (supported edges)
and 12 in. (intermediate framing; see footnote b).

$v_{\text{allowable}} = 180$ lb/ft.

Adjusted allowable shear (using SGAF = 0.93 from the previous part and using the 40% increase allowed by the code):

$$v_{\text{allowable}} = (0.93)(1.4)(180 \text{ lb/ft}) = \mathbf{234 \text{ lb/ft}} > v_d = 43 \text{ lb/ft} \quad \mathbf{OK}$$

The allowable shear is much greater than the actual shear, but the $\frac{3}{8}$-in. panel was the minimum required for the N–S direction.

Chord forces (Figure 6.27). In the N–S direction, A and B are chord members. The axial forces in chord members A and B are

$$T = C = \frac{M_{\text{N-S}}}{b_{\text{N-S}}} = \frac{(99{,}000 \text{ ft-lb})}{32 \text{ ft}} = \mathbf{3100 \text{ lb tension or compression}}$$

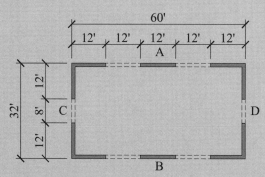

FIGURE 6.27 Diaphragm chords.

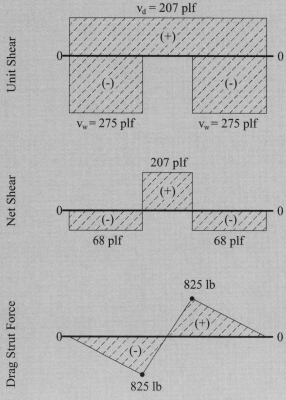

FIGURE 6.28 Unit shear and drag strut force diagrams, N–S direction.

where b_{N-S} is the distance between chords A and B.

In the E–W direction, C and D are chord members. The axial forces in chord members C and D are

$$T = C = \frac{M_{E-W}}{b_{E-W}} = \frac{(20{,}480 \text{ ft-lb})}{60 \text{ ft}} = \textbf{342 lb tension or compression}$$

where b_{E-W} is the distance between chords C and D.

Drag Struts. In the N–S direction, C and D are drag struts. The unit shear applied in the horizontal diaphragm in the N–S direction is $v_d = 206.3$ lb/ft. The unit shear in the N–S shear walls is

$$v_w = \frac{V_{N-S}}{\Sigma L_w} = \frac{6{,}600 \text{ lb}}{12 \text{ ft} + 12 \text{ ft}} = \textbf{275 lb/ft}$$

Note: In drawing the unit shears diagram (Figure 6.28), the horizontal diaphragm unit shears are plotted as positive and the shear wall unit shears are plotted as negative. The combined unit shears are a summation of the unit shear in the diaphragm and the unit shear in the walls. The drag strut force diagram is obtained by summing the area under the combined unit shear diagram.

For N–S wind, the maximum drag strut force in C and D is the largest force in the drag strut force diagram which is 825 lb (tension or compression). In the E–W direction, A and B are drag struts. The unit shear applied in the horizontal diaphragm in the E–W direction is $v_d = 42.7$ lb/ft. The unit shear in the E–W shear walls is

$$v_w = \frac{V_{E-W}}{\Sigma L_w} = \frac{2560 \text{ lb}}{12 \text{ ft} + 12 \text{ ft} + 12 \text{ ft}}\ \textbf{71.1 lb/ft}$$

Following the same procedure as that used in the construction of N–S unit shear, combined unit shear and drag strut force diagrams are shown in Figure 6.29.

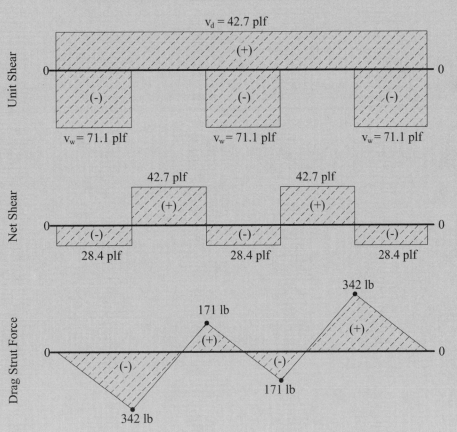

FIGURE 6.29 Unit shear and drag strut force diagrams, E–W direction.

For E–W wind, the maximum drag strut force in A and B is the largest force in the drag strut force diagram which is 342 lb (tension or compression).

The chord and drag strut forces are summarized in Table 6.4. The maximum force (tension and compression) for which the top plates should be designed is 3100 lb.

TABLE 6.4 Summary of Chord and Drag Strut Forces (lb), Example 6.4

Side	N–S Wind		E–W Wind	
	Chord	Drag Strut	Chord	Drag Strut
A	3100	—	—	342
B	3100	—	—	342
C	—	825	342	—
D	—	825	342	—

Design of top plates. Try double 2 × 6 top plates (same size as the wall studs given). As a result of the splicing of the double plates, only one 2 × 6 plate is effective in resisting the tension loads, whereas both 2 × 6 members are effective in resisting the axial compression force in the top plates (Figure 6.30).

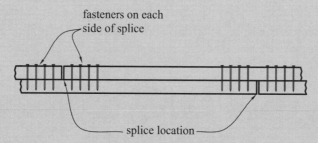

FIGURE 6-30 Top plate splice detail.

Design for tension. The maximum applied tension force T_{max} = **3100 lb** (see the summary of chord/drag strut forces). For the stress grade, assume hem-fir stud grade. Since a 2 × 6 is dimension lumber, use NDS-S Table 4A for design values. From the table the following design stress values are obtained:

$$F_t = 400 \text{ psi}$$

$$C_D(\text{wind}) = 1.6$$

$$C_M = 1.0$$

$$C_t = 1.0$$

$$C_F(F_t) = 1.0$$

$$C_i = 1.0 \text{ (assumed)}$$

$$A_g = 8.25 \text{ in}^2 \text{ (for one 2 × 6)}$$

The allowable tension stress

$$F'_t = F_t C_D C_M C_t C_F C_i$$

$$= (400)(1.6)(1.0)(1.0)(1.0)(1.0) = 640 \text{ psi}$$

The allowable tension force in the top plates

$$T_{\text{allowable}} = F'_t A_g = (640 \text{ psi})(8.25 \text{ in}^2) = \mathbf{5280 \text{ lb}} > T_{\text{applied}} = 3100 \text{ lb} \quad \mathbf{OK}$$

Design for compression. The top plate is fully braced about both axes of bending. It is braced by the stud wall for y–y or weak-axis bending, and it is braced by the roof or floor diaphragm for x–x or strong-axis bending, Therefore, the column stability factor $C_P = 1.0$. The maximum applied compression force, $P_{\max} = \mathbf{3100 \text{ lb}}$ (see the summary of chord/drag strut forces). For the stress grade, assume hem-fir stud grade. Since a 2 × 6 is dimension lumber, use NDSS Table 4A for design values. From the table the following design stress values are obtained:

$$F_c = 800 \text{ psi}$$

$$C_D(\text{wind}) = 1.6$$

$$C_M = 1.0$$

$$C_t = 1.0$$

$$C_F(F_c) = 1.0$$

$$C_i = 1.0 \text{ (assumed)}$$

$$C_P = 1.0$$

$$A_g = 16.5 \text{ in}^2 \text{ (for two 2 × 6's)}$$

The allowable compression stress is

$$F'_c = F_c C_D C_M C_t C_F C_i C_P$$
$$= (800)(1.6)(1.0)(1.0)(1.0)(1.0)(1.0) = 1280 \text{ psi}$$

The allowable compression force in the top plates is

$$P_{\text{allowable}} = F'_c A_g = (1280 \text{ psi})(16.5 \text{ in}^2) = \mathbf{21{,}120 \text{ lb}} > P_{\text{applied}} = 3100 \text{ lb} \quad \mathbf{OK}$$

Use two 2 × 6's hem-fir stud grade for the top plates.

Note: See Chapter 9 for design of the lap splice connection.

EXAMPLE 6.5

Design of a Horizontal Diaphragm with Partial Wood Blocking

Considering the diaphragm loading shown in Fiugre 6.31, determine the extent for which wood blocking is required assuming the floor framing is similar to that shown for Example 6.4.

Solution: For load case 1 and assuming a blocked diaphragm, the following is obtained from IBC Table 2306.3.1:

$\frac{3}{8}$-in.-thick C–DX plywood, Structural I.

8d common nails.

2× blocking required.

Nails spaced at 6 in. (all edges)

and 12 in. (intermediate framing, see footnote b)

Base $v_{\text{allowable}} = 270$ lb/ft.

SGAF = 0.93 (from Example 6.4). The adjusted allowable shear

$$v_{\text{allowable}} = (0.93)(270 \text{ lb/ft})(1.40) = \mathbf{351 \text{ lb/ft}} > v_d = 338 \text{ lb/ft} \quad \mathbf{OK}$$

Extent of wood blocking. Wood blocking is provided only in the zones where the unblocked diaphragm allowable unit shear is less than the unit shear applied. The zone on the shear diagram in which the unit shear is less than the unblocked diaphragm allowable unit shear (i.e., 312 lb/ft calculated in Example 6-4) is determined using similar triangles (Figure 6.31).

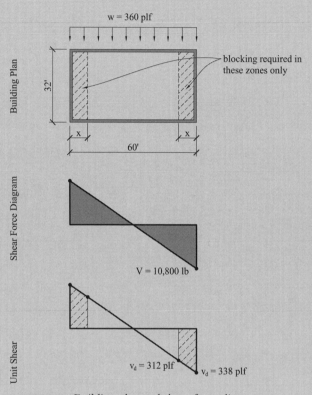

FIGURE 6.31 Building plan and shear force diagram.

$$\frac{338 \text{ lb/ft}}{30 \text{ ft}} = \frac{312 \text{ lb/ft}}{30 \text{ ft} - x} \quad x = 2.31 \text{ ft, say 4 ft (since joist spacing is 24 inches)}$$

Therefore, the horizontal diaphragm needs to be blocked in the N–S direction only in the zones within the first 2 ft from the shear walls (hatched area in Figure 6.31).

Thus far, we have considered only buildings with shear walls along the exterior faces only, which results in single-span simply supported horizontal diaphragms. When a building becomes too long with respect to its width, the diaphragm aspect ratio may exceed the 4:1 maximum allowed by the IBC, in which case interior shear walls will be required. With interior shear walls, the horizontal diaphragms are considered as separate simply supported diaphragms that span between the shear walls. We now consider an example building with both interior and exterior shear walls.

EXAMPLE 6.6

Horizontal Diaphragms with Interior and Exterior Shear Walls

Determine the forces in the chords and drag struts and the diaphragm unit shears at the roof level for a building with the roof plan shown in Figure 6.32. Consider only the N–S wind loads given.

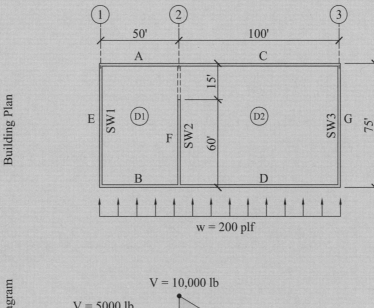

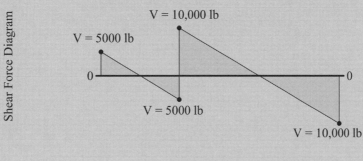

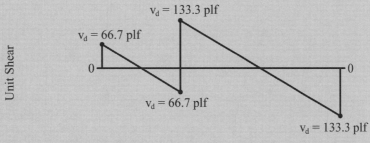

FIGURE 6.32 Building plan.

Solution:

Diaphragm D1. For the N–S wind, A and B are chords; E and F are drag struts.

$$\text{Maximum moment } M_{D1} = \frac{w_{D1} L^2}{8} = \frac{(200)(50)^2}{8} = 62{,}500 \text{ ft-lb}$$

The axial tension and compression force in chords A and B is

$$T = C = \frac{M_{D1}}{b_{D1}} = \frac{62{,}500 \text{ ft-lb}}{75 \text{ ft}} = \textbf{833 lb tension or compression}$$

$$v_{d1} = \frac{V_1}{b} = \frac{5000 \text{ lb}}{75 \text{ ft}} = \textbf{66.7 lb/ft} \text{ (unit shear in the diaphragm along line 1)}$$

$$v_{d2(\text{left})} = \frac{V_2}{b} = \frac{5000 \text{ lb}}{75 \text{ ft}} = \textbf{66.7 lb/ft} \text{ (unit shear in the diaphragm along line 2)}$$

Diaphragm D2. For the N–S wind, C and D are chords; F and G are drag struts.

$$\text{Maximum moment } M_{D2} = \frac{w_{D2} L^2}{8} = \frac{(200)(100)^2}{8} = 250{,}000 \text{ ft-lb}$$

The axial tension and compression force in chords A and B is

$$T = C = \frac{M_{D2}}{b_{D2}} = \frac{250{,}000 \text{ ft-lb}}{75 \text{ ft}} = \textbf{3334 lb tension or compression}$$

$$v_{d2(\text{right})} = \frac{V_2}{b} = \frac{10{,}000 \text{ lb}}{75 \text{ ft}} = \textbf{133.3 lb/ft} \text{ (unit shear in the diaphragm along line 2)}$$

$$v_{d3} = \frac{V_3}{b} = \frac{10{,}000 \text{ lb}}{75 \text{ ft}} = \textbf{133.3 lb/ft} \text{ (unit shear in the diaphragm along line 3)}$$

Drag strut forces (N–S wind):

1. The force in drag strut on line 1 = 0 since there are no wall openings along line 1 and therefore there is no force to be "dragged."
2. The force in drag strut on line 3 = 0 since there are no wall openings along line 3 and therefore there is no force to be dragged.
3. To calculate the force in the drag strut on line 2 (i.e., at the interior wall), the unit shear, net shear, and drag strut force diagrams must be constructed. The total diaphragm unit shear along line 2 is

$$v_{d(\text{total})} = v_{d(\text{left})} + v_{d(\text{right})}$$

$$= 66.7 \text{ lb/ft} + 133.3 \text{ lb/ft} = \textbf{200 lb/ft}$$

The unit shear in the shear wall along line 2 is

$$v_{w2} = \frac{V_2}{\Sigma L_w} = \frac{10{,}000 \text{ lb} + 5000 \text{ lb}}{60 \text{ ft}} = \textbf{250 lb/ft}$$

Following the procedure used in Example 6.4, we construct the unit shear, combined unit shear, and drag strut force diagrams (Figure 6.33). The chord and drag strut forces are summarized in Table 6.5. The maximum force in the drag strut on line 2 = 3000 lb.

TABLE 6.5 Summary of Chord and Drag Strut Forces (lb), Example 6.6

Chords	N–S Wind Chord Force	Drag Strut	N–S Wind Drag Strut Force
A	833	A	—
B	833	B	—
C	3334	C	—
D	3334	D	—
E	—	E	0
F	—	F	3000
G	—	G	0

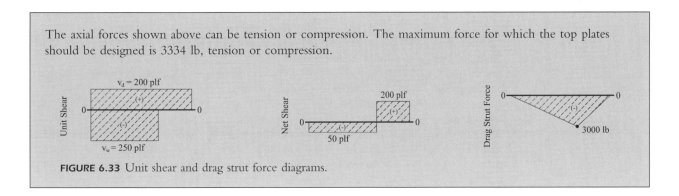

The axial forces shown above can be tension or compression. The maximum force for which the top plates should be designed is 3334 lb, tension or compression.

FIGURE 6.33 Unit shear and drag strut force diagrams.

Nonrectangular Diaphragms

The examples presented thus far have generally assumed rectangular diaphragms. In practice it is common to encounter several types of nonrectangular diaphragms. Although there are countless possibilities, some design considerations of nonrectangular diaphragms are summarized in Figure 6.34. The aim is to avoid the noncompatibility of the lateral deflections in nonrectangular diaphragms by using drag struts and collectors. The drag struts effectively convert a nonrectangular diaphragm into several rectangular diaphragm segments, thus ensuring the compatibility of the lateral deflections of the various diaphragm segments.

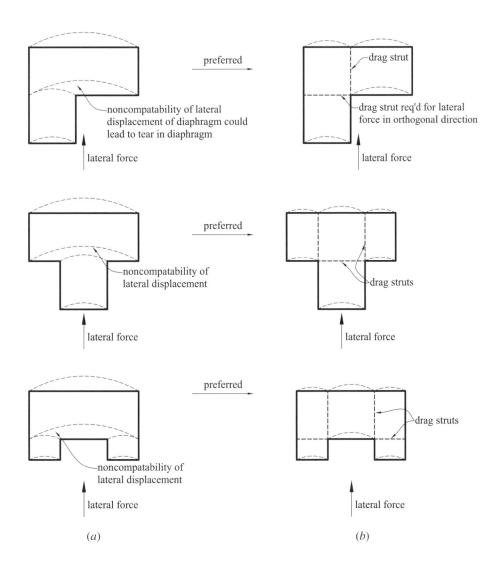

FIGURE 6.34 Bending action of nonrectangular diaphragms: (*a*) without drag struts; (*b*) with drag struts.

REFERENCES

1. ANSI/AF&PA, (2001), *Allowable Stress Design Manual for Engineered Wood Construction,* American Forest & Paper Association, Washington, DC.
2. ANSI/AF&PA, (2005), *National Design Specification Supplement: Design Values for Wood Construction,* American Forest & Paper Association, Washington, DC.
3. APA (2002), *Construction and Industrial Plywood,* PS 1-95, APA—The Engineered Wood Association, Tacoma, WA.
4. APA (2001), *APA Quality Assurance Policies for Structural-Use Panels Qualified to PRP-108,* APA—The Engineered Wood Association, Tacoma, WA.
5. APA (2002), *Performance Standard for Wood-Based Structural-Use Panels,* PS 2-92, APA—The Engineered Wood Association, Tacoma, WA.
6. APA (2001), *Engineered Wood Construction Guide,* APA—The Engineered Wood Association, Tacoma, WA.
7. APA (2004), *Diaphragms and Shear Walls: Design/Construction Guide,* APA—The Engineered Wood Association, Tacoma, WA.
8. APA (1998), *Plywood Design Specification,* APA—The Engineered Wood Association, Tacoma, WA.
9. ATC (1981), *Guidelines for the Design of Horizontal Diaphragms,* ATC-7, Applied Technology Council, Redwood City, CA.
10. Faherty, Keith F., and Williamson, Thomas G. (1989), *Wood Engineering and Construction,* McGraw-Hill, New York.
11. Halperin, Don A., and Bible, G. Thomas (1994), *Principles of Timber Design for Architects and Builders,* Wiley, New York.
12. ICC (2006), *International Building Code,* International Code Council, Washington, DC.
13. Stalnaker, Judith J., and Harris, Earnest C. (1997), *Structural Design in Wood,* Chapman & Hall, London.
14. Breyer, Donald et al. (2003), *Design of Wood Structures—ASD,* 5th ed., McGraw-Hill, New York.
15. Tarpy, Thomas S., Thomas, David J., and Soltis, Lawrence A. (1985), Continuous timber diaphragms, *J. Struct. Div. ASCE* 111(5):992–1002.
16. ANSI/American Forest & Paper Association, (2001), *Wood Structural Panels Supplement,* AF&PA, Washington, DC.

PROBLEMS

6.1 For a total roof load of 60 psf on 2× framing members at 24 in. o.c., determine an appropriate plywood thickness and span rating assuming panels with edge support and panels oriented perpendicular to the framing. Specify the minimum fasteners and spacing.

6.2 Repeat Problem 6.1 for panels oriented parallel to the supports.

6.3 For a total floor load of 75 psf on 2× floor framing at 24 in. o.c., determine an appropriate plywood thickness and span rating assuming panels with edge support and panels oriented perpendicular to the framing. Specify the minimum fasteners and spacing.

6.4 A one-story building is shown in Figure 6.35. The roof panels are $\frac{15}{32}$-in.-thick CD–X with a span rating of 32/16. The wind load is 25 psf. Assume that the diaphragm is unblocked and that the framing members are spruce-pine-fir. Determine the following for the transverse direction:

   (a) Unit shear in the diaphragms
   (b) Required fasteners
   (c) Maximum chord forces
   (d) Maximum drag strut forces (construct unit shear, net shear, and drag strut force diagrams)

6.5 Repeat Problem 6.4 for the longitudinal direction.

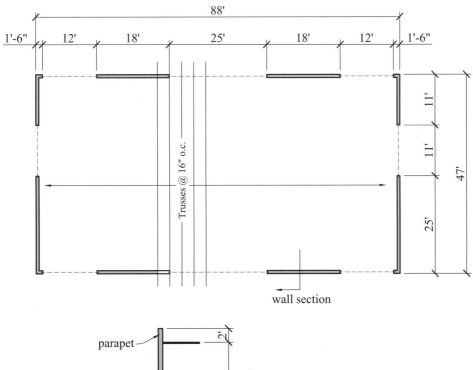
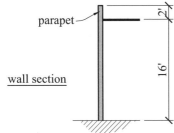

FIGURE 6.35

chapter *seven*

Vertical Diaphragms under Lateral Loads (Shear Walls)

7.1 INTRODUCTION

The design of vertical diaphragms or shear walls is one of the most critical elements in wood-framed structures. The lateral forces due to wind or seismic loads are transmitted from the horizontal diaphragm to the shear walls, which then transfer these loads to the foundation. The most common sheathing materials used are either plywood or gypsum wallboard (GWB).

The lateral load distribution to vertical diaphragms or shear walls is a function of the relative rigidity of the horizontal roof and floor diaphragms. Horizontal diaphragms can be classified as *flexible* or *rigid,* and most wood buildings can be assumed to have flexible diaphragms. Therefore, lateral wind loads are distributed to the shear walls based on the tributary area of the shear wall relative to the vertical wall surface on which the wind acts, while seismic loads are distributed based on the plan tributary area of the shear walls.

The capacity of a plywood shear wall is greater than that of a shear wall with GWB, especially when comparing their resistance to seismic loads. A typical wood-framed structure has plywood sheathing on the exterior walls only, with the remaining interior walls sheathed with GWB. If a building design requires the use of plywood sheathing on the interior walls, the plywood is typically covered with GWB for aesthetic purposes, so the designer must consider the economic impact of using interior plywood shear walls versus strengthening the exterior plywood shear walls.

Wall sheathing must also act to transfer the horizontal wind load to the wall studs on the exterior walls. IBC Section 1607.13 does require that interior walls be adequate to resist a lateral load of 5 psf, but this lateral pressure is typically small enough not to affect the design of the interior wall sheathing or studs.

Wall Sheathing Types

Although several types of wall sheathing are available for use in shear walls, in this book we focus on plywood sheathing only. There are two common types of plywood wall sheathing; the first is *exposed plywood panel siding* (e.g., APA 303–rated siding). These panels are manufactured with various surface textures and perform a dual purpose as a structural panel as well as being the exterior surface of the building. Several proprietary products are available that are recognized by the IBC for use as shear walls. Exposed plywood siding can either be fastened directly to wall studs or over a layer of $\frac{1}{2}$- or $\frac{5}{8}$-in.-thick GWB to meet fire rating requirements. IBC Table 2308.9.3(2) specifies the minimum thickness for exposed plywood as $\frac{3}{8}$ in. for studs spaced 16 in. o.c. and $\frac{1}{2}$ in. for studs spaced 24 in. o.c.

The second type is *wood structural panel wall sheathing*. This is commonly used on exterior walls applied directly to the wall framing with a finish material applied to the plywood. IBC Table 2308.9.3(3) specifies the minimum thickness for plywood not exposed to the weather.

The span rating for wall panels is usually indicated by one number, either 16 or 24 in. (see Figure 7.1). This number refers to the maximum wall stud spacing. Panels rated for wall sheathing are performance tested, with the short direction spanning across the supports (i.e., the weaker direction); therefore, the wall panels can be placed either horizontally or vertically. Loads normal to the wall panels typically do not control the thickness of the sheathing, rather the unit shear in the wall panels due to lateral loads will usually determine the panel thickness.

Plywood as a Shear Wall

When lateral loads act parallel to the face of the wall, the plywood wall sheathing acts as a shear wall. Shear walls are the main elements used to resist lateral loads due to wind and earthquake in wood-framed buildings. They act as vertical cantilevers that are considered fixed at the ground level (see Figure 7.2).

The shear wall diaphragm (i.e., the plywood) resists the shear, while the shear wall chords (i.e., the vertical members at the ends of the shear wall) resist the moments caused by the lateral forces; this moment creates a tension and compression force couple in the shear wall chords. IBC Section 2305.1.2.1 specifies that all of the plywood panel edges be fastened to a framing member, which may require blocking between the wall studs, depending on the wall sheathing orientation. The same code section also specifies that the plywood be attached to the framing members with fasteners spaced not less than $\frac{3}{8}$ in. from a panel edge, a minimum of 6 in. o.c. along the panel edges and 12 in. o.c. in the field—that is, along the intermediate framing members (Figure 7.3). It is recommended that the location of the nails from the edge of the stud framing should be about the same as the distance of the nail from the panel edge. Table 6-1 from the *Wood Structural Panels Supplement* [15] indicates that the minimum nail size should be 6d for wood panels $\frac{1}{2}$ in. and thinner and 8d nails for thicker panels.

The capacity of a shear wall to resist lateral loads is a function of several items; the plywood thickness, fastener size, and fastener layout. The lateral load capacity of wood shear wall panels is largely controlled by the load capacity of the fasteners. The plywood itself will rarely fail in shear because of the relatively high shear strength of the wall sheathing compared to the fastener capacity. The failure mode for shear walls includes one or more of the following scenarios:

1. Edge tearing of the diaphragm as the nails bear against it.
2. The pull-through of the nails from the wall sheathing or failure of the nails in bending, leaving the panels laterally unsupported, and thus making the panels susceptible to buckling.
3. Longitudinal splitting failure of the wood framing especially for larger nail sizes.

The capacities of several combinations of plywood thickness and fasteners are listed in IBC Table 2306.4.1 (see Figures 7.15 and 7.17). It is important to note that the shear capacities listed in this table can be increased by 40% when designing for wind loads (IBC Section 2306.4.1) because the shear capacities in IBC Tables 2306.3.1 and 2306.4.1 were derived based on a safety factor of 2.8, whereas the ASCE 7 wind loads are based on a safety factor of 2.0. [8]. These shear capacities apply to plywood fastened to one side of the studs. For plywood fastened to both sides of the studs, the shear capacity is doubled if the sheathing material and fasteners are the same on each side. When the shear capacity is not equal on both sides, the total capacity is the larger of either twice the smaller capacity, or the capacity of the stronger side.

In some cases, such as in exterior walls, plywood is applied on the exterior face and gypsum wall board (GWB) is applied on the interior face. While the GWB and fasteners have some capacity to act as a shear wall to resist lateral loads (see Ref. 9), it is usually not recommended that GWB and plywood be used in combination as a shear wall since the response modification factor *R* is much greater for wood than for GWB (see ASCE 7 Table 12.2.1). The *R* value is indirectly proportional to the seismic load. However, IBC Section 2305.3.8 does permit direct summation of the shear capacities for wood panels used in combination with GWB for resisting wind loads only. This is not permitted for seismic loads. So for the same building, the allowable

FIGURE 7.1 Wall siding grade stamp. (Courtesy of APA–The Engineered Wood Association, Tacoma, WA.)

218 | STRUCTURAL WOOD DESIGN

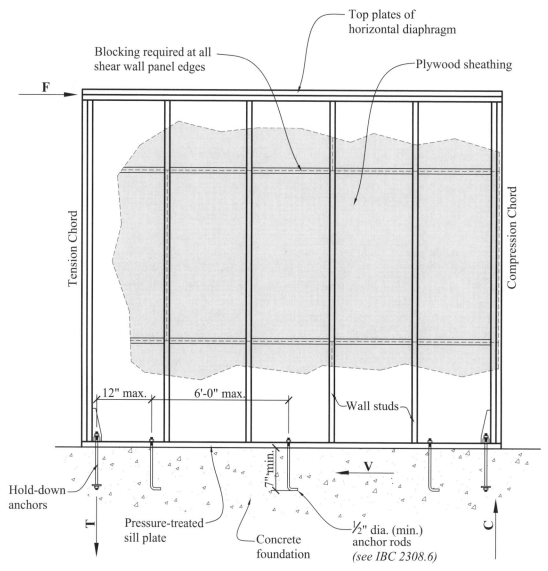

FIGURE 7.2 Shear wall as a vertical cantilever.

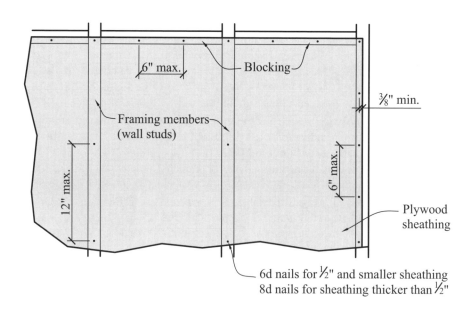

FIGURE 7.3 Typical minimum fastening for wall panels.

shear capacity of a wall could be much smaller for seismic design than for wind design given the discussion above and the 40% increase allowed for wind design. This difference in shear capacities for wind and seismic design is illustrated in Example 7.3.

Hold-down anchors are used to resist uplift tension, and sill anchors (or anchor rods) are used to resist sliding. Both are usually anchored into the concrete foundation. IBC Sections 2308.3.3 and 2308.6 specify the minimum spacing and dimensions for the anchor rods.

The *load path* for wind is as follows: Wind acts perpendicular to wall surfaces, resulting in vertical bending of these walls. The lateral reaction from the vertical bending of the walls is then transmitted to the diaphragms above and below the walls. The roof and floor diaphragms then transfer the lateral reactions to the vertical lateral force–resisting systems (i.e., the vertical diaphragms or shear walls), and the shear walls transfer the lateral loads safely to the ground or foundation.

For seismic loads, the ground motion results in the back-and-forth lateral displacement of the base of the building, causing a shear force at the base of the building. This base displacement causes lateral displacements of the diaphragms in the direction opposite to the base displacement. These lateral displacements result in equivalent lateral forces at the various diaphragm levels of the building. The equivalent lateral forces are then transmitted through the roof and floor diaphragms to the vertical lateral force–resisting systems (i.e., the vertical diaphragms or shear walls), and the shear walls transfer the lateral loads safely to the ground or foundation (Figure 7.4).

7.2 SHEAR WALL ANALYSIS

There are three methods given in the IBC for shear wall analysis (IBC Section 2305.3):

1. *Segmented approach.* The full height segments between window and door openings are assumed to be the only walls resisting lateral loads; thus, the stiffness contribution of the walls above and below the window or door openings is neglected. In this approach the hold-down anchors are located at the ends of each wall segment. The shear wall forces calculated using this approach are usually the largest of the three approaches. This is the method adopted in this book and is the most commonly used method in practice.

2. *Perforated shear wall approach.* All the walls, including those above and below window and door openings, are assumed to contribute to the lateral load-carrying capacity of the shear walls. In this approach, the hold-down anchors are located at the extreme ends of the *overall* shear wall; thus, each shear wall line will typically have only two hold-down anchors. The shear wall forces calculated using this approach are usually smaller than those calculated using the first method, but the edges of the wall openings must be reinforced to transfer the lateral forces. This method is sometimes used in buildings where the length and number of shear walls are limited.

3. *Force transfer method.* The walls are assumed to behave as a coupled shear wall system. In this case, the hold-down anchors are also located at the extreme ends of the entire shear wall; thus, each shear wall line will typically have only two hold-down anchors. To ensure coupled shear wall action, horizontal straps or ties would have to be added to tie the wall segments below and above the window and door openings to the wall segments on both sides of the opening. The shear wall forces calculated using this approach are usually the smallest of the three methods.

Shear Wall Aspect Ratios

Several IBC aspect ratio requirements pertain to shearwalls. IBC Section 2305.3.4 states that the height-to-width aspect ratio must be less than 3.5 to resist wind loads and less than 2.0 to resist seismic loads. However, Table 2305.3.4 indicates that the aspect ratio can be increased to 3.5

FIGURE 7.4 Typical load path for lateral loads.

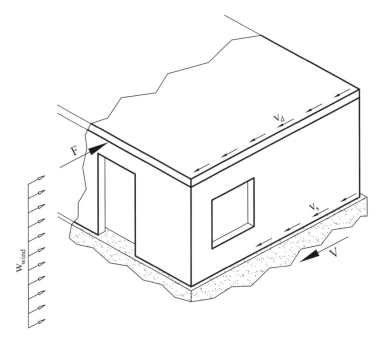

F = lateral force at the horizontal diaphragm
    for wind loads, F is the resultant from the tributary wind load at the diaphragm
    for seismic loads, F is calculated (e.g. Equivalent Lateral Force)
V = base shear
v_d = unit shear in the horizontal diaphragm
v_s = unit shear in the vertical diaphragm (shearwall)

for seismic loads provided the allowable shear obtained from IBC Table 2306.4.1 is multiplied by $2w/h$, where w is the length of the shear wall and h is the height of the shear wall (Figure 7.5).

Multistory Shear Wall Aspect Ratio

When calculating the shear wall aspect ratio in accordance with IBC requirements, the value used for H is the cumulative height from the base of the wall in question to the roof. Although the wall layout for most multistory wood-framed buildings is virtually the same for each floor (and thus shear wall lengths are the same for each floor), Figure 7.7 indicates the most economical use of shear walls for a typical wall elevation.

Shear Wall Overturning Analysis

To analyze a shear wall for the various load effects, a free-body diagram must first be drawn. Shown in Figure 7.8 is a typical shear wall along the exterior wall of a building with tributary lateral and gravity loads applied. From Figure 7.8,

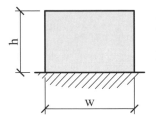

For wind loads: h/w < 3.5

For seismic loads: h/w < 2.0

FIGURE 7.5 Shear wall aspect ratios.

EXAMPLE 7.1

Definition of a Shear Wall

Determine which wall panels can be used as a shear wall in the building elevation shown in Figure 7.6.

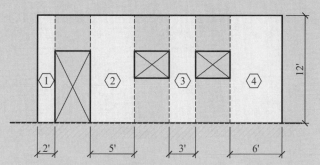

FIGURE 7.6 Shear wall aspect ratios.

Solution:

$$\text{Wall 1:} \quad \frac{h}{w} = \frac{12 \text{ ft}}{2 \text{ ft}} = 6 > 3.5$$

Therefore, wall 1 does not meet code for both wind and seismic designs, and will be neglected for shear action.

$$\text{Wall 2:} \quad \frac{h}{w} = \frac{12 \text{ ft}}{5 \text{ ft}} = 2.4 < 3.5$$

Therefore, wall 2 meets code for wind loads and also meets code for seismic loads provided that the allowable unit shear from IBC Table 2306.4.1 is multiplied by $2w/h = (2)(5 \text{ ft})/12 \text{ ft}$, or 0.83.

$$\text{Wall 3:} \quad \frac{h}{w} = \frac{12 \text{ ft}}{3 \text{ ft}} = 4 > 3.5$$

Therefore, wall 3 does not meet code for both wind and seismic designs, and will be neglected for shear action.

$$\text{Wall 4:} \quad \frac{h}{w} = \frac{12 \text{ ft}}{6 \text{ ft}} = 2 < 3.5$$

Therefore, wall 4 meets code for both wind and seismic designs.
Use Wall 2 and Wall 4 only to resist lateral loads.

w = length of the shear wall
$L_{H(1)}$ = length of the header (left side)
$L_{H(2)}$ = length of the header (right side)
H_1 = first-floor height
H_2 = second-floor height
F_{roof} = force applied to the shear wall at the roof level (either wind or seismic)
F_{2nd} = force applied to the shear wall at the second floor (either wind or seismic)
W_{2nd} = self-weight of the shear wall at the second floor
W_{1st} = self-weight of the shear wall at the first floor
R_{roof} = force due to uniform load on the shear wall at the roof
     = $w_{\text{roof}} w$
R_{2nd} = force due to uniform load on the shear wall at the second floor

FIGURE 7.7
Multistory shear wall aspect ratios.

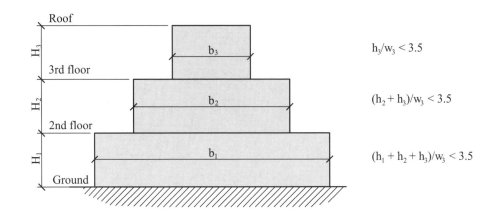

$h_3/w_3 < 3.5$

$(h_2 + h_3)/w_3 < 3.5$

$(h_1 + h_2 + h_3)/w_3 < 3.5$

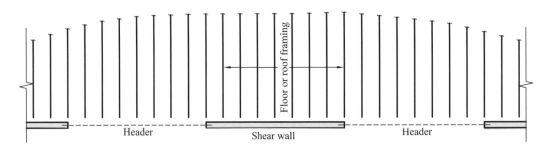

Floor or Roof Framing Plan

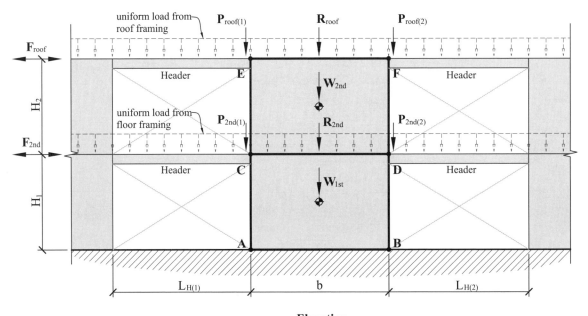

Elevation

FIGURE 7.8 Loads to a typical shear wall.

$$= w_{2nd} w$$

P_{roof} = force due to uniform load on the header at the roof

$$= \frac{w_{roof} L_{H1}}{2} \text{ or } \frac{w_{roof} L_{H2}}{2}$$

P_{2nd} = force due to uniform load on the header at the second floor

$$= \frac{w_{2nd} L_{H1}}{2} \text{ or } \frac{w_{2nd} L_{H2}}{2}$$

The uniform loads are resolved into resultant forces to simplify the overturning analysis. The uniform loads along the wall (i.e., the load from the roof or floor framing and the self-weight of the wall) act at the centroid of the wall, or at a distance of $w/2$ from the end of the shear wall.

The lateral wind or seismic loads acting on the shear wall can act in either direction. If the reactions from the left and right headers are equal, the overturning analysis has to be performed with the lateral load in only one direction. If the header reactions are not equal, the overturning analysis must be performed with the lateral applied in both directions, as will be shown later.

The loads that act on the roof level that result in the applied forces R_{roof} and P_{roof} can be some combination of dead, roof live, snow, or wind loads, depending on the governing load combination. Similarly, the loads that act on the second floor that result in the applied forces R_{2nd} and P_{2nd} can be some combination of dead and floor live loads. Figure 7.8 is somewhat generic but is typical for most shear walls. It is important to note that it is possible to have other gravity loads acting on the shear wall (such as a header from another beam framing perpendicular into the shear wall). These loads must also be accounted for in the overturning analysis.

The free-body diagram shown is specific for a two-story structure; however, it can be seen by inspection that this analysis can be used for a building with any number of stories. For simplicity, a two-story example is used here. The overturning analysis must also be performed at each level for multistory structures. This will require another isolated free-body diagram for each level, as shown in Figure 7.9 for the two-story structure. From Figure 7.9

T_1 and C_1 = shear wall **tension chord** and **compression chord** forces at the ground floor level

T_2 and C_2 = shear wall **tension chord** and **compression chord** forces at the second floor

> *Note:* For calculating the T_1 and T_2 chord forces, the *full* dead loads on the shear wall are used, whereas for C_1 and C_2 chord forces, only the *tributary* gravity loads apply; therefore, only the gravity loads in the compression chords are used.

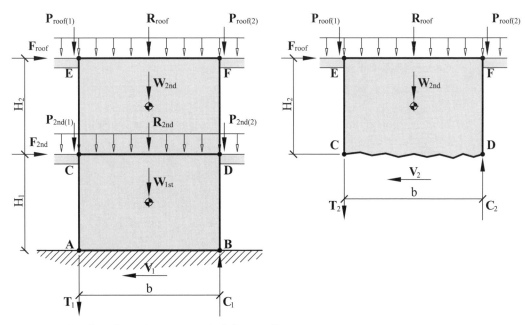

FIGURE 7.9 Loads and reactions to a typical shear wall.

V_1 = base shear at the ground level (= $F_{roof} + F_{2nd}$)
V_2 = base shear at the second floor (= F_{roof})

The lateral forces in this example are shown acting in one direction to show the resulting tension and compression chord forces.

The overturning moment can be calculated at each level as follows. At the second floor summing moments about point C or D yields

$$OM_2 = F_{roof}H_2 \tag{7.1}$$

At the ground level, summing moments about point A or B yields

$$OM_1 = [F_{roof}(H_2 + H_1)] + (F_{2nd}H_1) \tag{7.2}$$

The gravity loads prevent the shear wall from overturning or toppling over by creating a resisting moment that opposes the overturning moment. The magnitude of this *resisting moment* is the summation of moments about the same point that the overturning occurs. The applied loads in the overturning analysis are to be in accordance with the code-required load combinations, which are shown below for reference.

1. $D + F$ (IBC Equation 16-8)
2. $D + H + F + L + T$ (IBC Equation 16-9)
3. $D + H + F + (L_r \text{ or } S \text{ or } R)$ (IBC Equation 16-10)
4. $D + H + F + 0.75(L + T) + 0.75(L_r \text{ or } S \text{ or } R)$ (IBC Equation 16-11)
5. $D + H + F + (W \text{ or } 0.7E)$ (IBC Equation 16-12)
6. $D + H + F + 0.75(W \text{ or } 0.7E) + 0.75L + 0.75(L_r \text{ or } S \text{ or } R)$ (IBC Equation 16-13)
7. $0.6D + W + H$ (IBC Equation 16-14)
8. $0.6D + 0.7E + H$ (IBC Equation 16-15)

It can be seen by inspection that the first four combinations are eliminated for shear wall analysis and design since they do not include any lateral load terms. It can also be seen by inspection that the combination $D + H + F + 0.75(W \text{ or } 0.7E) + 0.75L + 0.75(L_r \text{ or } S \text{ or } R)$ will produce the worst-case compression chord force, and the combinations $0.6D + W + H$ or $0.6D + 0.7E + H$ will produce the worst-case tension chord forces. Since these combinations all include either wind or seismic loads, the load duration factor for the member designs will be $C_D = 1.6$. For simplicity, the following terms will not be considered since they do not typically apply to shear walls: H (earth pressure), F (fluid pressure), and T (temperature and shrinkage change).

Shear Wall Chord Forces: Tension Case

The controlling load combination for the tension chord will be either $0.6D + W$ or $0.6D + 0.7E$. These combinations are used for the tension case since the goal is to maximize the value in the tension chord to determine what the worst-case value is, and this is done by having the least amount of gravity load present in the model. The only gravity load effect present that resists overturning for the tension case is the dead load. Given that the dead load is multiplied by a 0.6 term, it is implied that there is a factor of safety of 1.67 (or 1/0.6) against overturning.

When calculating the maximum tension chord force (either T_1 or T_2 in Figures 7.8 and 7.9), the point at which to sum moments must first be determined. For the tension case, the following rules apply:

1. When the header reactions are equal, the summation of moments can be taken about any point (e.g., A or B at the base, or point C or D at the second level in Figure 7.8 or 7.9).
2. When the header reactions are not equal, the summation of moments should be taken about a point that lies on the same vertical line as the *larger* header reaction.

When the header reactions are not equal, it should be observed that the gravity load is minimized (and thus the tension chord force is maximized) when the summation of moments is taken about a point under the larger header reaction.

Wind Load Combination

Knowing that the governing load combination for wind loads is $0.6D + W$, the value T_1 or T_2 can be calculated from the free-body diagram in accordance with the following equilibrium equations by summing moments about the compression chord.

$$OM_2 - 0.6RM_{D2} - (T_2 w) = 0$$

Solving for T_2 yields

$$T_2 = \frac{OM_2 - 0.6RM_{D2}}{w} \quad (7.3)$$

where OM_2 = overturning moment at the second level [see equation (7.1)]
RM_{D2} = summation of the dead load resisting moments at the second level

Similarly, an equation for T_1 can also be developed:

$$T_1 = \frac{OM_1 - 0.6RM_{D1}}{w} \quad (7.4)$$

Seismic Load Combination

The governing load combination for seismic loads is $0.6D + 0.7E$. Section 12.4.2 of ASCE 7 defines the term E as follows:

$$E = \rho Q_E + 0.2S_{DS}D \text{ or } E = \rho Q_E - 0.2S_{DS}D$$

where ρ = redundancy coefficient (see Section 12.3.4 of ASCE 7)

> *Note:* When using the simplified procedure to calculate seismic loads, $\rho = 1.0$ for seismic design categories A, B, and C. For seismic design categories D, E, and F, the reader is referred to Section 12.3.4 of ASCE 7.

Q_E = horizontal effect of the seismic loads
S_{DS} = design spectral response acceleration (see Section 11.4.4 of ASCE 7)
D = dead load

By inspection it can be seen that the equation $E = \rho Q_E - 0.2S_{DS}D$ will govern for the design of the shear wall tension chord since the dead load effect is minimized, while $E = \rho Q_E + 0.2S_{DS}D$ will govern for the design of the compression chord.

The value of ρ can either be 1.0 or 1.3 and the value of S_{DS} can vary from 0.1 in low-seismic hazard areas to over 3.0 in a high-seismic hazard areas.

For the tension chord, $E = \rho Q_E - 0.2S_{DS}D$

Substituting this in the governing load combination $(0.6D + 0.7E)$ yields

$0.6D + 0.7(\rho Q_E - 0.2S_{DS}D) = (0.6 - 0.14S_{DS})D + 0.7\rho Q_E$ governing load combination

The value T_1 or T_2 can now be determined for the seismic load case as follows:

$$T_2 = \frac{(0.7\rho)OM_2 - (0.6 - 0.14S_{DS})RM_{D2}}{w} \quad (7.5)$$

$$T_1 = \frac{(0.7\rho)OM_1 - (0.6 - 0.14S_{DS})RM_{D1}}{w} \quad (7.6)$$

Note: If T_1 or $T_2 \leq 0$, this implies no uplift on tension chord
≥ 0, this implies uplift on tension chord

Shear Wall Chord Forces: Compression Case

The controlling load combination for the compression chord is $D + 0.75(W \text{ or } 0.7E) + 0.75L + 0.75(L_r \text{ or } S \text{ or } R)$. The code allows the effects from two or more transient loads to be reduced by a factor of 0.75. Therefore, the governing load combination can be rewritten as follows:

$$D + 0.75[W + L + (L_r \text{ or } S \text{ or } R)] \quad \text{for the wind load case}$$
$$D + 0.75[0.7E + L + (L_r \text{ or } S \text{ or } R)] \quad \text{for the seismic load case}$$

The term that accounts for the rain load effect R is ignored in this example since it typically does not control for most wood buildings. Also, the *larger* of the roof live load L_r and the snow load S should be used in the overturning analysis.

The general equations for determining the worst-case compression chord force can be developed similar to the equations developed earlier for the tension chord. The equations developed for the compression case will have to include an expanded form of the resisting moment RM term to include both a dead load and transient load term since different numerical coefficients will apply to each term. Recall that the tension case involved only dead-load effects, whereas the compression case will include other gravity load effects since the goal is to maximize the compression chord force. It should be recalled that only the *tributary* gravity loads on the compression chord are used to calculate the compression chord forces.

When calculating the maximum compression chord force (either C_1 or C_2 in Figures 7.8 and 7.9), the point about which to sum moments must first be determined. For the compression case, the following rules apply:

1. When the header reactions are equal, the summation of moments can be taken about any point (e.g., A or B at the base, or point C or D at the second level in Figure 7.8 or 7.9).
2. When the header reactions are not equal, the summation of moments should be taken about a point that lies on the same vertical line as the *smaller* header reaction.

When the header reactions are not equal, it should be observed that the gravity load is maximized (and thus the compression chord force is maximized) when the summation of moments is about a point under the smaller header reaction.

Wind Load Combination

The value of C_1 or C_2 can be calculated from the free-body diagram in accordance with the following equilibrium equations by summing moments about the tension chord at each level:

$$0.75 \text{OM}_2 + \text{RM}_{D2} + 0.75 \text{RM}_{T2} - C_2 w = 0$$

Solving for C_2 yields

$$C_2 = \frac{0.75 \text{OM}_2 + \text{RM}_{D2} + 0.75 \text{RM}_{T2}}{w} \quad (7.7)$$

where OM_2 = overturning moment at the second level [see equation (7.1)]
RM_{D2} = summation of the tributary dead load resisting moments at the second level
RM_{T2} = summation of the tributary transient load resisting moments at the second level

Similarly, an equation for C_1 can also be developed by summing moments about point A:

$$C_1 = \frac{0.75 \text{OM}_1 + \text{RM}_{D1} + 0.75 \text{RM}_{T1}}{w} \quad (7.8)$$

Seismic Load Combination

From the seismic load effect for the tension case, we can write

$$E = \rho Q_E + 0.2 S_{DS} D$$

Combining this with the governing load combination yields

$$(1.0 + 0.105 S_{DS}) D + 0.75 [0.7 \rho Q_E + L + (L_r \text{ or } S \text{ or } R)]$$

or

$(1.0 + 0.105S_{DS})D + 0.525\rho Q_E + 0.75L + 0.75(L_r \text{ or } S \text{ or } R)$ governing load combination

The value of C_1 or C_2 for the seismic case can now be calculated as follows:

$$C_2 = \frac{0.525\rho OM_2 + (1.0 + 0.105S_{DS})RM_{D2} + 0.75RM_{T2}}{w} \quad (7.9)$$

$$C_1 = \frac{0.525\rho OM_1 + (1.0 + 0.105S_{DS})RM_{D1} + 0.75RM_{T1}}{w} \quad (7.10)$$

where OM_1 and OM_2 = overturning moment caused by the horizontal seismic force, Q_E

In calculating the resisting moments (RM_{D1}, RM_{D2}, RM_{T1}, RM_{T2}) in equations (7.7) to (7.10), it should be noted that the full P loads or header reactions (see Figure 7.10) and the tributary R and W (i.e., distributed loads) loads should be used since the compression chords only support distributed gravity loads that are tributary to it, in addition to the concentrated header reactions or P loads. The tributary width of the compression chord is one-half of the stud spacing. Thus,

$$R_{trib} = \frac{R(\frac{1}{2} \times \text{stud spacing})}{w}$$

$$W_{trib} = \frac{W(\frac{1}{2} \times \text{stud spacing})}{w}$$

Recall that R and W are the total loads on the whole shear wall, while the terms R_{trib} and W_{trib} represent the amount of gravity load that is tributary to the compression chords. Since these loads are located at the compression chord (i.e., at the end of the wall), the moment arm relative to the tension chord for the terms R_{trib} and W_{trib} is equal to the length of the wall w and should be used accordingly in the resisting moment (RM_{D1}, RM_{D2}, RM_{T1}, RM_{T2}) calculations. The resisting moment about the tension chord for the R_{trib} and W_{trib} terms can be written as follows:

$$RM(R_{trib}) = \frac{R \times (\frac{1}{2} \times \text{stud spacing})}{w} \times w = R(\frac{1}{2} \times \text{stud spacing})$$

$$RM(W_{trib}) = \frac{W \times (\frac{1}{2} \times \text{stud spacing})}{w} \times w = W \times (\frac{1}{2} \times \text{stud spacing})$$

7.3 SHEAR WALL DESIGN PROCEDURE

The procedure for designing shear walls is as follows:

1. Calculate the lateral forces at each level (wind and seismic loads) and the gravity loads that act on the shear wall. Draw free-body diagrams similar to those shown in Figures 7.8 and 7.9.
2. Design the shear wall for the unit shear in the shear wall at each level. Give the full specification for the shear wall as follows:
   (a) Plywood grade and thickness
   (b) Nail size and penetration
   (c) Nail spacing (at edges and field)
   Recall that blocking is required at all panel edges.
3. Calculate the shear wall chord forces (T_1, T_2, C_1, C_2, etc.). Design the chords for these forces. Recall that the weak axis of the wall studs are braced by the wall sheathing, and the unbraced length of the chord for bending about the strong axis is equivalent to the floor-to-floor height.
4. Design the connections:
   (a) Hold-down anchors to resist T_1 or T_2
   (b) Sill anchors to resist the base shear V_1
   (c) Between the horizontal and vertical diaphragms to ensure a continuous load path

EXAMPLE 7.2
Design of Shear Walls

Design shear walls SW1 and SW2 for the two-story building shown in Figure 7.10 considering wind loads only. Assume a floor-to-floor height of 10 ft. Hem-fir lumber framing is used and assume a lateral wind load of 20 psf. See Figures 7.11 and 7.12 for additional information.

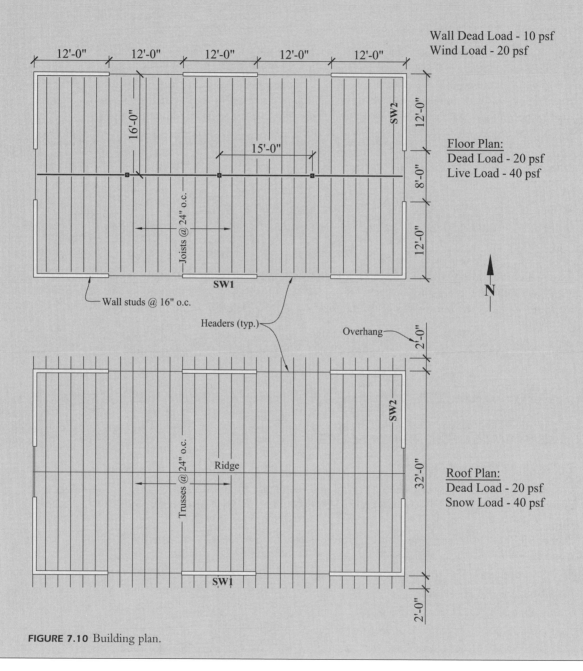

FIGURE 7.10 Building plan.

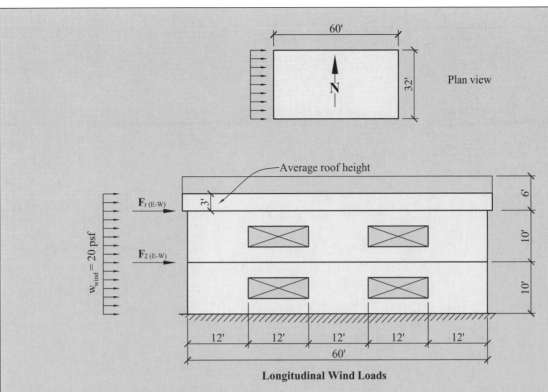

FIGURE 7.11 Longitudinal elevation.

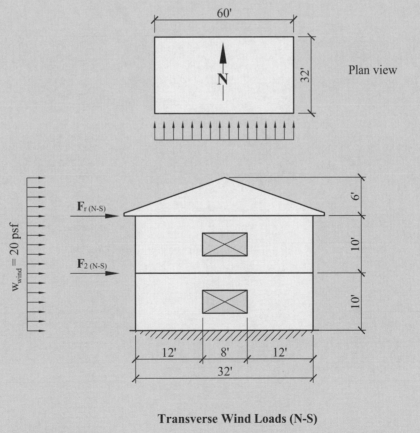

FIGURE 7.12 Transverse elevation.

Solution:
  East–West wind:

$$F_{r(E-W)} = (20 \text{ psf})\left(\frac{6 \text{ ft}}{2} + \frac{10 \text{ ft}}{2}\right)(32 \text{ ft}) = \mathbf{5120 \text{ lb}}$$

$$F_{2(E-W)} = (20 \text{ psf})\left(\frac{10 \text{ ft}}{2} + \frac{10 \text{ ft}}{2}\right)(32 \text{ ft}) = \mathbf{6400 \text{ lb}}$$

These values represent the total force at each level. They will have to be distributed to each shear wall in the E–W direction.

  North–South wind:

$$F_{r(N-S)} = (20 \text{ psf})\left(6 \text{ ft} + \frac{10 \text{ ft}}{2}\right)(60 \text{ ft}) = \mathbf{13{,}200 \text{ lb}}$$

$$F_{2(N-S)} = (20 \text{ psf})\left(\frac{10 \text{ ft}}{2} + \frac{10 \text{ ft}}{2}\right)(60 \text{ ft}) = \mathbf{12{,}000 \text{ lb}}$$

These values represent the total force at each level. They will have to be distributed to each shear wall in the N–S direction.

  Lateral loads on the E–W shear wall (SW1). Research has shown that wood diaphragms are typically classified as *flexible diaphragms* (see further discussion in Chapter 6 and Ref. 13); thus, the amount of lateral load on any individual shear wall is proportional to the tributary area of the shear wall in question. For this example, the lateral load on SW1 is proportional to the length of the wall relative to the other east–west shear walls. We must first check that SW1 meets the aspect ratio requirements in IBC Section 2305.3.3:

$$\frac{h}{w} \leq 3.5 \text{ (2.0 for seismic loads)}$$

$$\frac{10 \text{ ft} + 10 \text{ ft}}{12 \text{ ft}} = \mathbf{1.67 \leq 2.0}$$

Therefore, SW1 meets aspect ratio requirements. There are six shear walls, each 12 ft long, in the E–W direction, so the lateral load to SW1 at each level is

$$F_r = \frac{\text{length of SW1}}{\Sigma \text{ all shear wall lengths}} F_{r(E-W)}$$

$$= \frac{12 \text{ ft}}{(6)(12 \text{ ft})}(5120)$$

$$= \mathbf{854 \text{ lb}}$$

$$F_2 = \frac{\text{length of SW1}}{\Sigma \text{ all shear wall lengths}} F_{2(E-W)}$$

$$= \frac{12 \text{ ft}}{(6)(12 \text{ ft})}(6400)$$

$$= \mathbf{1067 \text{ lb}}$$

  Gravity loads on the E–W shear wall (SW1). At the roof level:

Direct load on wall:

$$R_D = (20 \text{ psf})\left(\frac{32 \text{ ft}}{2} + 2 \text{ ft}\right)(12 \text{ ft}) = \mathbf{4320 \text{ lb}}$$
$$\phantom{R_D = (20 \text{ psf})}\underbrace{\phantom{\left(\frac{32 \text{ ft}}{2} + 2 \text{ ft}\right)}}_{\text{(TW + overhang)}}\underbrace{\phantom{(12 \text{ ft})}}_{\text{(wall length)}}$$

$$R_S = (40 \text{ psf})\left(\frac{32 \text{ ft}}{2} + 2 \text{ ft}\right)(12 \text{ ft}) = \mathbf{8640 \text{ lb}}$$

$$W_D = (10 \text{ psf})(10 \text{ ft})(12 \text{ ft}) = \mathbf{1200 \text{ lb}}$$
$$\underbrace{\phantom{(10 \text{ psf})(10 \text{ ft})}}_{\text{(wall height)}}\underbrace{\phantom{(12 \text{ ft})}}_{\text{(wall length)}}$$

Reaction from headers (both ends equal):

$$P_D = (20 \text{ psf})\left(\frac{32 \text{ ft}}{2} + 2 \text{ ft}\right)\left(\frac{12 \text{ ft}}{2}\right) = \mathbf{2160 \text{ lb}}$$
$$\phantom{P_D = (20 \text{ psf})\left(\frac{32 \text{ ft}}{2} + 2 \text{ ft}\right)}\underbrace{\phantom{\left(\frac{12 \text{ ft}}{2}\right)}}_{\text{(half of header length)}}$$

$$P_S = (40 \text{ psf})\left(\frac{32 \text{ ft}}{2} + 2 \text{ ft}\right)\left(\frac{12 \text{ ft}}{2}\right) = \mathbf{4320 \text{ lb}}$$

At the second floor:

Direct load on wall:

$$R_D = (20 \text{ psf})\left(\frac{16 \text{ ft}}{2}\right)(12 \text{ ft}) = \mathbf{1920 \text{ lb}}$$
$$\phantom{R_D = (20 \text{ psf})}\underbrace{\phantom{\left(\frac{16 \text{ ft}}{2}\right)}}_{\text{(TW)}}\underbrace{\phantom{(12 \text{ ft})}}_{\text{(wall length)}}$$

$$R_L = (40 \text{ psf})\left(\frac{16 \text{ ft}}{2}\right)(12 \text{ ft}) = \mathbf{3840 \text{ lb}}$$

$$W_D = (10 \text{ psf})(10 \text{ ft})(12 \text{ ft}) = \mathbf{1200 \text{ lb}}$$
$$\underbrace{\phantom{(10 \text{ psf})(10 \text{ ft})}}_{\text{(wall height)}}\underbrace{\phantom{(12 \text{ ft})}}_{\text{(wall length)}}$$

Reaction from headers (both ends equal):

$$P_D = (20 \text{ psf})\left(\frac{16 \text{ ft}}{2}\right)\left(\frac{12 \text{ ft}}{2}\right) = \mathbf{960 \text{ lb}}$$
$$\phantom{P_D = (20 \text{ psf})\left(\frac{16 \text{ ft}}{2}\right)}\underbrace{\phantom{\left(\frac{12 \text{ ft}}{2}\right)}}_{\text{(half of header length)}}$$

$$P_L = (40 \text{ psf})\left(\frac{16 \text{ ft}}{2}\right)\left(\frac{12 \text{ ft}}{2}\right) = \mathbf{1920 \text{ lb}}$$

See Figure 7.13.

Lateral loads on the N–S shear wall (SW2). The lateral load on SW2 is proportional to the length of the wall relative to the other N–S shear walls. We must first check that SW2 meets the aspect ratio requirements in IBC Section 2305.3.3:

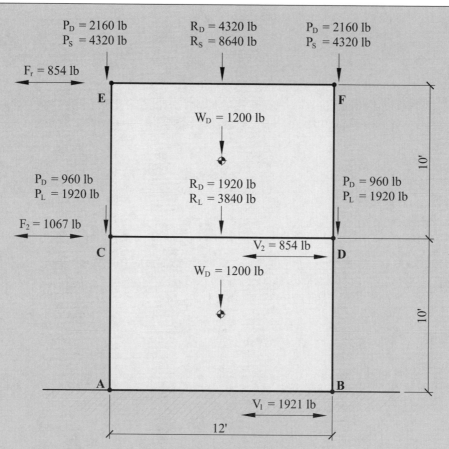

FIGURE 7.13 Free-body diagram of SW1.

$$\frac{h}{w} \leq 3.5 \ (2.0 \text{ for seismic loads})$$

$$\frac{10 \text{ ft} + 10 \text{ ft}}{12 \text{ ft}} = 1.67 \leq 2.0$$

Therefore, SW2 meets aspect ratio requirements. There are four shear walls, each 12 ft long, in the N–S direction, so the lateral load to SW2 at each level is

$$F_r = \frac{\text{length of SW2}}{\Sigma \text{ all shear wall lengths}} F_{r(N-S)}$$

$$= \frac{12 \text{ ft}}{(4)(12 \text{ ft})} (13{,}200)$$

$$= \mathbf{3300 \ lb}$$

$$F_2 = \frac{\text{length of SW2}}{\Sigma \text{ of all shear wall lengths}} F_{2(N-S)}$$

$$= \frac{12 \text{ ft}}{(4)(12 \text{ ft})} (12{,}000)$$

$$= \mathbf{3000 \ lb}$$

Gravity loads on the E–W shear wall (SW2). At the roof level:

Direct load on wall:

$$R_D = (20 \text{ psf}) \underbrace{\left(\frac{2 \text{ ft}}{2}\right) (12 \text{ ft})}_{\text{(truss spacing)(wall length)}} = \mathbf{240 \text{ lb}}$$

$$R_S = (40 \text{ psf}) \left(\frac{2 \text{ ft}}{2}\right) (12 \text{ ft}) = \mathbf{480 \text{ lb}}$$

$$W_D = (10 \text{ psf}) \underbrace{(10 \text{ ft}) (12 \text{ ft})}_{\text{(wall height)(wall length)}} = \mathbf{1200 \text{ lb}}$$

Reaction from left hand header ($P_{(1)}$):

$$P_D = \left(20 \text{ psf} \times \underbrace{\frac{2 \text{ ft}}{2}}_{\text{truss spacing}}\right) \times \frac{8 \text{ ft}}{2} = 80 \text{ lb}$$

$$P_S = \left(40 \text{ psf} \times \underbrace{\frac{2 \text{ ft}}{2}}_{\text{truss spacing}}\right) \times \frac{8 \text{ ft}}{2} = 160 \text{ lb}$$

Reaction from right hand header ($P_{(2)}$): Since there is no header on the right hand side of SW2,

$$P_D = 0$$
$$P_S = 0$$

At the second floor:

Direct load on wall:

$$R_D = (20 \text{ psf}) \underbrace{\left(\frac{2 \text{ ft}}{2}\right) (12 \text{ ft})}_{\text{(joist spacing)(wall length)}} = \mathbf{240 \text{ lb}}$$

$$R_L = (40 \text{ psf}) \left(\frac{2 \text{ ft}}{2}\right) (12 \text{ ft}) = \mathbf{480 \text{ lb}}$$

$$W_D = (10 \text{ psf}) \underbrace{(10 \text{ ft}) (12 \text{ ft})}_{\text{(wall height)(wall length)}} = \mathbf{1200 \text{ lb}}$$

Reaction from left end header:

$$\text{Tributary area of header reaction} = \frac{\overbrace{(16 \text{ ft}/2 + 16 \text{ ft}/2)}^{\text{(trib. width)}} \overbrace{(15 \text{ ft}/2)}^{\text{(half of girder length)}}}{2} = 60 \text{ ft}^2$$

$$P_{D1} = (20 \text{ psf})(60 \text{ ft}^2) = \mathbf{1200 \text{ lb}}$$

$$P_{L1} = (40 \text{ psf})(60 \text{ ft}^2) = \mathbf{2400 \text{ lb}}$$

Reaction from right end header: There is no header on the right end; therefore,

$$P_{D2} = \mathbf{0}$$
$$P_{L2} = \mathbf{0}$$

See Figure 7.14.

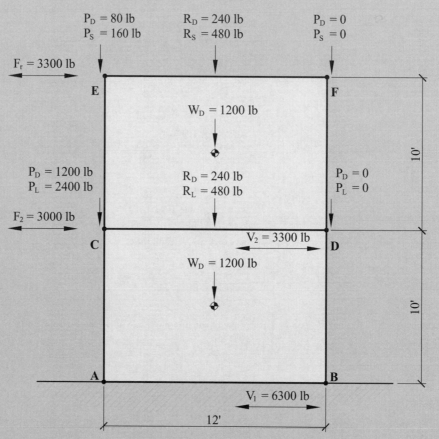

FIGURE 7.14 Free-body diagram of SW2.

Design of wall sheathing for SW1. Unit shears:

$$v_2 = \frac{F_r}{\text{wall length}} = \frac{854 \text{ lb}}{12 \text{ ft}} = \textbf{72 lb/ft}$$

$$v_1 = \frac{F_r + F_2}{\text{wall length}} = \frac{(854 \text{ lb} + 1067 \text{ lb})}{12 \text{ ft}} = \textbf{160 lb/ft}$$

The shear wall–fastening pattern could be different for each level, but both values are relatively low enough such that a minimum plywood thickness and fastening pattern will be required. Plywood $\frac{3}{8}$ in. thick is selected as a minimum practical size to allow flexibility in architectural finishes [cf. IBC Tables 2308.9.3(2) and 2308.9.3(3)]. From IBC Table 2306.4.1 (see Figure 7.15), select $\frac{3}{8}$-in.-thick Structural I CD–X, 8d nails at 6 in. o.c. (panel edges) and 12 in. o.c. (panel field), minimum penetration = $1\frac{3}{8}$ in., blocking required. $v_{\text{cap}} = 230$ plf.

Panel Grade	Minimum Nominal Panel Thickness (in.)	Minimum Fastener Penetration in Framing (in.)	PANELS APPLIED DIRECT TO FRAMING				
			NAIL (common or galvanized box) or staple size	Fastener spacing at panel edges (in.)			
				6	4	3	2
Structural I Sheathing	5/16	1 1/4	6d	200	300	390	510
		1	1 1/2" 16 Gage	165	245	325	415
	3/8	1 3/8	8d	**230**	360	460	610
		1	1 1/2" 16 Gage	155	235	315	400
	7/16	1 3/8	8d	255	395	505	670
		1					

FIGURE 7.15 Partial view of IBC Table 2306.4.1 (SW1). (Adapted from Ref. 8.)

This value is valid only for wall framing with Douglas fir-larch or southern pine. For wall framing with other wood species, the values must be reduced in accordance with footnote (a). From NDS Table 11.3.2A, the specific gravity of hem-fir wood species is

$$\gamma_{\text{hem-fir}} = 0.43$$

Adjusted shear capacity. The specific gravity adjustment factor from footnote (a) of IBC Table 2306.4.1 is given as

$$\text{SGAF} = 1 - (0.5 - G)$$

$$\text{SGAF}_{\text{hem-fir}} = 1 - (0.5 - 0.43) = 0.93$$

$$v_{\text{cap}} = (230 \text{ plf})(0.93) = 213 \text{ plf}$$

Applying the 40% increase in capacity (per IBC Section 2306.4.1 for wind loads only) gives us

$$v_{\text{cap}} = (213 \text{ plf})(1.4) = \mathbf{299 \text{ plf}}$$

Comparing the actual with the allowable values, we have

$$v_1 = 160 \text{ plf} < v_{\text{cap}} = 299 \text{ plf} \quad \text{OK}$$

Tension chord force, SW1. Isolating the upper level of the shear wall, and solving for T_2 from the free-body diagram by summing moments about point D (Figure 7.16) yields

$$\text{OM}_2 = (854 \text{ lb})(10 \text{ ft}) = 8540 \text{ ft-lb}$$

$$\text{RM}_{D2} = [(2160)(12)] + \left[(4320 + 1200)\left(\frac{12 \text{ ft}}{2}\right)\right] = 59{,}040 \text{ ft-lb}$$

$$T_2 = \frac{\text{OM}_2 - 0.6\text{RM}_{D2}}{w} \quad \text{[from equation (7-3)]}$$

$$= \frac{(8540) - (0.6)(59{,}040)}{12 \text{ ft}} = \mathbf{-2240 \text{ lb}} \text{ (the minus sign indicates no net uplift)}$$

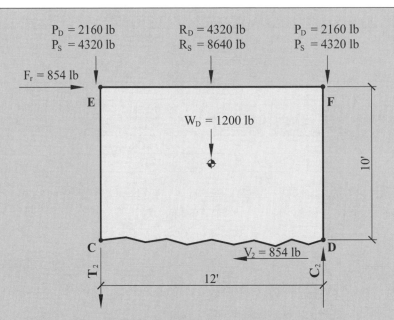

FIGURE 7.16 Tension chord force, SW1.

By referring back to Figure 7.13 for the complete loads, the maximum chord force at the ground level can be determined by summing moments about point B:

$$\text{OM}_1 = [(854)(10 \text{ ft} + 10 \text{ ft})] + [(1067)(10 \text{ ft})] = 27{,}750 \text{ ft-lb}$$

$$\text{RM}_{D1} = [(2160 + 960)(12 \text{ ft})] + \left[(4320 + 1920 + 1200 + 1200)\left(\frac{12 \text{ ft}}{2}\right)\right] = 89{,}280 \text{ ft-lb}$$

$$T_1 = \frac{\text{OM}_1 - 0.6\text{RM}_{D1}}{w} \text{ [from equation (7-4)]}$$

$$T_1 = \frac{(27{,}750) - (0.6)(89{,}280)}{12 \text{ ft}} = -\mathbf{2152 \text{ lb}} \text{ (the minus sign indicates no net uplift)}$$

Compression chord force, SW1. Referring back to Figure 7.16, and solving for C_2 from the free-body diagram by summing moments about point C, and accounting for the R and W loads that are *tributary* to the compression chord (stud spacing is 16 in. or 1.33 ft; see Figure 7.10):

$$\text{RM}_{T2} = [(4320)(12 \text{ ft})] + \left[(8640)\left(\frac{1.33 \text{ ft}}{2}\right)\right] = 57{,}600 \text{ ft-lb}$$

$$\text{OM}_2 = 8540 \text{ ft-lb (from tension chord force calculations)}$$

$$\text{RM}_{D2} = [(2160)(12 \text{ ft})] + \left[(4320 + 1200)\left(\frac{1.33 \text{ ft}}{2}\right)\right] = 29{,}600 \text{ ft-lb}$$

$$C_2 = \frac{0.75\text{OM}_2 + \text{RM}_{D2} + 0.75\text{RM}_{T2}}{w} \text{ [from equation (7.7)]}$$

$$= \frac{(0.75)(8540) + 29{,}600 + (0.75)(57{,}600)}{12 \text{ ft}} = \mathbf{6601 \text{ lb}}$$

By referring back to Figure 7.13 for the complete loads, the maximum chord force at the ground level C_1 can be determined by summing moments about point A:

$$\text{RM}_{T1} = [(4320 + 1920)(12 \text{ ft})] + \left[(8640 + 3840)\left(\frac{1.33 \text{ ft}}{2}\right)\right] = 83{,}200 \text{ ft-lb}$$

$$\text{OM}_1 = 27{,}750 \text{ ft-lb (from tension chord force calculations)}$$

$$\text{RM}_{D1} = [(2160 + 960)(12 \text{ ft})]$$

$$+ \left[(4320 + 1920 + 1200 + 1200)\left(\frac{1.33 \text{ ft}}{2}\right)\right] = 43{,}200 \text{ ft-lb}$$

$$C_1 = \frac{0.75\text{OM}_1 + \text{RM}_{D1} + 0.75\text{RM}_{T1}}{w} \text{ [from equation (7.8)]}$$

$$= \frac{(0.75)(27{,}750) + 43{,}200 + (0.75)(83{,}200)}{12 \text{ ft}} = \mathbf{10{,}535 \text{ lb}}$$

Design of wall sheathing for SW2. Unit shears:

$$v_2 = \frac{F_r}{\text{wall length}} = \frac{3300 \text{ lb}}{12 \text{ ft}} = \mathbf{275 \text{ lb/ft}}$$

$$v_1 = \frac{F_r + F_2}{\text{wall length}} = \frac{3300 \text{ lb} + 3000 \text{ lb}}{12 \text{ ft}} = \mathbf{525 \text{ lb/ft}}$$

The shear wall–fastening pattern could be different for each level, but for simplicity only the unit shear in the ground floor is considered.

From IBC Table 2306.4.1 (see Figure 7.17), select 15/32-in.-thick Structural I CD–X, 8d nails at 4 in. o.c. (panel edges) and 12 in. o.c. (panel field), minimum penetration = $1\frac{3}{8}$ in., blocking required. $\mathbf{\mathit{v}_{cap} = 430}$ **plf.**

This value is valid only for wall framing with Douglas fir-larch or southern pine because IBC Table 2306.4.1 is based on these species of wood. From the previous SW1 wall sheathing design, we obtained the specific gravity adjustment factor, SGAF, for hem-fir wood species as 0.93. Therefore,

			PANELS APPLIED DIRECT TO FRAMING				
Panel Grade	Minimum Nominal Panel Thickness (in.)	Minimum Fastener Penetration in Framing (in.)	NAIL (common or galvanized box) or staple size	Fastener spacing at panel edges (in.)			
				6	4	3	2
Structural I Sheathing	5/16	1¼	6d	200	300	390	510
		1	1½" 16 Gage	165	245	325	415
	3/8	1⅜	8d	230	360	460	610
		1	1½" 16 Gage	155	235	315	400
	7/16	1⅜	8d	255	395	505	670
		1	1½" 16 Gage	170	260	345	440
	15/32	1⅜	8d	280	**430**	550	730
		1	1½" 16 Gage	185	280	375	475
		1½	10d	340	510	665	870
	5/16 or ¼ [c]	1¼	6d	180	270	350	450
		1	1½" 16 Gage	145	220	295	375
		1¼					390

FIGURE 7.17 Partial view of IBC Table 2306.4.1 (SW2). (Adapted from Ref. 8.)

$$v_{cap} = (430 \text{ plf})(0.93) = 399 \text{ plf}$$

Applying the 40% increase in capacity (per IBC Section 2306.4.1) yields

$$v_{cap} = (399 \text{ plf})(1.4) = \mathbf{559 \text{ plf}}$$

Comparing the actual with the allowable values, we have

$$v_1 = 525 \text{ plf} < v_{cap} = 559 \text{ plf} \quad \text{OK}$$

Tension Chord Force, SW2. Isolating the upper level of the shear wall and solving for T_2 from the free-body diagram by summing moments about point C (Figure 7.18) yields

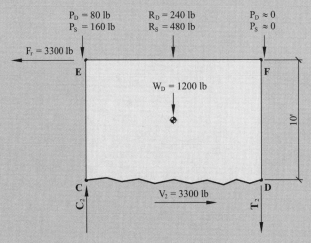

FIGURE 7.18 Tension chord force, SW2.

$$OM_2 = (3300)(10 \text{ ft}) = 33,000 \text{ ft-lb}$$

$$RM_{D2} = \left[(240 + 1200)\left(\frac{12 \text{ ft}}{2}\right)\right] = 8640 \text{ ft-lb}$$

$$T_2 = \frac{OM_2 - 0.6 RM_{D2}}{w} \text{ [from equation (7-3)]}$$

$$= \frac{(33,000) - (0.6)(8640)}{12 \text{ ft}} = \mathbf{2318 \text{ lb}} \text{ (net uplift)}$$

By referring back to Figure 7.14 for the complete loads, the maximum chord force in wall SW2 at the ground level can be determined by taking a summation of moments about point A (since gravity loads are minimized; see Section 7.2):

$$OM_1 = [(3300)(10 \text{ ft} + 10 \text{ ft})] + [(3000)(10 \text{ ft})] = 96,000 \text{ ft-lb}$$

$$RM_{D1} = \left[(240 + 240 + 1200 + 1200)\left(\frac{12 \text{ ft}}{2}\right)\right] = 17,280 \text{ ft-lb}$$

$$T_1 = \frac{OM_1 - 0.6 RM_{D1}}{w} \text{ [from equation (7.4)]}$$

$$= \frac{(96,000) - (0.6)(17,280)}{12 \text{ ft}} = \mathbf{7136 \text{ lb}} \text{ (net uplift)}$$

Compression chord force, SW2. Referring back to Figure 7.18, and solving for C_2 from the free-body diagram by summing moments about point D (yields higher forces than summing moments about C) and accounting for the R and W loads that are tributary to the compression chord (stud spacing is 16 in. or 1.33 ft; see Figure 7.10); we obtain

$$RM_{T2} = \left[(480)\left(\frac{1.33 \text{ ft}}{2}\right)\right] + 160(12 \text{ ft}) = 2240 \text{ ft-lb}$$

$$OM_2 = 33{,}000 \text{ ft-lb (from tension chord force calculations)}$$

$$RM_{D2} = \left[(240 + 1200)\left(\frac{1.33 \text{ ft}}{2}\right)\right] + 80(12 \text{ ft}) = 1920 \text{ ft-lb}$$

$$C_2 = \frac{0.75 OM_2 + RM_{D2} + 0.75 RM_{T2}}{w} \text{ [from equation (7.7)]}$$

$$= \frac{(0.75)(33{,}000) + 1920 + (0.75)(2240)}{12 \text{ ft}} = \mathbf{2363 \text{ lb}}$$

By referring back to Figure 7.14 for the complete loads, the maximum chord force at the ground level can be determined by taking a summation of moments about point B (since the compression chord force is maximized):

$$RM_{T1} = [(2400 + 160)(12 \text{ ft})] + \left[(480 + 480)\left(\frac{1.33 \text{ ft}}{2}\right)\right] = 31{,}360 \text{ ft-lb}$$

$$OM_1 = 96{,}000 \text{ ft-lb (from tension chord force calculations)}$$

$$RM_{D1} = [(1200 + 80)(12 \text{ ft})] + \left[(240 + 240 + 1200 + 1200)\left(\frac{1.33 \text{ ft}}{2}\right)\right] = 17{,}280 \text{ ft-lb}$$

$$C_1 = \frac{0.75 OM_1 + RM_{D1} + 0.75 RM_{T1}}{w} \text{ [from equation (7.8)]}$$

$$= \frac{(0.75)(96{,}000) + 17{,}280 + (0.75)(31{,}360)}{12 \text{ ft}} = \mathbf{9400 \text{ lb}}$$

TABLE 7.1 Summary of Shear Wall Chord Design (lb), Example 7.2

	Level 2			Ground Level		
	V_2	T_2	C_2	V_1	T_1	C_1
SW1	854	−2,240[a]	6,601	1,921	−2,152[a]	10,535
SW2	3,300	2,318	2,363	6,300	7,136	9,400

[a]A negative value actually indicates net compressive force; theoretically, hold-down anchors would not be required.

The shear wall chord design is summarized in Table 7.1. The chord members can now be designed for the worst-case tension and compression forces (see examples in Chapter 8).

Connection design. A connection between the upper shear wall and the lower shear wall to resist the net uplift at the second floor is required. From the chord design,

$$T_{2\max} = 2318 \text{ lb}$$

A steel strap is selected to span across the floor framing and tie the upper shear wall chord to the lower shear wall chord (Figure 7.19). A preengineered connector could be selected from a manufacturer's catalog (using a load duration factor of 1.6 or 160%), which is common in practice. A design example using a steel strap is provided in Chapter 9.

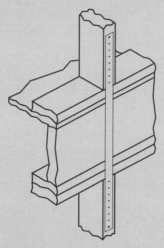

FIGURE 7.19 Hold-down strap.

A connection between the lower shear wall and the foundation to resist the net uplift at the ground floor is required. From the chord design, T_{1max} = 7136 lb. A preengineered connector (Figure 7.20) and the required embedment into the concrete foundation could be selected from a manufacturer's catalog (using a load duration factor of 1.6 or 160%), which is common in practice. The actual design of the hold-down anchor and embedment involves principles of steel and concrete design which is beyond the scope of this book. It should be noted that steel anchors loaded in tension should be deformed or hooked (i.e., not smooth) in order to engage the concrete properly.

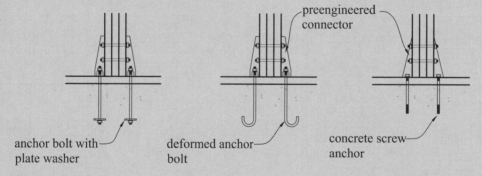

FIGURE 7.20 Hold-down anchor types.

The uplift connectors are typically provided at both ends of the shear wall since the lateral loads can act in the reverse direction. The tension chord members need to be checked for stresses due to the tension chord force, T_1 and T_2, in addition to the bending moment ($T_1 e$ and $T_2 e$) resulting from the eccentricity e of the hold-down anchors (see Figure 7.21). Use the *net area* of the chord member to account for the bolt holes required for the hold-down or tie-down anchors.

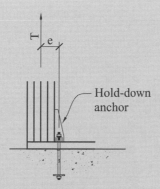

FIGURE 7.21 Hold-down anchor with eccentricity.

The sill anchors should also be designed to resist the base shear. Considering the worst-case base shear $V_1 = 6300$ lb, the allowable shear parallel to the grain in the sill plate is

$$Z' = ZC_D C_M C_t C_g C_\Delta C_{eg} C_{di} C_{tn}$$

$$C_D = 1.6 \text{ (wind)}$$

$$C_M = 1.0 \text{ (MC} \leq 19\%)$$

$$C_t = 1.0 \text{ } (T \leq 100°F)$$

$$C_g, C_\Delta, C_{eg}, C_{di}, C_{tn} = 1.0 \text{ (see Chapter 9 for design examples with } C_g, C_\Delta, C_{eg}, C_{di}, C_{tn},$$
$$\text{or NDS code Section 10.3)}$$

$$Z_\parallel = 590 \text{ lb (NDS Table 11E)}$$

The allowable shear per bolt is

$$Z' = ZC_D C_M C_t C_g C_\Delta C_{eg} C_{di} C_{tn}$$
$$= (590)(1.6)(1.0)(1.0)(1.0)(1.0)(1.0)(1.0) = 944 \text{ lb per bolt}$$

Therefore, the number of sill anchor bolts required = (6300 lb)/(944 lb/bolt) $\simeq$ 7 bolts. The maximum bolt spacing is

$$\frac{12\text{-ft wall length} - 1\text{-ft edge distance} - 1\text{-ft edge distance}}{7 \text{ bolts} - 1} = 1.67 \text{ ft} \approx 1 \text{ ft 8 in.}$$

Use $\frac{1}{2}$-in.-diameter anchor bolts at 1 ft 8 in. o.c. (see Figure 7-2 for a typical layout).

For ease of wall placement, a contractor may request that the sill anchor bolt holes be oversized. One way to do this, and still ensure a snug tight connection between the anchors and sill plate so that the sill anchors are able to transfer the lateral base shear, is to sleeve the bolt holes with metal tubing and fill the areas between the bolt and the inside face of the sleeve with expansive cement. [16]

The compression chord members also need to be checked for stresses due to the compression chord forces C_1 and C_2. When checking the stresses in the chord members at midheight, include buckling (i.e., include the C_p factor), and when checking the stresses at the base of the chords, use the *net area* of the chord member to account for the bolt holes required for the hold-down or tie-down anchors. Design the sill plate for compression perpendicular to the grain due to the compression chord forces C_1 and C_2.

> **EXAMPLE 7.3**
>
> *Difference in Shear Wall Capacity for Wind and Seismic Design*
>
> A 3-ft-long × 8-ft-tall shear wall panel has the following design parameters:
>
> $\frac{3}{8}$-in. plywood sheathing with 8d nails at 4 in. o.c. [edge nailing (EN)] and 12 in. o.c. [field nailing (FN)] on the outside face.
>
> $\frac{1}{2}$ in. Gypsum wall board (GWB) on the inside face of the wall. The GWB is unblocked and fastened per IBC Table 2306.4.5 with an allowable unit shear capacity of 100 plf.
>
> Determine the allowable unit shear capacity of the shear wall panel for both wind and seismic designs.
>
> *Solution:* Wall aspect ratio = 8 ft/3 ft = 2.67.
>
> < 3.5   OK for wind design
> > 2.0   Not OK for seismic design unless the allowable unit shear from IBC Table 2306.4.1 is multiplied by $2w/h$
>
> Allowable unit shear for $\frac{3}{8}$-in. plywood (8d nails at 4 in. EN; 12 in. o.c. FN) = 320 plf (IBC Table 2306.4.1). The allowable unit shear of GWB = 100 plf (IBC Table 2306.4.5).
>
> *Wind design:*
>
> - IBC Section 2306.4.1 allows a 40% increase in the unit shear capacities from IBC Table 2306.4.1.
> - IBC Section 2305.3.9 allows the addition of the unit shear capacity of walls of dissimilar materials when designing for wind loads.
>
> Total allowable unit shear of wall = (1.40)(320 plf) + 100 plf = 548 plf
>
> *Seismic design:*
>
> - Addition of unit shear capacities of dissimilar materials is not allowed by the IBC for seismic design. Consequently, only the plywood sheathing will be considered since it has the higher shear capacity.
> - Since the aspect ratio of 2.67 exceeds 2.0, the unit shear capacity from IBC Table 2306.4.1 must be multiplied by $2w/h$.
>
> Total allowable unit shear of wall = $\frac{(2)(3 \text{ ft})}{(8 \text{ ft})}$(320 plf) = 240 plf
>
> *Note:* From the results above, the allowable unit shear for the shear wall panel is more than twice as high for wind design as it is for seismic design. Although the wind load may have been the controlling load for this building, the shear capacity under seismic loads is quite low compared to the capacity under wind loads which indicates that checking this building and shear wall panel for seismic loads is also necessary. This example points to the need to compare not only the wind and seismic loads but also the respective shear capacities of the shear wall under both loading conditions. The importance of this difference in shear wall capacities has been highlighted in Ref. 14.

7.4 COMBINED SHEAR AND UPLIFT IN WALL SHEATHING

In high wind areas of the country, several hold-down anchors may be required between floors to resist the uplift due to wind loads, and this can often lead to increased cost and there may also be nailing interference because of the relatively large number of fasteners that may be required for these anchor devices. In lieu of providing hold-down straps between floors (see Figure 7.19) to resist uplift loads, the exterior sheathing can be fastened in such a way to create a vertical plywood panel splice across the floor to transfer the uplift force. Figure 7.22 illustrates two possible details that can be used for this condition. In both cases, nails are added to resist the uplift force in addition to the nails required for shear.

The detail shown in Figure 7.22a should be used where the rim joist is sawn lumber or any lumber that is susceptible to shrinkage. The exterior sheathing is pre-engineered and is generally dry when it is installed, whereas sawn lumber will have some water content at the time of installation and will shrink over time. The exterior sheathing and plywood shim should have a clear distance of about $\frac{1}{2}''$ at the ends to allow for shrinkage. The shim should also be made from the same material and have the same thickness and orientation as the exterior sheathing. Furthermore, the plywood shim acts to transfer the tension force since sawn lumber has very little capacity when loaded perpendicular to the grain. When an OSB or plywood rim joist is used, then a shim is not required since these products are not susceptible to shrinkage and have sufficient tension capacity when loaded in this manner.

The detail shown in Figure 7.22b should be used when the rim joist is a pre-engineered I-joist or LVL since shrinkage in sawn lumber would likely cause the exterior sheathing to buckle in this case.

Using the exterior sheathing to support both uplift and in-plane shear creates a situation where the sheathing and fasteners are subjected to combined loading. However, the shear capacity of a shear wall is controlled by the fasteners such that it is generally difficult to add enough fasteners so that the sheathing fails. Therefore, the design approach taken here will be two-fold: limit the tension in the sheathing due to uplift to a reasonably low value and add fasteners to resist the uplift in addition to the fasteners required for lateral shear.

The American Plywood Association (APA) [17] recommends that the uplift on the sheathing should not exceed 1,000 plf for $\frac{3}{8}$ inch panels and 1,500 plf for $\frac{7}{16}$ inch panels and larger. These values are approximated from the tensile capacity of the panels using the *Wood Structural Panels Supplement* [15] or the *Panel Design Specification* [5] as follows:

From Table 3.2—Wood Structural Panels Supplement (or Table 4A of the Panel Design Specification):

$$F_t A = 600 \text{ plf} \ (\text{for } \tfrac{3}{8} \text{ inch panels})$$
$$F_t A = 990 \text{ plf} \ (\text{for } \tfrac{7}{16} \text{ inch panels})$$

These values are applicable to panels installed horizontally, that is panels which have their long direction perpendicular to the wall studs. When the long axis of the panel is oriented parallel to the wall studs, the allowable tension is much higher, so a conservative (horizontal) orientation has been assumed for the panels. Applying a load duration factor of $C_D = 1.6$ to the design tension values above yields allowable uplift load on plywood panels as

$$F_t A = (1.6)(600 \text{ plf}) = 960 \text{ plf} \ (\text{for } \tfrac{3}{8} \text{ inch panels})$$
$$F_t A = (1.6)(990 \text{ plf}) = 1584 \text{ plf} \ (\text{for } \tfrac{7}{16} \text{ inch panels})$$

These values are reasonably close to the APA [16] recommended values of 1,000 plf and 1,500 plf for $\frac{3}{8}$ inch and $\frac{7}{16}$-inch panels, respectively. Panels thicker than $\frac{7}{16}$ inch are limited to 1,500 plf for uplift. It should be noted that these uplift limits are generally conservative for most wood framed buildings.

For the fasteners, the extra nails should be uniformly distributed to each wall stud or into the rim joist as shown in Figure 7.22. The design of fasteners is covered in greater detail in Chapter 8, but a brief overview of the fastener design is illustrated in the following example to show the full scope of the design requirements of a plywood vertical lap splice.

EXAMPLE 7.4

Combined Shear and Uplift in a Shear Wall

A $\frac{3}{8}$-inch exterior shear wall panel is subjected to a uniform uplift of 375 plf. The wall framing is 2x6 studs at 16 in. o.c. and the lumber is Hem-Fir No. 2. The floor framing is pre-engineered I-joists. Determine if the panel is adequate for combined shear and uplift and determine an appropriate nailing pattern if the exterior sheathing is used as a lap splice assuming that the splice detail shown in Figure 7.22 is used. Assume normal temperature and moisture conditions.

Solution:

The maximum recommended uplift is 1,000 plf for a $\frac{3}{8}$-inch panel which is greater than the applied uplift of 375 plf, so the panel is adequate in combined shear and uplift.

The allowable tension in the studs needs to be checked. From NDS Table 4A, the following is obtained for 2x6 Hem-Fir No. 2 studs:

$$F_t = 525 \text{ psi}$$
$$C_D = 1.6 \text{ (wind loads)}$$
$$C_i = 1.0$$
$$C_F = 1.3$$
$$A = 8.25 \text{ in.}^2$$
$$F'_t = F_t C_D C_M C_t C_F C_i$$
$$= 525 \times 1.6 \times 1.0 \times 1.0 \times 1.3 \times 1.0 = 1092 \text{ psi}$$
$$T_{max} = F'_t A = (1092 \text{ psi})(8.25 \text{ in.}^2) = 9009 \text{ lb.}$$

Using a stud spacing of 16 in., the uplift force on each stud is:

$$(375 \text{ plf})\left(\frac{16 \text{ in.}}{12}\right) = 500 \text{ lb} < T_{max} = 9009 \text{ lb.}$$

Therefore, the studs are adequate to resist the uplift force.

For the fasteners, 8d common nails will be assumed. From NDS Table 11Q using $t_s = \frac{3}{8}$ inch (i.e. plywood thickness), the allowable shear per nail is $Z = 66$ lb. Applying the load duration factor, which is the only adjustment factor that applies in this case per NDS Table 10.3.1, yields:

$$Z' = ZC_D = (66 \text{ lb})(1.6) = 105 \text{ lb/nail}$$

Using a tension force of 500 lb per stud, the required number of fasteners per stud is:

$$\frac{500}{105} = 4.73 \approx 5 \text{ nails each side of the splice}$$

From NDS Table L4: The length, L and the diameter of 8d common nails is,

$$L = 2.5 \text{ in.}$$
$$D = 0.131 \text{ in.}$$

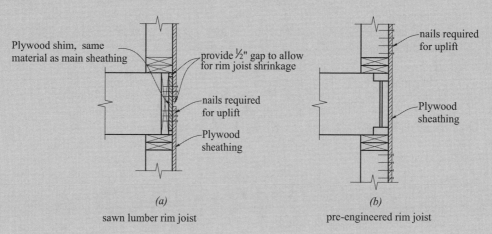

(a) sawn lumber rim joist

(b) pre-engineered rim joist

FIGURE 7.22 Uplift detail using exterior sheathing.

> Minimum spacing (*see Table 8-1 and Table 8-2*):
>
> $\quad\quad$ end distance = 15D $\quad\quad$ = 1.96 in. ⇒ **2.0 in.**
>
> $\quad\quad$ edge distance = 2.5D $\quad\quad$ = 0.33 in. ⇒ **0.75 in. provided, centered in stud**
>
> $\quad\quad$ center-center spacing = 15D = 1.96 in. ⇒ **2.0 in.**
>
> $\quad\quad$ minimum p = 10D $\quad\quad$ = 1.31 in. ⇒ **approx. 2 in. provided**
>
> From Figure 7-22*b*, 5 nails would be required into each stud above and below the floor framing. These nails would be in addition to the nails required for shear.

REFERENCES

1. ANSI/AF&PA, (2001), *Allowable Stress Design Manual for Engineered Wood Construction*, American Forest & Paper Association, Washington, DC.
2. ANSI/AF&PA, (2005), *National Design Specification Supplement: Design Values for Wood Construction*, American Forest and Paper Association, Washington, DC.
3. APA (2001), *Engineered Wood Construction Guide*, APA—The Engineered Wood Association, Tacoma, WA.
4. APA (2004), *Diaphragms and Shear Walls: Design/Construction Guide*, APA—The Engineered Wood Association, Tacoma, WA.
5. APA (1998), *Plywood Design Specification*, APA—The Engineered Wood Association, Tacoma, WA.
6. ASCE (2005), *Minimum Design Loads for Buildings and Other Structures*, ASCE-7, American Society of Civil Engineers, Reston, VA.
7. GA (1996), *Shear Values for Screw Application of Gypsum Board on Walls*, GA-229-96, Gypsum Association, Washington, DC.
8. ICC (2006), *International Building Code*, International Codes Council, Falls Church, VA.
9. Faherty, Keith F., and Williamson, Thomas G. (1989), *Wood Engineering and Construction*, McGraw-Hill, New York.
10. Halperin, Don A., and Bible, G. Thomas (1994), *Principles of Timber Design for Architects and Builders*, Wiley, New York.
11. Stalnaker, Judith J., and Harris, Earnest C. (1997), *Structural Design in Wood*, Chapman & Hall, London.
12. Breyer, Donald et al. (2003), *Design of Wood Structures—ASD*, 5th Ed., McGraw-Hill, New York.
13. Tarpy, Thomas S., Thomas, David J., and Soltis, Lawrence A. (1985), Continuous timber diaphragms, *J. Struct. Div. ASCE* 111(5): 992–1002.
14. Beebe, Karyn (2006), Enhanced lateral design with wood structural panels, *Struct. Eng.*, March.
15. ANSI/American Forest & Paper Association, (2001), *Wood Structural Panels Supplement*, AF&PA, Washington, DC.
16. Knight, Brian (2006), High Rise Wood Frame Construction—Cornerstone Condominiums, STRUCTURE, pp. 68–70, June.
17. APA—The Engineered Wood Association (2005), Using Wood Structural Panels for Combined Uplift and Shear Resistance, Tacoma, WA.

PROBLEMS

7.1 For the building plan shown in Figure 7.23 and based on the following given information:

$\quad$ Wall framing is 2 × 6 at 16 in. o.c., hem-fir, Select Structural.

$\quad$ Normal temperature and moisture conditions apply.

$\quad$ Roof dead load = **15 psf,** wall dead load = **10 psf,** snow load P_f = **40 psf.**

$\quad$ Floor-to-roof height = 18 ft.

$\quad$ Consider loads in the N–S direction only.

FIGURE 7.23 Building plan.

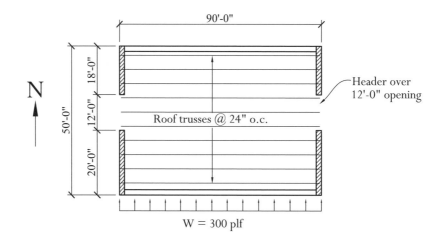

(a) For the 18- and 20-ft shear walls on the east and west face only, select the required plywood thickness, fastening, and edge support requirements. Assume that the plywood is applied directly to the framing and is on one side only. Give the full specification.

(b) Determine the maximum tension force in the shear wall chords of the 20-ft-long shear wall.

(c) Determine the maximum compression force in the shear wall chords of the 20-ft-long shear wall.

7.2 For the building plan shown in Figure 7.24 for a one-story building and based on the following given information:

Wall framing is 2 × 6 at 16 in. o.c., DF-L, No. 1.

Normal temperature and moisture conditions apply.

Roof dead load = **15 psf,** wall dead load = **10 psf,** roof live load = **20 psf.**

Floor-to-roof height = 18 ft.

Applied wind load = 250 plf to the diaphragm.

(a) For the typical 19-ft shear wall, select the required plywood thickness, fastening, and edge support requirements. Assume that the plywood is applied directly to the framing and is on two sides. Give the full specification.

(b) Determine the maximum tension force in the shear wall chords for the interior shear wall.

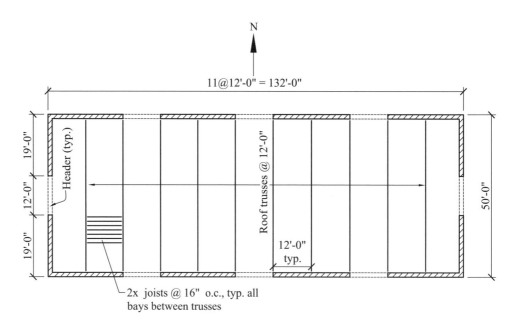

FIGURE 7.24 Building plan.

(c) Determine the maximum compression force in the shear wall chords for the interior shear wall.

(d) Select an appropriate anchor rod layout to resist the lateral loads applied.

7.3 Determine the allowable unit shear capacity for a shear wall panel under both wind and seismic loads assuming the following design parameters: exterior face of wall: $\frac{15}{32}$-in. plywood with 8d nails at 4 in. o.c. (EN); 12 in. o.c. (FN) interior face of wall; GWB with allowable unit shear of 175 plf.

7.4 Determine the allowable unit shear capacity for a shear wall panel under both wind and seismic loads assuming the following design parameters: exterior face of wall: $\frac{15}{32}$-in. plywood with 8d nails at 4 in. o.c. (EN); 12 in. o.c. (FN) interior face of wall: $\frac{3}{8}$-in. plywood with 8d nails at 4 in. o.c. (EN); 6 in. o.c. (FN).

7.5 Design a plywood lap splice between floors using sawn lumber with 2 × 6 wall studs at 16 in. o.c. and $\frac{7}{16}$ in. wall sheathing for a net uplift of 600 plf. Stud spacing is 16 in., and lumber is DF-L No. 1. Sketch your design.

chapter *eight*

CONNECTIONS

8.1 INTRODUCTION

Connections within any structure are usually very small relative to the supported members but they are a very critical component in the load path. Beams and other major structural members are only as strong as their connectors. In practice, the design of connections is oftentimes overlooked and is thus delegated by default to either someone preparing shop drawings or to the builder in the field. Although this process typically results in a structure with adequate connections, proper attention to the design and detailing of connections can help limit the amount of field-related problems.

In practice, the designer has three basic options for connections:

1. *Use a preengineered connector.* Several proprietary products are available, and each manufacturer usually makes available a catalog detailing each connector with a corresponding load capacity. The manufacturer usually also indicates the required placement of nails, screws, or bolts required for each connector. This is the least costly option from a design standpoint and is often the preferred method from the builder's standpoint.
2. *Use the design equations in the NDS code.* This can be a tedious process that is best carried out with the aid of a computer software. This method requires adequate detailing to ensure that the connectors are placed properly in the field.
3. *Use the connector capacity tables in the NDS code.* Several tables are provided in the NDS code for various connectors and conditions and are based on the equations in the NDS code. Occasionally, conservative assumptions must be made for loading conditions not contained in the tables, but they are not as tedious as using the equations directly.

Several types of connectors are available, but the following connector types will be covered in this book since they are the most commonly used: (1) bolts (Table L1), (2) lag screws (Table L2), (3) wood screws (Table L3), and (4) nails (Table L4). The tables indicated are found in the appendix of the NDS code [7]. These tables list the various dimensional characteristics of each connector type. Bolts have the highest strength but require the most labor. Conversely, nails generally have the lowest strength but require the least labor and equipment. Figure 8.1 illustrates each of these connector types.

Bolts are greater than $\frac{1}{4}$ in. in diameter and are installed completely through the connected members through predilled holes that are at least $\frac{1}{32}$ in. greater in diameter than the bolt, but less than $\frac{1}{16}$ in. (NDS code Section 11.1.2.2). Bolts are not permitted to be forcibly driven.

Lag screws are greater than $\frac{1}{4}$ in. in diameter and are installed into a wood member. Lag screws typically have a threaded and a smooth portion (see Figure 8.12). The NDS code gives the required diameter of clearance holes and lead holes for installing lag screws to prevent splitting of the wood member during construction (NDS code Section 11.1.3.2). The *clearance hole* is the area around the unthreaded shank and should have the same diameter and length as the unthreaded shank. The *lead hole* is the area around the threaded portion of the shank. The diameter of the lead hole is generally smaller than the diameter of the shank, depending on the specific gravity of the wood.

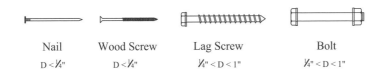

FIGURE 8.1 Fastener types.

Wood screws are similar to lag screws but are generally less than $\frac{1}{4}$ in. in diameter. The NDS code (Section 11.1.4.2) gives the required diameter of lead holes to prevent splitting.

Nails are generally less than $\frac{1}{4}$ in. in diameter and are forcibly driven by a hammer or other mechanical means. Nails can have numerous head and shank types, but the discussion of these is beyond the scope of this book. We limit our discussion to *common nails,* which are representative of most structural nails. Lead holes are not specifically required in the NDS code, but are allowed (see NDS code Section 11.1.5.3).

8.2 DESIGN STRENGTH

There are two basic types of loads that connections are designed to resist: lateral loads and withdrawal loads. As with the other member designs in this book, each connector or group of connectors has a base design value that has to be modified to determine the allowable or adjusted design values. Several factors have an effect on the load-carrying capacity of a connection. These factors can be seen by inspection of the adjustment factors indicated in Table 10.3.1 of the NDS code. These adjusted design values are

$$Z' = Z C_D C_M C_t C_g C_\Delta C_{eg} C_{di} C_{tn} \tag{8.1}$$

$$W' = W C_D C_M C_t C_{eg} C_{tn} \tag{8.2}$$

where Z = nominal lateral design value for a single fastener
Z' = adjusted lateral design value for a single fastener
W = nominal withdrawal design value for a single fastener
W' = adjusted withdrawal design value for a single fastener
C_D = load duration factor ($\leq$ 1.6, see Section 10.3.2 of the NDS code)
C_M = wet service factor (Table 10.3.3 of the NDS code)
C_t = temperature factor (Table 10.3.4 of the NDS code)
C_g = group action factor (Section 10.3.6 of the NDS code)
C_Δ = geometry factor (Section 11.5.1 of the NDS code)
C_{eg} = end grain factor (Section 11.5.2 of the NDS code)
C_{di} = diaphragm factor (Section 11.5.3 of the NDS code)
C_{tn} = toenail factor (Section 11.5.4 of the NDS code)

8.3 ADJUSTMENT FACTORS FOR CONNECTORS

Wood has a tendency to undergo volumetric changes over time, causing a change in the position of the connectors, resulting in a reduced load-carrying capacity. Other factors that affect the capacity of a connection are the direction of the load, the geometry of the connectors, and the location of the connector within the wood member. The adjustment factors indicated in Section 8.2 are intended to account for all of these variables. Each of these adjustment factors is now discussed in greater detail.

Load Duration Factor C_D

The load duration factor is found in Table 2.3.2 of the NDS code and is usually the same factor used for the connected member. Section 10.3.2 of the NDS code does limit the value of C_D to 1.6 since laboratory testing of connections at short durations is limited (i.e., impact loads).

Wet Service Factor C_M

Since wood has a tendency to lose water and shrink over time, the capacity of a connector depends on the moisture content of the wood at the time of fabrication and while in service.

Wood that is dry, or wood with a moisture content below 19% does not result in a reduced connection capacity and $C_M = 1.0$. For wood with a moisture content greater than 19% either at the time of fabrication or while in service, refer to Table 10.3.3 of the NDS code for the C_M factor.

Temperature Factor C_t

While the temperature range for most wood-framed buildings is such that the connection capacity does not have to be reduced, Table 10.3.4 of the NDS code gives the values for C_t for various temperature ranges.

Group Action Factor C_g

When a series of fasteners aligned in a row are loaded, the load distribution to each fastener is not equal. The group action factor accounts for this condition. The group action factor does not apply to smaller-diameter fasteners ($D < \frac{1}{4}$ in.). Note that a row is defined as a group of fasteners aligned and loaded parallel to the grain. The value of C_g is defined in Section 10.3.6 of the NDS code as follows (NDS code Equation 10.3–1):

$$C_g = \frac{m(1 - m^{2n})}{n[(1 + R_{EA}m^n)(1 + m) - 1 + m^{2n}]} \frac{1 + R_{EA}}{1 - m} \tag{8.3}$$

where n = number of fasteners in a row

$$R_{EA} = \text{the lesser of } \frac{E_s A_s}{E_m A_m} \text{ or } \frac{E_m A_m}{E_s A_s} \tag{8.4}$$

(see Figure 8.10 for main and side member description)

E_m = modulus of elasticity of the main member, psi
E_s = modulus of elasticity of the side member, psi
A_m = gross cross-sectional area of the main member, in^2
A_s = sum of the gross cross sectional areas of the side member, in^2

$$m = u - \sqrt{u^2 - 1} \tag{8.5}$$

$$u = 1 + \gamma \frac{s}{2}\left(\frac{1}{E_m A_m} + \frac{1}{E_s A_s}\right) \tag{8.6}$$

s = center-to-center spacing between adjacent fasteners in a row
γ = load/slip modulus for a connection, lb/in. $\tag{8.7}$
  = $180{,}000 D^{1.5}$ for dowel-type fasteners in wood-to-wood connections
  = $270{,}000 D^{1.5}$ for dowel-type fasteners in wood-to-metal connections
D = diameter of fastener, in.

It can be seen by inspection that the group action factor is a function of the number of fasteners, the stiffness of the main and side members, the spacing of the fasteners, the slip modulus, and the diameter of the fasteners. It can also be seen that the equation is rather tedious. The NDS code does give the value of C_g for various connection geometries in Tables 10.3.6A through 10.3.6D. Example 8.1 illustrates that use of the tables as a reasonable alternative to using the equation for most cases.

Fastener too close to the end of a member

Fasteners spaced too close togther

FIGURE 8.2 Splitting due to inadequate fastener spacing.

Geometry Factor C_Δ

Connectors with inadequate spacing have a tendency to split (see Figure 8.2); thus, to maximize the load-carrying capacity of a connection, the designer must consider the geometry of the connection. The geometry factor applies to all connectors greater than $\frac{1}{4}$ in. in diameter and is a function of the following: edge distance, end distance, center-to-center spacing, and row spacing. For connectors with $D < \frac{1}{4}$ in., $C_\Delta = 1.0$, but minimum spacing requirements must be met. For connectors with $D < \frac{1}{4}$ in., the NDS code indicates that the spacing "shall be sufficient to prevent splitting of the wood" (see Sections 11.1.4.7 and 11.1.5.6 of the NDS code). Although this is too vague for any direct application, there is sufficient guidance given in the NDS commentary (C11.1.4.7 and C11.1.5.6) relative to the spacing requirements for small-diameter connectors, as summarized in

Table 8.1. It can be seen from the table that having prebored holes reduces the spacing required, because the prebored hole reduces the possibility of splitting when the fastener is installed.

There is also a required minimum penetration into the main member for screws and nails that must be met (NDS code Section 11.1), as well as a required minimum penetration to achieve the full design value. The design value for a connector that has a penetration depth less than the full design penetration is multiplied by $p/8D$ for lag screws and $p/10D$ for wood screws or nails. The minimum penetration allowed by the NDS code is $4D$ for lag screws and $6D$ for wood screws and nails (see Table 8.2).

For larger-diameter connectors ($D > \frac{1}{4}$ in.), there are two options for spacing:

1. Meet the *base* spacing requirements for the full design value of the connector; for this case, $C_\Delta = 1.0$.
2. Provide a reduced spacing that is less than the base spacing required for the maximum design value, but greater than the code *minimum*. For this case,

$$C_\Delta = \frac{\text{spacing provided}}{\text{base spacing for maximum design value}}$$

The NDS code provides values for both the base spacing and minimum spacing for all of the following parameters: edge distance, end distance, center-to-center spacing, and row spacing. There is a unique value for C_Δ for each of these four spacing requirements. The C_Δ used for design is the smallest value obtained from these unique values. Table 8.3 summarizes the base spacing and minimum spacing requirements for larger-diameter connectors.

End Grain Factor C_{eg}

When fasteners are oriented parallel to the grain of a wood member, the connection capacity is reduced. This is the case when fasteners are inserted into the end grain of a member (see Figure 8.3). For withdrawal loads W, lag screws have an end grain factor of $C_{eg} = 0.75$. Wood screws and nails have virtually no capacity in withdrawal when loaded in the end grain of a member and thus are not permitted. For lateral loads Z, both nails and screws are adjusted by $C_{eg} = 0.67$.

Diaphragm Factor C_{di}

The capacity of a wood diaphragm is a function of the connector type and spacing and is covered in Chapter 6. The capacity of a diaphragm can be found in other references (such as IBC Table 2306.3.1); however, there may be conditions where the diaphragm capacity has to

TABLE 8.1 Recommended Minimum Spacing Requirements for Connectors with $D < \frac{1}{4}$ in.

		Minimum Spacing for Connection Configuration:		
	Type of Loading[a]	Wood Side Member Without Prebored Holes	Wood Side Member with Prebored Holes *or* Steel Side Plates Without Prebored Holes	Steel Side Plates with Prebored Holes
End distance	‖ to grain, tension	15D	10D	5D
	‖ to grain, compression	10D	5D	3D
Edge distance	Any	2.5D	2.5D	2.5D
Center-to-center spacing (pitch)	‖ to grain	15D	10D	5D
	⊥ to grain	10D	5D	2.5D
Row spacing (gage)	Staggered	2.5D	2.5D	2.5D
	In-line	5D	3D	2.5D

[a] ‖ = parallel to the grain
⊥ = perpendicular to the grain

TABLE 8.2 Minimum Penetration Values for Various Connectors

Connector	$p_{\min}$ (reduced)	$p_{\min}$ (full)
Lag screws	4D	8D
Wood screws	6D	10D
Nails	6D	10D

TABLE 8.3 Minimum Spacing Requirements for Connectors with $D > \frac{1}{4}$ in.[a]

		Minimum Spacing	
	Type of Loading	Reduced	Full (Base)
End distance	$\perp$ to grain	2D	4D
	$\parallel$ to grain, compression	2D	4D
	$\parallel$ to grain, tension (softwood)	3.5D	7D
	$\parallel$ to grain, tension (hardwood)	2.5D	5D
Edge distance	$\parallel$ to grain, $l/D \leq 6$	N/A	1.5D
	$\parallel$ to grain, $l/D > 6$	N/A	1.5D or $\frac{1}{2}$ row span (whichever is greater)
	$\perp$ to grain, loaded edge	N/A	4D
	$\perp$ to grain, unloaded edge	N/A	1.5D
Center-to-center spacing	$\parallel$ to grain	3D	4D
	$\perp$ to grain	N/A	3D
Row spacing	$\parallel$ to grain	N/A	1.5D
	$\perp$ to grain, $l/D \leq 2$	N/A	2.5D
	$\perp$ to grain, $2 < l/D \leq 6$	N/A	$(5l + 10D)/8$
	$\perp$ to grain, $l/D > 6$	N/A	5D

l/D (bolt slenderness) is the smaller of l_m/D or l_s/D.
l_m = length of the fastener in wood main member.
l_s = total length of fastener in wood side member(s).
D = diameter of the fastener.
l = length of the fastener.

FIGURE 8.3 End grain loading.

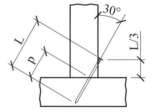

FIGURE 8.4 Toenail connection geometry.

be calculated directly. For this case the capacity of the connectors is multiplied by $C_{di} = 1.1$. This factor applies only to wood panels attached to framing members. It does not apply to other connectors in the load path, such as drag struts or chords.

Toenail Factor C_{tn}

Section 11.1.5.4 of the NDS code gives the geometric configuration of a toenail connection (see Figure 8.4). This is the optimal configuration to prevent the side member from splitting vertically. Toenail connections are typically used where the loads are minimal, such as a beam to sill plate, stud to sill plate, or blocking to a beam (see Figure 8.5). For toenailed connections loaded in withdrawal (W), $C_{tn} = 0.67$. Note that the wet service factor C_M for toenailed connections loaded in withdrawal does not apply. For lateral loads (Z), $C_{tn} = 0.83$.

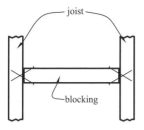
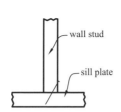
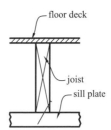

Blocking to joist   Stud to sill plate   Joist to sill plate

FIGURE 8.5 Typical toenail connections.

EXAMPLE 8.1

Group Action Factor—Wood-to-Wood Connection

With reference to Figure 8.6, determine the group action factor C_g. Verify the results from the appropriate NDS table.

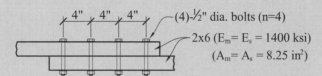

FIGURE 8.6 Group action factor: wood to wood.

Solution: From Equation 8.3:

$$C_g = \frac{m(1 - m^{2n})}{n[(1 + R_{EA}m^n)(1 + m) - 1 + m^{2n}]} \frac{1 + R_{EA}}{1 - m}$$

For the case at hand,

$$R_{EA} = \text{the lesser of } \frac{E_s A_s}{E_m A_m} \text{ or } \frac{E_m A_m}{E_s A_s} = 1.0$$

$$\gamma = 180{,}000 D^{1.5} = (180{,}000)(0.5^{1.5}) = 63{,}640 \text{ (see Section 8.3)}$$

$$u = 1 + \gamma \frac{s}{2}\left(\frac{1}{E_m A_m} + \frac{1}{E_s A_s}\right)$$

$$= 1 + (63{,}640)\left(\frac{4 \text{ in.}}{2}\right)\left[\frac{1}{(1.4 \times 10^6)(8.25)} + \frac{1}{(1.4 \times 10^6)(8.25)}\right] = 1.022$$

$$m = u - \sqrt{u^2 - 1}$$

$$= (1.022) - \sqrt{(1.022)^2 - 1} = 0.811$$

$n = 4$ (number of fasteners in a row)

Then

$$C_g = \left[\frac{(0.811)(1 - 0.811^{(2)(4)})}{(4)[(1 + (1.0)(0.811)^4)(1 + 0.811) - 1 + (0.811)^{(2)(4)}]}\right]\left(\frac{1 + 1.0}{1 - 0.811}\right) = 0.979$$

From NDS Table 10.3.6A with $A_s/A_m = 1.0$ and $A_s = 8.25 \text{ in}^2$, $C_g = 0.94$. This table is conservative for values of D less than 1 in., so the calculated value is consistent with the value obtained from the table. It is also within 5%, so the use of the NDS table for design is appropriate.

EXAMPLE 8.2

Group Action Factor—Wood-to-Steel Connection

With reference to Figure 8.7, determine the group action factor C_g. Verify the results from the appropriate NDS table.

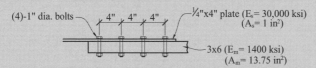

FIGURE 8.7 Group action factor: wood to steel.

Solution: From Equation 8.3:

$$C_g = \frac{m(1 - m^{2n})}{n[(1 + R_{EA}m^n)(1 + m) - 1 + m^{2n}]} \frac{1 + R_{EA}}{1 - m}$$

For the case at hand,

$$R_{EA} = \text{the lesser of } \frac{E_s A_s}{E_m A_m} \quad \text{or} \quad \frac{E_m A_m}{E_s A_s}$$

$$= \frac{(30)(0.25 \times 4)}{(1.4)(13.75)} = 1.56 \quad \text{or} \quad \frac{(1.4)(13.75)}{(30)(0.25 \times 4)} = 0.642$$

$$\gamma = 270{,}000 D^{1.5} = (270{,}000)(1^{1.5}) = 270{,}000 \quad \text{(see Section 8.3)}$$

$$u = 1 + \gamma \frac{s}{2}\left(\frac{1}{E_m A_m} + \frac{1}{E_s A_s}\right)$$

$$= 1 + (270{,}000)\left(\frac{4 \text{ in.}}{2}\right)\left[\frac{1}{(1.4 \times 10^6)(13.75)} + \frac{1}{(30 \times 10^6)(0.25 \times 4)}\right] = 1.046$$

$$m = u - \sqrt{u^2 - 1}$$

$$= (1.046) - \sqrt{(1.046)^2 - 1} = 0.739$$

$n = 4$ (number of fasteners in a row)

Then

$$C_g = \frac{(0.739)(1 - 0.739^{(2)(4)})}{(4)[(1 + (0.642)(0.739)^4)(1 + 0.739) - 1 + 0.739^{(2)(4)}]} \frac{1 + 0.642}{1 - 0.739} = 0.912$$

From NDS Table 10.3.6C with $A_m/A_s = 13.75$ and $A_m = 13.75$ in^2, $C_g = 0.91$ (by interpolation), thus verifying the value from the equation.

EXAMPLE 8.3

Geometry Factor for a Nailed Connection

For the lap splice connections shown in Figure 8.8, it has been found that 12-10d nails are required. Determine if the fastener layout is adequate for the full design value. Assume that the holes are not prebored.

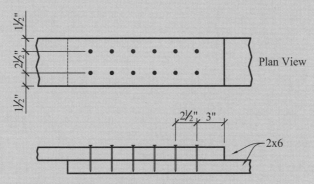

FIGURE 8.8 Nailed lap splice connection.

Solution: Recall that $C_\Delta = 1.0$ for nailed connections; thus, to obtain the full design value for the fasteners, the following minimum spacings are required (see Table 8.1):

$$\text{end distance: } 15D$$

$$\text{edge distance: } 2.5D$$

$$\text{center-to-center spacing: } 15D$$

$$\text{row spacing: } 5D$$

For 10d nails, $D = 0.148$ in., $L = 3$ in. (Table L4 NDS code). Thus,

end distance: $(15)(0.148 \text{ in.}) = 2.2$ in. < 3 in. OK

edge distance: $(2.5)(0.148 \text{ in.}) = 0.37$ in. < 1.5 in. OK

center-to-center spacing: $(15)(0.148 \text{ in.}) = 2.2$ in. < 2.5 in. OK

row spacing: $(5)(0.148 \text{ in.}) = 0.75$ in. < 2.5 in. OK

minimum penetration: $p = 10D = (10)(0.148 \text{ in.}) = 1.48$ in. < 1.5 in. OK

The fastener layout is adequate.

EXAMPLE 8.4

Geometry Factor for a Bolted Connection

For the lap splice connections shown in Figure 8.9, determine the geometry factor for the connection shown. The lumber is Douglas fir-larch.

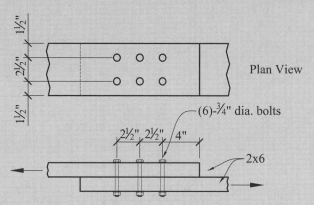

FIGURE 8.9 Bolted lap splice connection.

Solution: l/D = smaller of:

$$\frac{l_m}{D} = \frac{1.5}{0.75} = 2$$

$$\frac{l_s}{D} = \frac{1.5}{0.75} = 2 \text{ (same result; use } l/D = 2)$$

The following distances are provided:

end distance: 4 in.

edge distance: 1.5 in.

center-to-center spacing: 2.5 in.

row spacing: 2.5 in.

The following minimum spacings are required (see Table 8.3):

	Minimum	Full
End distance	3.5D	7D
Edge distance	N/A	1.5D
Center-to-center spacing	3D	4D
Row spacing	N/A	1.5D

thus:

	Minimum		Full
End distance	(3.5)(0.75 in.) = 2.63 in.	to	(7)(0.75 in.) = 5.25 in.
Edge distance	N/A		(1.5)(0.75 in.) = 1.13 in.
Center-to-center spacing	(3)(0.75 in.) = 2.25 in.	to	(4)(0.75 in.) = 3 in.
Row spacing	N/A		(1.5)(0.75 in.) = 1.13 in.

> By inspection, the spacings provided are all greater than the minimum value required. For the parameters that do not have a minimum value, the full or base spacing must be provided. The geometry factor must now be calculated for the other distance parameters:
>
> $$C_\Delta = \frac{\text{spacing provided}}{\text{base spacing}} \qquad (8.8)$$
>
> end distance: $C_\Delta = 4\text{ in.}/5.25\text{ in.} = 0.76$
>
> edge distance: $C_\Delta = 1.0$ (full or base spacing must be provided)
>
> center-to-center spacing: $C_\Delta = 2.5\text{ in.}/3\text{ in.} = 0.83$
>
> row spacing: $C_\Delta = 1.0$ (full or base spacing must be provided)
>
> The lowest value for C_Δ controls; therefore, **$C_\Delta = 0.76$.**

8.4 BASE DESIGN VALUES: LATERALLY LOADED CONNECTORS

In Section 11.3 of the NDS code, a *yield limit model* is presented to determine the nominal lateral design value for a single fastener, Z, based on the following assumptions:

1. The faces of the connected members are in contact.
2. The load acts perpendicular to the axis of the dowel or fastener.
3. Edge distance, end distance, center-to-center spacing, and row spacing requirements are met (either base values or minimum values).
4. The minimum fastener penetration requirement is met.

The nominal lateral design value Z using the yield limit model is a function of the following factors:

- The mode of failure (see Figure 8.10)
- The diameter of the fastener, D
- The bearing length of the fastener, l
- The wood species (i.e., specific gravity G)
- The direction of loading relative to the grain orientation, θ
- The yield strength of the fastener in bending, F_{yb}
- The dowel bearing strength, F_e

There are four basic modes of failure for connectors and the analysis of each mode leads to a corresponding set of equations for determining the design load capacity. Figure 8.10 illustrates each of these modes in single and double shear.

1. *Mode I* is a bearing failure such that the connector crushes the wood member it is in contact with. This failure can occur in either the main member (mode I_m) or in the side member (mode I_s).
2. *Mode II* is a failure such that the connector pivots within the wood members (i.e., the fastener does not yield). There is localized crushing of the wood fibers at the faces of the wood members. This type of failure occurs only in single shear connections.
3. *Mode III* is a combined failure of the connector such that a plastic hinge forms in the connector in conjunction with a bearing failure in either the main member (mode III_m) or in the side member (mode III_s). This type of failure occurs only in single shear connections.
4. *Mode IV* is a failure of the connector such that two plastic hinges form (one in each shear plane), with limited localized crushing of the wood fibers near the shear plane(s).

FIGURE 8.10
Connector failure modes. (Adapted from Ref. 7, Figure I1. Courtesy of the American Forest & Paper Association, Washington, DC.)

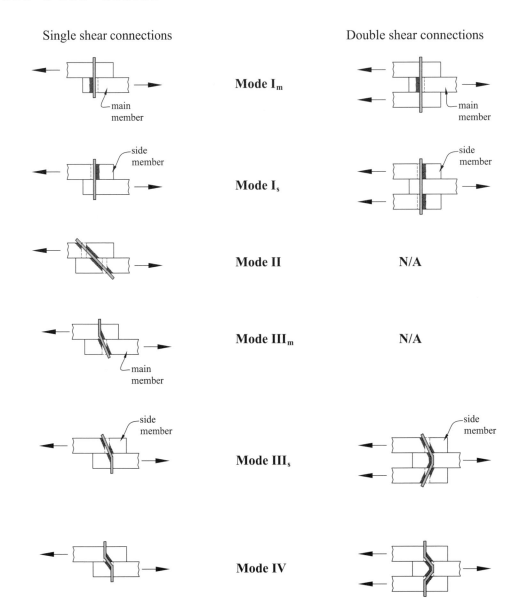

Note: In two-member connections, the main member is usually the member that receives the 'pointed' end; in three-member connections, the main member is the middle member.

Since the failure is either crushing of the wood or yielding of the fastener, the dowel bearing strength F_e and the yield strength of the fastener F_{yb} are the most critical factors in determining Z. The dowel bearing strength is a function of the fastener diameter, the wood species, and the direction of the loading. Values for F_e are calculated as follows:

$$\text{For } D < \tfrac{1}{4} \text{ in.:} \quad F_e = 16{,}600 G^{1.84} \text{ (psi)} \tag{8.9}$$

$$\text{For } D \geq \tfrac{1}{4} \text{ in.:} \quad F_{e\|} = 11{,}200 G \text{ (psi)} \tag{8.10}$$

$$\text{For } D \geq \tfrac{1}{4} \text{ in.:} \quad F_{e\perp} = \frac{6100 G^{1.45}}{\sqrt{D}} \text{ (psi)} \tag{8.11}$$

where D = fastener diameter, in.
G = specific gravity of wood (Table 11.3.2A NDS code)

TABLE 8.4 Yield Strength in Bending of Various Fasteners

Fastener Type	Diameter (in.)	F_{yb} (psi)
Bolt, lag screw	$D > 0.375$	45,000
Common, box, or	$0.344 < D < 0.375$	45,000
sinker nail; lag screw;	$0.273 < D < 0.344$	60,000
wood screw (low- to	$0.236 < D < 0.273$	70,000
medium-carbon steel)	$0.177 < D < 0.236$	80,000
	$0.142 < D < 0.177$	90,000
	$0.099 < D < 0.142$	100,000

Values for F_e are given for various dowel sizes and wood species in the NDS Table 11.3.2. It can be seen that load direction relative to the grain of the wood is not a factor for small-diameter fasteners (less than $\frac{1}{4}$ in.). For larger-diameter fasteners, the dowel bearing strength is constant when loaded parallel to the grain but decreases as the fastener size increases when loaded perpendicular to the grain. The dowel bearing strength is also directly proportional to the density of the wood.

The yield strength of the fastener in bending, F_{yb}, is determined from tests performed in accordance with ASTM F1575 (small-diameter fasteners) and ASTM F606 (large-diameter fasteners). When published data are not available for a specific fastener, the designer should contact the manufacturer to determine the fastener yield strength. Typical values for F_{yb} are given in Table I1 of the NDS code, which is summarized in Table 8.4.

When fasteners are loaded at an angle to the grain (see Figure 8.11), the Hankinson formula shall be used to determine the dowel bearing strength:

$$F_{e\theta} = \frac{F_{e\parallel} F_{e\perp}}{F_{e\parallel} \sin^2\theta + F_{e\perp} \cos^2\theta} \tag{8.12}$$

When connections are made with steel side plates, Section I.2 of the NDS code explains that the nominal bearing strength of steel is $2.4F_u$ for hot-rolled steel and $2.2F_u$ for cold-formed steel. Applying a load duration factor of 1.6, the dowel bearing strength on the steel is as follows:

$$F_e = 1.5F_u \text{ (hot-rolled steel)}$$

$$F_e = 1.375F_u \text{ (cold-formed steel)}$$

where F_u = ultimate tensile strength
= 58,000 psi for ASTM A36 steel, $\frac{1}{4}$ in. and thicker
= 45,000 psi, 3 gage and thinner ASTM A653, grade 33

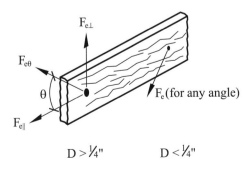

FIGURE 8.11 Fasteners loaded at various angles to the grain.

When fasteners are embedded in concrete, such as anchor bolts for a sill plate (see Chapter 9), the dowel bearing strength can be assumed to be 7500 psi for concrete with a compressive strength of at least $f'_c = 2500$ psi. Table 11E of the NDS code gives lateral design values Z for various wood species and bolt sizes. Note that this table assumes a minimum embedment of 6 in. into the concrete. Having discussed the controlling factors in determining the lateral design value Z for a single fastener, we can now look at the equations to calculate Z. Each connection must be analyzed for each failure mode (see Figure 8.10). The NDS code gives a yield limit equation to determine Z, each corresponding to a specific mode of failure. The smallest value determined from these equations is the controlling value to use for design. The equations are as follows:

Yield Mode	Single Shear	Double Shear	
I_m	$Z = \dfrac{Dl_m F_{em}}{R_d}$	$Z = \dfrac{Dl_m F_{em}}{R_d}$	(8.13)
I_s	$Z = \dfrac{Dl_s F_{es}}{R_d}$	$Z = \dfrac{2Dl_s F_{es}}{R_d}$	(8.14)
II	$Z = \dfrac{k_1 Dl_s F_{es}}{R_d}$		(8.15)
III_m	$Z = \dfrac{k_2 Dl_m F_{em}}{(1 + 2R_e)R_d}$		
III_s	$Z = \dfrac{k_3 Dl_s F_{em}}{(2 + R_e)R_d}$	$Z = \dfrac{2k_3 Dl_s F_{em}}{(2 + R_e)R_d}$	(8.16)
IV	$Z = \dfrac{D^2}{R_d}\sqrt{\dfrac{2F_{em}F_{yb}}{3(1 + R_e)}}$	$Z = \dfrac{2D^2}{R_d}\sqrt{\dfrac{2F_{em}F_{yb}}{3(1 + R_e)}}$	(8.17)

where

$$k_1 = \frac{\sqrt{R_e + 2R_e^2(1 + R_t + R_t^2) + R_t^2 R_e^3} - R_e(1 + R_t)}{(1 + R_e)} \tag{8.18}$$

$$k_2 = -1 + \sqrt{2(1 + R_e) + \frac{2F_{yb}(1 + 2R_e)D^2}{3F_{em}l_m^2}} \tag{8.19}$$

$$k_3 = -1 + \sqrt{\frac{2(1 + R_e)}{R_e} + \frac{2F_{yb}(2 + R_e)D^2}{3F_{em}l_s^2}} \tag{8.20}$$

D = diameter of connector, in.
F_{yb} = dowel bending strength, psi (see NDS code Table I1)
R_d = reduction term (NDS code, Table 11.3.1B or see Table 8.5)
$K_\theta = 1 + (0.25)(\theta/90)$
$R_e = \dfrac{F_{em}}{F_{es}}$
$R_t = \dfrac{l_m}{l_s}$
l_m = main member dowel length, in.
l_s = side member dowel length, in.
F_{em} = main member dowel bearing strength, psi
F_{es} = side member dowel bearing strength, psi

TABLE 8.5 Reduction Factor, R_d

Fastener Size (in.)	Yield Mode	R_d
$0.25 \leq D < 1$	I_m, I_s	$4K_\theta$
	II	$3.6K_\theta$
	III_m, III_s, IV	$3.2K_\theta$
$0.17 < D < 0.25$	I_m, I_s, II, III_m, III_s, IV	$10D + 0.5$
$D \leq 0.17$	I_m, I_s, II, III_m, III_s, IV	2.2

F_{em} or F_{es} = F_e for $D \leq \frac{1}{4}$ in.
= $F_{e\parallel}$ or $F_{e\perp}$ for applied loads either parallel or perpendicular to the grain
= $F_{e\theta}$ for applied loads at an angle to the grain
= 87,000 psi for ASTM A36 steel side plates (1.5 Fu)
= 61,850 psi ASTM A653, grade 33 steel side plates (1.375 Fu)
= 7500 psi (fastener embedded into concrete)

For lag screws, the diameter D is typically taken as the root diameter D_r or the reduced body diameter at the threads (see Figure 8.12). Section 11.3.4 of the NDS code stipulates that the main member dowel length l_m does not include the length of the tapered tip when the penetration depth into the main member is less than 10D. When the penetration is greater than 10D, then l_m would include the length of the tapered tip.

As an alternative to solving for Z using the yield limit equations for lag screws, Tables 11J and 11K are provided in the NDS code. These tables are for lag screws in single shear (wood-to-wood or wood-to-steel) since lag screws are not typically loaded in double shear. The values in this table are valid when the penetration into the main member is at least 8D. When the penetration is such that $4D < p < 8D$, the value for Z in Tables 11J or 11K is adjusted by $p/8D$. Note that NDS Tables 11A through 11J uses the following notations:

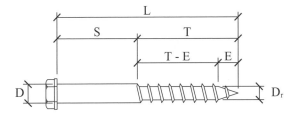

D = diameter
D_r = root diameter
L = overall length
S = unthreaded shank length
T = thread length
E = tapered tip length

FIGURE 8.12 Elements of a lag screw.

$Z_{\parallel}$ = lateral design shear strength of a fastener in single shear with the *side* member loaded *parallel* to the grain and the *main* member loaded *parallel* to the grain

$Z_\perp$ = Lateral design shear strength of a fastener in single shear with the *side* member loaded *perpendicular* to the grain and the *main* member loaded *perpendicular* to the grain

$Z_{m\perp}$ = Lateral design shear strength of a fastener in single shear with the *side* member loaded *parallel* to the grain and the *main* member loaded *perpendicular* to the grain

$Z_{s\perp}$ = Lateral design shear strength of a fastener in single shear with the *side* member loaded *perpendicular* to grain and the *main* member loaded *parallel* to the grain

The following examples illustrate the calculation of Z for various loading conditions.

EXAMPLE 8.5

Laterally Loaded Nail, Single Shear

For the connection shown in Figure 8.13, determine the nominal lateral design value Z and compare the results from the appropriate table in the NDS code. Lumber is spruce-pine-fir for the main and side members.

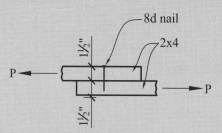

FIGURE 8.13 Laterally loaded nail.

Solution:

$G = 0.42$ (NDS code Table 11.3.2A)

$D = 0.131$ in. (NDS code Table L4)

$l = 2.5$ in. (NDS code Table L4)

$F_{yb} = 100{,}000$ psi (Table 8.4)

$F_{em} = F_{es} = 16{,}600 G^{1.84}$

$\quad = (16{,}600)(0.42)^{1.84} = 3364$ psi (agrees with Table 11.3.2 of the NDS code)

$R_e = \dfrac{F_{em}}{F_{es}} = \dfrac{3364}{3364} = 1.0$

$R_t = \dfrac{l_m}{l_s} = \dfrac{2.5 \text{ in.} - 1.5 \text{ in.}}{1.5 \text{ in.}} = 0.667$

$R_d = 2.2$

$k_1 = \dfrac{\sqrt{R_e + 2R_e^2(1 + R_t + R_t^2) + R_t^2 R_e^3} - R_e(1 + R_t)}{1 + R_e}$

$\quad = \dfrac{\sqrt{1 + (2)(1)^2[1 + 0.667 + (0.667)^2] + (0.667)^2(1)^3} - (1)(1 + 0.667)}{1 + 1} = \mathbf{0.357}$

$k_2 = -1 + \sqrt{2(1 + R_e) + \dfrac{2F_{yb}(1 + 2R_e)D^2}{3F_{em} l_m^2}}$

$\quad = -1 + \sqrt{(2)(1 + 1) + \dfrac{(2)(100{,}000)[1 + (2)(1)](0.131)^2}{(3)(3364)(1)^2}} = \mathbf{1.24}$

$k_3 = -1 + \sqrt{\dfrac{2(1 + R_e)}{R_e} + \dfrac{2F_{yb}(2 + R_e)D^2}{3F_{em} l_s^2}}$

$\quad = -1 + \sqrt{\dfrac{2(1 + 1)}{(1)} + \dfrac{(2)(100{,}000)(2 + 1)(0.131)^2}{(3)(3364)(1.5)^2}} = \mathbf{1.11}$

Calculate Z for each mode of failure.

Mode I_m:

$$Z = \frac{Dl_m F_{em}}{R_d} = \frac{(0.131)(1)(3364)}{2.2} = \mathbf{200\ lb}$$

Mode I_s:

$$Z = \frac{Dl_s F_{es}}{R_d} = \frac{(0.131)(1.5)(3364)}{2.2} = \mathbf{300\ lb}$$

Mode II:

$$Z = \frac{k_1 l_s F_{es}}{R_d} = \frac{(0.357)(0.131)(1.5)(3364)}{2.2} = \mathbf{107\ lb}$$

Mode III_m:

$$Z = \frac{k_2 D l_m F_{em}}{(1 + 2R_e)R_d} = \frac{(1.24)(0.131)(1)(3364)}{[1 + (2)(1)](2.2)} = \mathbf{82\ lb}$$

Mode III_s:

$$Z = \frac{k_3 D l_s F_{em}}{(2 + R_e)R_d} = \frac{(1.11)(0.131)(1.5)(3364)}{(2 + 1)(2.2)} = \mathbf{111\ lb}$$

Mode IV:

$$Z = \frac{D^2}{R_d}\sqrt{\frac{2F_{em}F_{yb}}{3(1 + R_e)}} = \frac{(0.131)^2}{2.2}\sqrt{\frac{(2)(3364)(100,000)}{3(1 + 1)}} = \mathbf{82\ lb}$$

The lowest value was 82 lb (modes III_m and IV) therefore, $Z = 82$ lb, which agrees with Table 11N of the NDS code.

EXAMPLE 8.6

Laterally Loaded Bolt, Double Shear

For the connection shown in Figure 8.14, determine the nominal lateral design value $Z_\perp$ and compare the results from the appropriate table in the NDS code. Lumber is hem-fir for the main member, and the steel plates are $\frac{1}{4}$ in. A36.

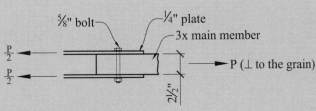

FIGURE 8.14 Laterally loaded bolt.

Solution:

$$G = 0.43 \text{ (NDS code, Table 11.3.2A)}$$

$$D = 0.625 \text{ in.}$$

$$l_m = 2.5 \text{ in. (3}\times\text{ main member)}$$

$$F_{yb} = 45{,}000 \text{ psi (Table 8.4)}$$

$$F_{em} = F_{e\perp} = \frac{6100 G^{1.45}}{\sqrt{D}} = \frac{(6100)(0.43)^{1.45}}{\sqrt{0.625}}$$

$$= 2269 \text{ psi (agrees with Table 11.3.2 of the NDS code)}$$

$$F_{es} = 87{,}000 \text{ (A36 side plates)}$$

$$R_e = \frac{F_{em}}{F_{es}} = \frac{2269}{87{,}000} = 0.026$$

$$R_t = \frac{l_m}{l_s} = \frac{2.5 \text{ in.}}{0.25 \text{ in.}} = 10$$

$$\theta = 90°$$

$$K_\theta = 1 + (0.25)(\theta/90) = 1 + (0.25)(90/90) = 1.25$$

$$k_3 = -1 + \sqrt{\frac{2(1 + R_e)}{R_e} + \frac{2F_{yb}(2 + R_e)D^2}{3F_{em}l_s^2}}$$

$$= -1 + \sqrt{\frac{(2)(1 + 0.026)}{0.026} + \frac{(2)(45{,}000)(2 + 0.026)(0.625)^2}{(3)(2269)(0.25)^2}} = 14.7$$

Calculate Z for each mode of failure.

Mode I_m ($R_d = 4K_\theta$):

$$Z = \frac{Dl_m F_{em}}{R_d} = \frac{(0.625)(2.5)(2269)}{(4)(1.25)} = \mathbf{709 \text{ lb}}$$

Mode I_s ($R_d = 4K_\theta$):

$$Z = \frac{2Dl_s F_{es}}{R_d} = \frac{(2)(0.625)(0.25)(87{,}000)}{(4)(1.25)} = \mathbf{5437 \text{ lb}}$$

Mode II: Not applicable

Mode III_m: Not applicable

Mode III_s ($R_d = 3.2K_\theta$):

$$Z = \frac{2k_3 Dl_s F_{em}}{(2 + R_e)R_d} = \frac{(2)(14.7)(0.625)(0.25)(2269)}{(2 + 0.026)(3.2)(1.25)} = \mathbf{1285 \text{ lb}}$$

Mode IV:

$$Z = \frac{2D^2}{R_d}\sqrt{\frac{2F_{em}F_{yb}}{3(1 + R_e)}} = \frac{(2)(0.625)^2}{(3.2)(1.25)}\sqrt{\frac{(2)(2269)(45{,}000)}{3(1 + 0.026)}} = \mathbf{1591 \text{ lb}}$$

The lowest value was 709 lb (mode I_m); therefore, $Z_\perp = 709$ lb, which agrees with Table 11G of the NDS code.

EXAMPLE 8.7

Laterally Loaded Lag Screw Perpendicular to the Grain

A $\frac{3}{8}$-in. lag screw is connected to a Douglas fir-larch column perpendicular to the grain with a 10-gage side plate (ASTM A653 grade 33) (Figure 8.15). Determine the nominal lateral design value Z, and compare the results from the appropriate table in the NDS code. Assume that the penetration into the main member is at least $10D$.

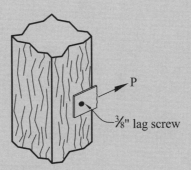

FIGURE 8.15 Laterally loaded lag screw.

Solution:

$$G = 0.50 \text{ (NDS code Table 11.3.2A)}$$

$$D = 0.375 \text{ in.}$$

$$D_r = 0.265 \text{ in. (Table L2)}$$

$$l_m = 10D = (10)(0.265) = 2.65 \text{ in.}$$

$$l_s = 0.134 \text{ in. (10 gage)}$$

$$F_{yb} = 45{,}000 \text{ psi (Table 8.4)}$$

$$F_{em} = F_{e\perp} = \frac{6100 G^{1.45}}{\sqrt{D}} = \frac{(6100)(0.50)^{1.45}}{\sqrt{0.375}} = 3646 \text{ psi}$$

$$F_{es} = 61{,}850 \text{ psi (ASTM A653, grade 33)}$$

$$R_e = \frac{F_{em}}{F_{es}} = \frac{3646}{61{,}850} = 0.059$$

$$R_t = \frac{l_m}{l_s} = \frac{2.65 \text{ in.}}{0.134 \text{ in.}} = 19.78$$

$$\theta = 90°$$

$$K_\theta = 1 + 0.25(\theta/90) = 1 + 0.25(90/90) = 1.25$$

$$k_1 = \frac{\sqrt{R_e + 2R_e^2(1 + R_t + R_t^2) + R_t^2 R_e^3} - R_e(1 + R_t)}{1 + R_e}$$

$$= \frac{\sqrt{0.059 + (2)(0.059)^2[1 + 19.78 + (19.78)^2] + (19.78)^2(0.059)^3} - (0.059)(1 + 19.78)}{1 + 0.059}$$

$$= \mathbf{0.480}$$

$$k_2 = -1 + \sqrt{2(1 + R_e) + \frac{2F_{yb}(1 + 2R_e)D^2}{3F_{em}l_m^2}}$$

$$= -1 + \sqrt{2(1 + 0.059) + \frac{(2)(45,000)[1 + (2)(0.059)](0.265)^2}{(3)(3646)(2.65)^2}} = 0.487$$

$$k_3 = -1 + \sqrt{\frac{2(1 + R_e)}{R_e} + \frac{2F_{yb}(2 + R_e)D^2}{3F_{em}l_s^2}}$$

$$= -1 + \sqrt{\frac{2(1 + 0.059)}{0.059} + \frac{(2)(45,000)(2 + 0.059)(0.265)^2}{(3)(3646)(0.134)^2}} = 9.11$$

Calculate Z for each mode of failure.

Mode I_m ($R_d = 4K_\theta$):

$$Z = \frac{Dl_m F_{em}}{R_d} = \frac{(0.265)(2.65)(3646)}{(4)(1.25)} = 512 \text{ lb}$$

Mode I_s ($R_d = 4K_\theta$):

$$Z = \frac{Dl_s F_{es}}{R_d} = \frac{(0.265)(0.134)(61,850)}{(4)(1.25)} = 439 \text{ lb}$$

Mode II ($R_d = 3.6K_\theta$):

$$Z = \frac{k_1 Dl_s F_{es}}{R_d} = \frac{(0.480)(0.265)(0.134)(61,850)}{(3.6)(1.25)} = 234 \text{ lb}$$

Mode III_m ($R_d = 3.2K_\theta$):

$$Z = \frac{k_2 Dl_m F_{em}}{(1 + 2R_e)R_d} = \frac{(0.487)(0.265)(2.65)(3646)}{(1 + (2)(0.059))(3.2)(1.25)} = 279 \text{ lb}$$

Mode III_s ($R_d = 3.2K_\theta$):

$$Z = \frac{k_3 Dl_s F_{em}}{(2 + R_e)R_d} = \frac{(9.11)(0.265)(0.134)(3646)}{(2 + 0.059)(3.2)(1.25)} = 143 \text{ lb}$$

Mode IV ($R_d = 3.2K_\theta$):

$$Z = \frac{D^2}{R_d}\sqrt{\frac{2F_{em}F_{yb}}{3(1 + R_e)}} = \frac{(0.265)^2}{(3.2)(1.25)}\sqrt{\frac{(2)(3646)(45,000)}{3(1 + 0.059)}} = 178 \text{ lb}$$

The lowest value was 143 lb (mode III_s); therefore, $Z_\perp = 143$ lb, which agrees with Table 11K of the NDS code ($Z_\perp = 140$ lb).

EXAMPLE 8.8

Knee Brace Connection

For the knee brace connection shown in Figure 8.16, design the connectors assuming 16d common nails. The wood species is spruce-pine-fir and is exposed to the weather. Loads are due to lateral wind.

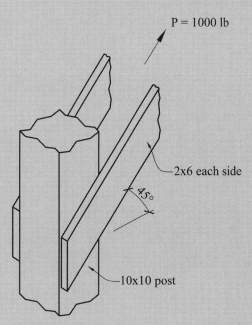

FIGURE 8.16 Knee brace connection.

Solution: From NDS Table L-4: $D = 0.162$ in., $L = 3\frac{1}{2}$ in. Minimum distances (from Table 8.1):

$$\text{end distance: } 15D$$

$$\text{edge distance: } 2.5D$$

$$\text{center-to-center spacing: } 15D$$

$$\text{row spacing: } 5D$$

Thus

$$\text{end distance: } (15)(0.162 \text{ in.}) = 2.43 \text{ in.}$$

$$\text{edge distance: } (2.5)(0.162 \text{ in.}) = 0.405 \text{ in.}$$

$$\text{center-to-center spacing: } (15)(0.162 \text{ in.}) = 2.43 \text{ in.}$$

$$\text{row spacing: } (5)(0.162 \text{ in.}) = 0.81 \text{ in.}$$

$$\text{minimum penetration: } p = 10D = (10)(0.162 \text{ in.}) = 1.62 \text{ in.}$$

From NDS Table 11N, $Z = 120$ lb. From NDS Table 10.3.3, $C_M = 0.7$. All other C factors $= 1.0$.

$$Z' = ZC_D C_M C_t C_g C_\Delta C_{eg} C_{di} C_{tn}$$

$$= (120)(1.6)(0.7)(1.0)(1.0)(1.0)(1.0)(1.0) = 134 \text{ lb}$$

The number of nails required

$$N_{req'd} = \frac{1000}{134} = 7.5 \text{ nails} \simeq 8 \text{ nails}$$

Use 4-16d nails on each side of the splice. See Figure 8.17.

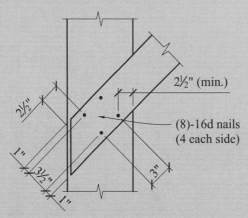

FIGURE 8.17 Nailed knee brace connection detail.

8.5 BASE DESIGN VALUES: CONNECTORS LOADED IN WITHDRAWAL

In *withdrawal loading* a fastener has a load applied parallel to its length such that the load stresses the fastener in tension and tends to pull the fastener out of the main member in which it is embedded. From a practical standpoint, wood screws and lag screws are the preferred connector for withdrawal loading because the threads create a more positive connection to the wood. Lag screws are permitted to be installed on either the side grain or the end grain when loaded in withdrawal. Recall that the end grain factor applies for lag screws loaded in withdrawal ($C_{eg} = 0.75$). Wood screws and nails are not permitted to be loaded in withdrawal from the end grain of wood.

Section 11.2 of the NDS code gives the withdrawal design values W for various connectors as follows:

$$W = \begin{cases} 1800 G^{3/2} D^{3/4} & \text{(lag screws)} \quad (8.21) \\ 2850 G^2 D & \text{(wood screws)} \quad (8.22) \\ 1380 G^{5/2} D & \text{(nails)} \quad (8.23) \end{cases}$$

where W = withdrawal design value, lb (per inch of penetration)
G = specific gravity (see NDS code Table 11.3.2A)
D = fastener diameter, in.

The value for W calculated from the equations above represents the withdrawal design value per inch of penetration into the main member, per fastener. NDS code Tables 11.2A, 11.2B, and 11.2C give tabulated values for W. The fastener could also fail by yielding in tension and should be checked for this condition, but wood connections are not usually controlled by the fastener strength in pure tension.

EXAMPLE 8.9

Fasteners Loaded in Withdrawal

With reference to Figure 8.18, determine the adjusted withdrawal values W' (P_{max}) for the following connector types: (a) $\frac{3}{8}$-in. lag screw; (b) no. 12 wood screw; (c) 16d common nail. Normal temperature and moisture conditions apply and $C_D = 1.6$ (wind). Lumber is hem-fir.

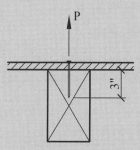

FIGURE 8.18 Fasteners loaded in withdrawal.

Solution: From NDS code Table 11.3.2A, $G = 0.43$ (hem-fir).

(a) $\frac{3}{8}$-in. lag screw:

$$W = 1800 G^{3/2} D^{3/4}$$
$$= (1800)(0.43)^{3/2}(0.375)^{3/4} = 243 \text{ lb/in.} \quad \text{(agrees with NDS code Table 11.2A)}$$

Adjusted for penetration:

$$W = (3 \text{ in.})(243 \text{ lb/in.}) = 729 \text{ lb}$$
$$P_{max} = W' = W C_D C_M C_t C_{eg} C_{tn}$$
$$= (729)(1.6)(1.0)(1.0)(1.0)(1.0) = \mathbf{1166 \text{ lb}}$$

(b) *no. 12 wood screw.* From Table L3 of the NDS code:

$$D = 0.216 \text{ in.}$$
$$W = 2850 G^2 D$$
$$= (2850)(0.43)^2(0.216) = 114 \text{ lb/in.} \quad \text{(agrees with Table 11.2B of the NDS code)}$$

Adjusted for penetration:

$$W = (3 \text{ in.})(114 \text{ lb/in.}) = 342 \text{ lb}$$
$$P_{max} = W' = W C_D C_M C_t C_{eg} C_{tn}$$
$$= (342)(1.6)(1.0)(1.0)(1.0)(1.0) = \mathbf{546 \text{ lb}}$$

(c) *16d common nail.* From Table L3 of the NDS code:

$$D = 0.162 \text{ in.}$$
$$W = 1380 G^{5/2} D$$
$$= (1380)(0.43)^{5/2}(0.162) = 27 \text{ lb/in.} \quad \text{(agrees with Table 11.2C of the NDS code)}$$

Adjusted for penetration:

$$W = (3 \text{ in.})(27 \text{ lb/in.}) = 81 \text{ lb}$$
$$P_{max} = W' = W C_D C_M C_t C_{eg} C_{tn}$$
$$= (81)(1.6)(1.0)(1.0)(1.0) = \textbf{130 lb}$$

8.6 COMBINED LATERAL AND WITHDRAWAL LOADS

With reference to Figure 8.19, there are conditions in which connectors are subjected to lateral and withdrawal load simultaneously. For this case, the NDS code specifies the following interaction equations for combined loading:

$$Z'_\alpha = \begin{cases} \dfrac{W'pZ'}{W'p \cos^2\alpha + Z' \sin^2\alpha} & \text{for lag screws or wood screws} \quad (8.24) \\[2ex] \dfrac{W'pZ'}{W'p \cos\alpha + Z' \sin\alpha} & \text{for nails} \quad (8.25) \end{cases}$$

where Z'_α = adjusted resulting design value under combined loading
W' = adjusted withdrawal design value
p = length of penetration into the main member, in.
Z' = adjusted lateral design value
α = angle between the wood surface and the direction of load applied

The value Z'_α would then be compared to the resultant load P shown in Figure 8.19.

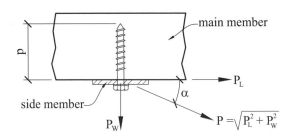

FIGURE 8.19
Combined lateral and withdrawal loading.

EXAMPLE 8.10

Fastener Loaded by Lateral and Withdrawal Loads

A connection similar to that shown in Figure 8.19 requires a no. 14 wood screws under combined lateral and withdrawal loads. Determine P_{max} when $\alpha = 30°$ and a penetration, $p = 1.5$ in. Lumber is hem-fir. The side plate is 10 gage (ASTM A653 grade 33) and the loads are $D + S$. Assume that normal temperature and moisture conditions apply.

Solution:

$$D = 0.242 \text{ in. (NDS code Table L3)}$$
$$W = 127 \text{ lb/in. (NDS code Table 11.2B)}$$
$$Z = 160 \text{ lb (NDS code Table 11N)}$$
$$p = 1.5 \text{ in. (greater than } 6D \text{ but less than } 10D)$$

Therefore,

$$Z = (160 \text{ lb})(p/10D) = (160)[1.5/(10)(0.242)] = \textbf{99 lb}$$
$$W' = W C_D C_M C_t C_{eg} C_{tn}$$
$$= (127)(1.15)(1.0)(1.0)(1.0)(1.0) = \textbf{146 lb}$$
$$C_g = 1.0 \ (D < \tfrac{1}{4} \text{ in.})$$

$C_\Delta = 1.0$, but minimum spacing must be provided for $D < \tfrac{1}{4}$ in.

From Table 8.1, required spacings (assume no prebored holes):

$$\text{end distance: } 10D$$
$$\text{edge distance: } 2.5D$$
$$\text{center-to-center spacing: } 10D$$
$$\text{row spacing: } 3D$$

Therefore,

$$\text{end distance: } (10)(0.242 \text{ in.}) = 2.42 \text{ in.}$$
$$\text{edge distance: } (2.5)(0.242 \text{ in.}) = 0.605 \text{ in.}$$
$$\text{center-to-center spacing: } (10)(0.242 \text{ in.}) = 2.42 \text{ in.}$$
$$\text{row spacing: } (3)(0.242 \text{ in.}) = 0.726 \text{ in.}$$

$$Z' = Z C_D C_M C_t C_g C_\Delta C_{eg} C_{di} C_{tn}$$
$$= (99)(1.15)(1.0)(1.0)(1.0)(1.0)(1.0)(1.0)(1.0) = \textbf{113 lb}$$
$$Z'_\alpha = \frac{W' p Z'}{W' p \cos^2\alpha + Z' \sin^2\alpha} = \frac{(146)(1.5)(113)}{(146)(1.5)\cos^2 30 + (113)\sin^2 30} = \textbf{128 lb}$$

$P_{max} = \textbf{128 lb}$

EXAMPLE 8.11

Diaphragm Fastener Loaded by Lateral and Withdrawal Loads

For the diaphragm connection shown in Figure 8.20, determine the capacity of the 10d toenail for (a) lateral loads and (b) withdrawal loads. Loads are due to wind and normal temperature and moisture conditions apply. The lumber is DF-L.

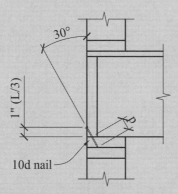

FIGURE 8.20 Diaphragm connection detail.

Solution:

$$D = 0.148 \text{ in. (NDS code Table L3)}$$

$$L = 3 \text{ in. (NDS code Table L3)}$$

$$p = 3 \text{ in.} - \frac{(L/3)}{6530} = 1.84 \text{ in.} > 10d = 1.48 \text{ in.} \quad \text{OK}$$

(a) From NDS Table 11N, $Z = 118$ lb (side member thickness = 1.5 in.):

$$Z' = ZC_D C_M C_t C_g C_\Delta C_{eg} C_{di} C_{tn}$$

$$C_D = 1.6 \text{ (wind)}$$

$$C_{di} = 1.1 \text{ (NDS code Section 11.5.3)}$$

$$C_{tn} = 0.83 \text{ (NDS code Section 11.5.4.2)}$$

All other C factors = 1.0.

$$Z' = (118)(1.6)(1.0)(1.0)(1.0)(1.0)(1.1)(0.83) = \textbf{172 lb}$$

(b) From NDS Table 11.2C, $W = 36$ lb/in. ($G = 0.50$, see NDS Table 11.3.2A); $p = 1.73$ in. from part (a); therefore,

$$W = (36 \text{ lb/in.})(1.73 \text{ in.}) = 62 \text{ lb}$$

$$W' = WC_D C_M C_t C_{eg} C_{tn}$$

$$C_D = 1.6 \text{ (wind)}$$

$$C_{tn} = 0.67 \text{ (NDS Section 11.5.4.1)}$$

All other C factors = 1.0 ($C_M = 1.0$ for toenail connections; see NDS code Section 11.5.4).

$$W' = (62)(1.6)(1.0)(1.0)(1.0)(0.67) = \textbf{66 lb}$$

8.7 PREENGINEERED CONNECTORS

There are a wide variety of preengineered connectors available for wood construction. Most manufacturers will readily supply a catalog of their connectors with corresponding load-carrying capacity as well as installation requirements. The selection of a manufacturer usually depends on regional availability or the builder's preference. For most cases, the load tables and connector details from the different manufacturers are very similar. A generic load table is shown in Figure 8.21. A brief explanation of some of the elements of the table follows.

1. The connector designation is usually based on the geometric properties of the connector. "JH" could stand for "joist hanger" and "214" would represent the joist that the connector is intended to hold.
2. The dimensions are important for detailing considerations.
3. The manufacturer usually lists the lumber that should be used with each connector. For other species of lumber, a reduction factor is usually applied.
4. The percentages noted corresponds to the load duration factor. 100% or full load duration corresponds to $C_D = 1.0$ (or floor live load). 160% corresponds to $C_D = 1.6$ (wind or seismic loads).

Figure 8.22 illustrates other preengineered connector types and where they are typically used in a wood structure.

8.8 PRACTICAL CONSIDERATIONS

Wood members within a structure will experience dimensional changes over time due mainly to the drying of the wood. The ambient humidity, which can change with the seasons, can also contribute to the shrinking and swelling of wood members. Because of this, wood connections should be detailed to account for dimensional changes. In general, wood connections should be detailed such that the stresses that can cause splitting are avoided.

Beam connections should preferably be detailed such that the load is transferred in bearing. Figure 8.23 show details that are not recommended because the connectors induce stresses perpendicular to the grain and could cause splitting. Figure 8.24 indicates preferred connections in which loads are transferred in bearing. Note that bolts (or other connectors) are still provided within the beam to resist longitudinal forces and for stability. Many preengineered joist hangers employ this concept.

Tension splice connections, which might occur in the bottom chord of a truss, should be detailed with vertically slotted holes to allow for dimensional changes. If shear forces have to be

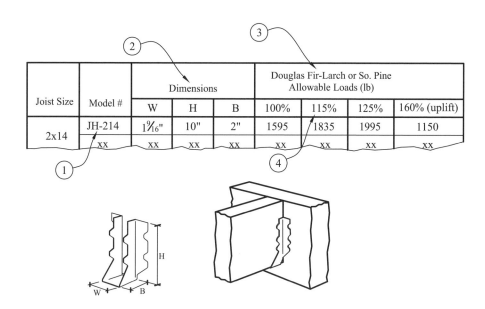

Joist Size	Model #	Dimensions			Douglas Fir-Larch or So. Pine Allowable Loads (lb)			
		W	H	B	100%	115%	125%	160% (uplift)
2x14	JH-214	1 9/16"	10"	2"	1595	1835	1995	1150
	xx	xx	xx	xx	xx	xx	xx	xx

FIGURE 8.21 Preengineered connector load table.

FIGURE 8.22
Preengineered connector types.

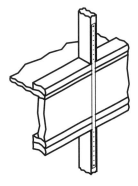
Shear wall strap

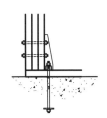
Hold-down anchor

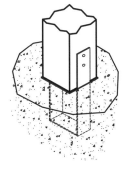
Column base

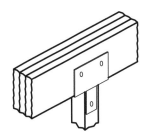
Column cap

Sloped joist hanger

FIGURE 8.23
Nonpreferred beam connections.

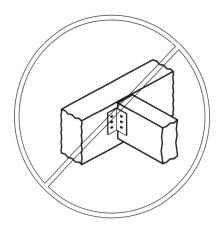
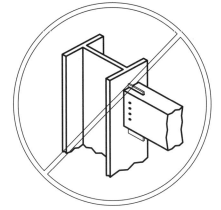

FIGURE 8.24
Preferred beam connections.

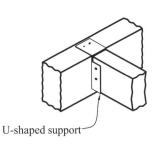
U-shaped support

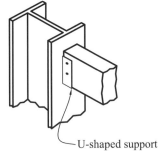
U-shaped support

transferred across the splice, a combination of slotted and round holes can be used. Figure 8.25 shows preferred tension splice details.

Truss connections should be detailed such that the member centerlines intersect at a common work point (Figure 8.26). This will ensure that the axial loads in the truss members do not induce shear and bending stresses in the members.

Base connections for wall studs and posts should be kept sufficiently high above the concrete foundation to avoid moisture that can cause decay. Figure 8.27 indicates base connections that

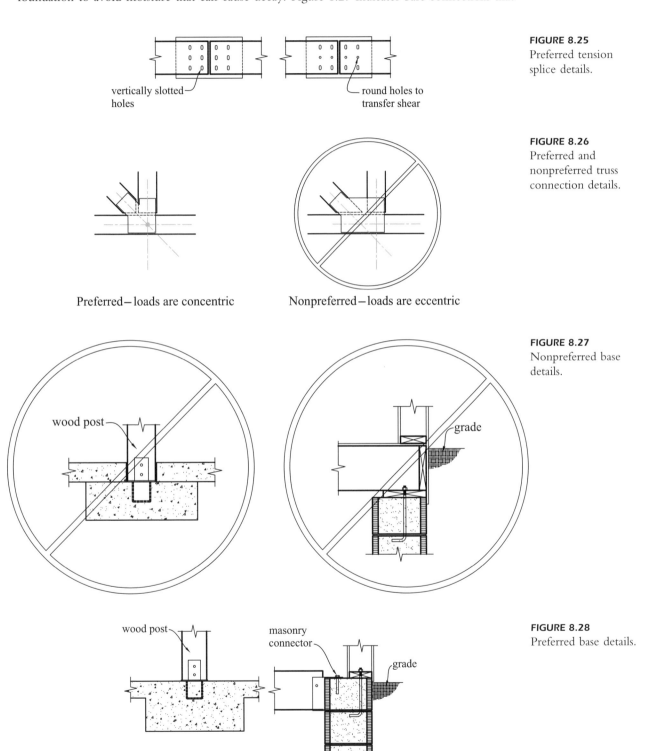

FIGURE 8.25
Preferred tension splice details.

FIGURE 8.26
Preferred and nonpreferred truss connection details.

FIGURE 8.27
Nonpreferred base details.

FIGURE 8.28
Preferred base details.

are not recommended because they allow the wood and connectors to be exposed to direct moisture. The preferred details for these conditions are shown in Figure 8.28.

REFERENCES

1. Willenbrock, Jack H., Manbeck, Harvey B., and Suchar, Michael G. (1998), *Residential Building Design and Construction,* Prentice Hall, Upper Saddle River, NJ.
2. Ambrose, James (1994), *Simplified Design of Wood Structures,* 5th ed., Wiley, New York.
3. Faherty, Keith F., and Williamson, Thomas G. (1989), *Wood Engineering and Construction,* McGraw-Hill, New York.
4. Halperin, Don A., and Bible, G. Thomas (1994), *Principles of Timber Design for Architects and Builders,* Wiley, New York.
5. ICC (2006), *International Building Code,* International Code Council, Washington, DC.
6. Stalnaker, Judith J., and Harris, Earnest C. (1997), *Structural Design in Wood,* Chapman & Hall, London.
7. ANSI/AF&PA (2005), *National Design Specification for Wood Construction,* American Forest & Paper Association, Washington, DC.
8. Breyer, Donald et al. (2003), *Design of Wood Structures—ASD,* 5th ed., McGraw-Hill, New York.
9. NAHB (2000), *Residential Structural Design Guide— 2000,* National Association of Home Builders Research Center, Upper Marlboro, MD.
10. Newman, Morton (1995), *Design and Construction of Wood-Framed Buildings,* McGraw-Hill, New York.
11. Kang, Kaffee (1998), *Graphic Guide to Frame Construction— Student Edition,* Prentice Hall, Upper Saddle River, NJ.
12. Stefano, Jim (2004) Detailing timber connections, *STRUCTURE,* pp. 15–17, February.

PROBLEMS

8.1 Calculate the nominal withdrawal design values W for the following and compare the results with the appropriate table in the NDS code: (a) common nails: 8d, 10d, 12d; (b) wood screws: nos. 10, 12, 14; (c) lag screws: $\frac{1}{2}$ in., $\frac{3}{4}$ in. The lumber is Douglas Fir-Larch.

8.2 Calculate the group action factor for the connection shown in Figure 8.29, compare the results with NDS code Table 10.3.6A, and explain why the results are different. The lumber is hem-fir Select Structural.

8.3 For the following connectors, calculate the dowel bearing strength parallel to the grain ($F_{e\parallel}$), perpendicular to the grain ($F_{e\perp}$), and at an angle of 30° to the grain ($F_{e\theta}$). The lumber is spruce-pine-fir: (a) 16d common nail; (b) wood screws: nos. 14 and 18; (c) bolts: $\frac{5}{8}$ in., 1 in. Compare the results with Table 11.3.2 of the NDS code.

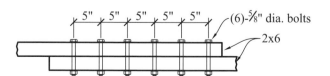

FIGURE 8.29 Problem 8-2.

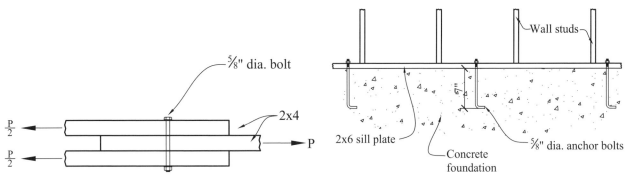

FIGURE 8.30 Problem 8-4.   **FIGURE 8.31** Problem 8.5.

8.4 Using the yield limit equations, determine the nominal lateral design value Z for the connection shown in Figure 8.30. Compare the results with the appropriate table in the NDS code. The lumber is Douglas fir-larch.

8.5 For the sill plate connection shown in Figure 8.31, calculate the design shear capacity Z' parallel to the grain. The concrete strength is $f'_c = 3000$ psi and the lumber is spruce-pine-fir. Loads are dues to seismic (E), and normal temperature and moisture conditions apply.

8.6 Design a 2×6 lap splice connection using (a) wood screws and (b) bolts. The load applied is 1800 lb due to wind. The lumber is southern pine.

chapter *nine*

BUILDING DESIGN CASE STUDY

9.1 INTRODUCTION

The design of a simple wood building structure (see Figure 9.1) is presented in this chapter to help the reader tie together the various structural concepts presented in earlier chapters. This design case study building is a wood-framed office building located in Rochester, New York. It consists of roof trusses, plywood roof sheathing, stud walls with plywood sheathing, plywood floor sheathing, floor joists, a glulam floor girder, wood columns supporting the floor beam, and stairs leading from the ground floor to the second floor. The plan dimensions of the building are 30 ft × 48 ft and the floor-to-floor height is 12 ft. The exterior walls have $\frac{5}{8}$-in. gypsum wall board on the inside and $\frac{5}{8}$-in. wood shingle on plywood sheathing on the outside. The ceiling is $\frac{5}{8}$-in. gypsum wall board and the second floor is finished with $\frac{7}{8}$-in.-thick hardwood flooring. The roofing is asphalt shingles. Spruce-pine-fir wood species is specified for all sawn-lumber members and Hem-Fir for the glulam girder. The design aids presented in Appendix B are used for selection of the ground-floor column, exterior wall stud, and the floor joists. Note that for any given project, the designer should verify what species and grade of lumber and sheathing are available where the building is to be constructed.

Checklist of Items to Be Designed
The following is a checklist of all the items to be designed for this case study building.

Sheathing for Floor, Roof, and Walls Initially, select the roof and floor sheathing grade, minimum thickness, span rating, and edge support requirements for *gravity* loads based on the IBC tables. The plywood grade and thickness obtained for the gravity load condition should then be checked to determine if it is adequate to resist the *lateral* loads when the roof and floor sheathing act as horizontal diaphragms.

Roof Truss The truss is designed for the case of dead load plus snow load, and structural analysis software is used to analyze the roof truss for the member forces. The computer results can be verified using a hand calculation method such as the method of sections or the method of joints.

Stud Walls All the IBC load cases have to be considered and 2 × 6 studs are assumed.

Floor Joist, Glulam Floor Girder, and Columns The critical load case is dead load plus floor live load.

Roof Diaphragm Select the nailing and edge support requirements. Calculate the chord and drag strut forces and design the roof diaphragm chord and drag strut (i.e., the top plates).

Floor Diaphragm Select the nailing and edge support requirements. Calculate the chord and drag strut forces and design the floor diaphragm chord and drag strut (i.e., the top plates).

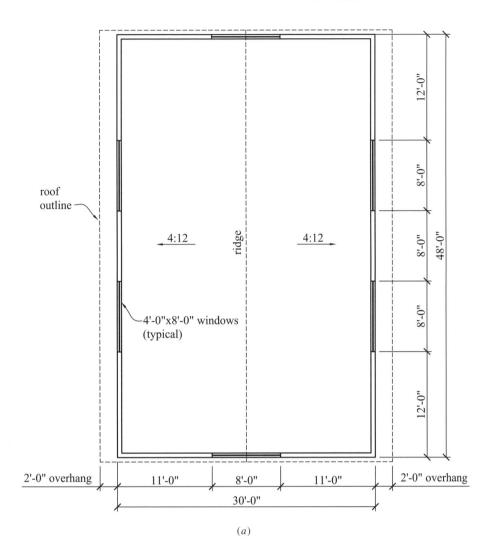

FIGURE 9.1 Case study building: (*a*) roof plan; (*b*) second floor plan; (*c*) ground floor plan; (*d*) possible roof truss profiles.

Shear Walls Analyze and design the most critical panel in *each direction*, checking the tension and compression chords; indicate whether hold-down devices are needed, and specify the vertical diaphragm nailing requirements.

Lintel/Headers Design lintels and headers at door and window openings for gravity loads from above (i.e., dead plus snow loads for lintels/headers below the roof, and dead plus floor live load for lintels below the second floor).

9.2 GRAVITY LOADS

In this section we calculate the roof dead and live loads, the floor dead and live loads, and the wall dead load (i.e., the self-weight of the wall).

Roof Dead Load

To calculate the roof dead load, we assume roof trusses spaced at 2 ft o.c., plywood roof sheathing, asphalt shingle roofing, and $\frac{5}{8}$-in. gypsum wall board ceiling. Other dead loads that will be included are the weight of mechanical and electrical fixtures, insulation and possible reroofing, assuming that the existing roof will be left in place during the reroofing operation. The dead loads are obtained as follows using the dead load tables in Appendix A:

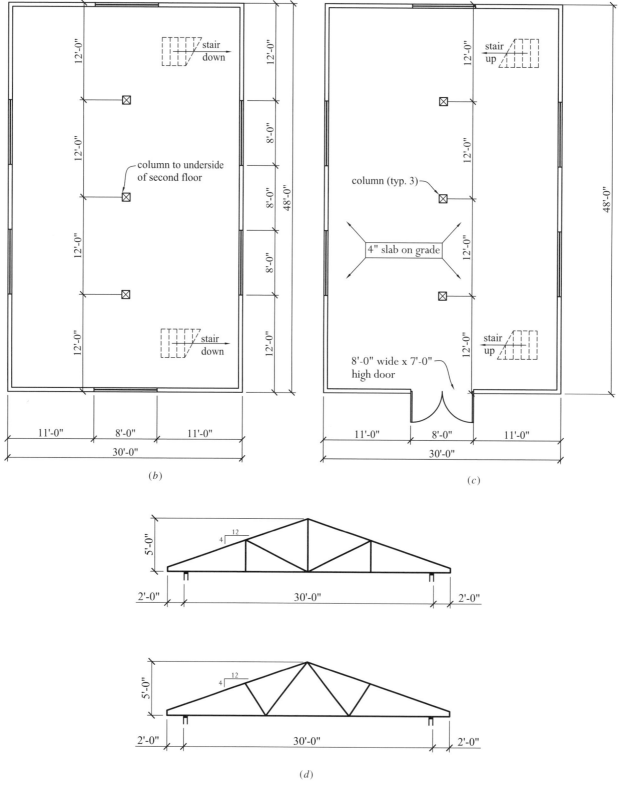

FIGURE 9.1 (*Continued*)

Roofing ($\frac{5}{16}$ in. asphalt shingles)	= 2.5 psf
Reroofing (i.e., future added roof)	= 2.5 psf (assumed)
$\frac{3}{8}$-in. plywood sheathing (= 0.4 psf/$\frac{1}{8}$ in. × .3)	= 1.2 psf
Framing (truss span = 30 ft; see Appendix A)	= 5.0 psf
1 in. fiber glass insulation (supported by the bottom chord)	= 1.1 psf
$\frac{5}{8}$-in. gypsum ceiling attached to the truss bottom chord (5 psf/in × $\frac{5}{8}$ in)	= 3.2 psf
Mechanical and electrical (supported off the bottom chord)	= 5.0 psf
Total roof dead load D_{roof}	= 20.5 psf of sloped roof area

The dead load supported by the roof truss bottom chord includes the insulation, the gypsum ceiling, and the mechanical and electrical loads. Therefore, the total dead load on the bottom chord is 9.3 psf of horizontal plan area, while the dead load supported by the truss top chord is 11.2 psf. Note that the load on the top chord is in psf of sloped roof area and must be converted to psf of horizontal plan area before combining it with the loads on the truss.

The total roof dead load in psf of the *horizontal plan area* is given as

$$w_{\text{DL}} \text{ (top chord)} = (11.2 \text{ psf}) \left(\frac{15.8 \text{ ft}}{15 \text{ ft}}\right) \approx 12 \text{ psf}$$

$$w_{\text{DL}} \text{ (bottom chord)} \approx 10 \text{ psf}$$

It should be noted that for calculating the roof wind uplift load, the dead load must not be overestimated because it is not conservative to do so. For example, the 2.5 psf reroofing dead load should not be considered for this case since reroofing is a future event and may not be there for quite a number of years. Therefore, for uplift load calculations, only the dead loads that are likely to be on the structure during a wind event should be used to calculate the net uplift wind pressure. For this building, the applicable roof dead load is calculated as

$$w_{\text{DL}} \text{ (for uplift calculations only)} = (11.2 \text{ psf} - 2.5 \text{ psf}) \left(\frac{15.8 \text{ ft}}{15 \text{ ft}}\right) + 9.3 \text{ psf} = 18.5 \text{ psf}$$

Roof Live Load

The snow load S and roof live load L_r acting on the roof of this building will now be calculated.

Roof Live Load L_r From Section 2.4, since $F = 4$, the roof slope factor is obtained as $R_2 = 1.0$.

The tributary area of the roof truss = 2 ft × 30 ft = 60 ft^2 < 200 ft^2; therefore, $R_1 = 1.0$.
The roof live load on the truss $L_r = 20 R_1 R_2 = (20)(1.0)(1.0) = 20$ psf.
12 psf ≤ L_r ≤ 20 psf   OK
The tributary roof area for a typical exterior stud = (2 ft)(30 ft/2) = 30 ft^2 < 200 ft^2; therefore, $R_1 = 1.0$.
The roof live load on the exterior wall stud $L_r = 20 R_1 R_2 = (20)(1.0)(1.0) = 20$ psf.
12 psf ≤ L_r ≤ 20 psf   OK

Snow Load The roof slope (4-in-12) implies that $\tan \theta = 4/12$; therefore, $\theta = 18.43°$. Using Figure 1608.2 from the New York State version of the IBC, we obtain the ground snow load P_g for Rochester, New York to be 50 psf. The reader should note that for this building location, the national snow map in IBC Figure 1608.2 or ASCE 7 Figure 7-1 shows a ground snow load value of 40 psf, but the New York State snow map indicates a ground snow load value of 50 psf. This higher value will be used for the design of this building.

Assuming a "partially" exposed roof and a building site with terrain category C, we obtain the exposure coefficient $C_e = 1.0$ (ASCE 7 Table 7-2).

The thermal factor for a heated building $C_t = 1.0$ (ASCE 7 Table 7-3).

The importance factor $I = 1.0$ (IBC Table 1604.5).

The slope factor $C_s = 1.0$ (ASCE Figure 7-2 with roof slope $\theta = 18.43°$ and a warm roof).

The flat roof snow load $P_f = 0.7 C_e C_t I P_g = (0.7)(1.0)(1.0)(1.0)(50) = 35$ psf.

Thus, the design roof snow load, $S = P_s = C_s P_f = (1.0)(35) = 35$ psf.

The design total load in psf of the *horizontal plan area* is obtained using the IBC load combinations (in Section 2.1). Where downward wind loads on the roof are not critical (this will be checked later), as is often the case, the controlling load combination, with the deal load modified to account for the roof slope, is given by Equation 2.1 as

$$w_{TL} = D \frac{L_1}{L_2} + (L_r \text{ or } S \text{ or } R) \qquad \text{psf of the } horizontal\ plan\ area$$

Since the roof live load L_r (20 psf) is less that the design snow load S (35 psf), the snow load is more critical and will be used in calculating the total roof load.

L_1 = sloped length of roof truss top chord = 15.8 ft

L_2 = horizontal projected length of roof truss top chord = 15 ft

The total load on the top chord and bottom chords of the roof truss will be calculated separately.

$w_{TL}(\textbf{top chord}) = 12$ psf $+ 35$ psf

$= \textbf{47 psf}$ of the horizontal plan area

$w_{TL}(\textbf{bottom chord}) \approx \textbf{10 psf}$ of the horizontal plan area (this is from the ceiling load only)

The total load on the roof truss $w_{TL} = 47 + 10 = \textbf{57 psf}$ of the horizontal plan area.

The tributary width for a typical interior truss = 2 ft.

The total load in pounds per horizontal linear foot (lb/ft) on the roof truss is given as

$$w_{TL}(\text{lb/ft}) = w_{TL}(\text{psf})(\text{tributary width})$$

$$= (57 \text{ psf})(2 \text{ ft}) = \textbf{114 lb/ft}$$

The maximum shear force in the roof truss is

$$V_{max} = w_{TL} \frac{L_2}{2} = (114)\left(\frac{30 \text{ ft}}{2}\right) = 1710 \text{ lb}$$

The total load in pounds per horizontal linear foot (lb/ft) on the top chord and bottom chords are calculated as

$$w_{TL}(\text{top chord}) = 47 \text{ psf} \times 2 \text{ ft} = 94 \text{ lb/ft}$$

$$w_{TL}(\text{bottom chord}) = 10 \text{ psf} \times 2 \text{ ft} = 20 \text{ lb/ft}$$

Floor Dead Load

Floor covering/finishes (assume $\frac{7}{8}$-in. hardwood)	= 4.0 psf
1-in. plywood sheathing (0.4 psf/$\frac{1}{8}$ in. $\times$ 8)	= 3.2 psf
Framing (assume 2 $\times$ 14 at 12 in. o.c.)	= 4.4 psf
$\frac{5}{8}$-in. gypsum ceiling (= 5 psf/in. $\times \frac{5}{8}$ in.)	= 3.2 psf
Mechanical and electrical	= 5.0 psf

Partitions (assumed)	= 20.0 psf
Total floor dead load D_{floor}	≈ 40 psf

Note that the minimum value allowed for partitions per IBC Section 1607.5 or ASCE 7 Section 4.2.2 is 15 psf.

Floor Live Load

We neglect floor live-load reduction for this design example. It should be noted that for light-framed wood structures, the effect of floor live-load reduction is usually minimal. For office buildings, the floor live load $L = 50$ psf (load is higher in corridor areas). The live-load reduction factor for the various structural elements of this building will be determined in the design sections for these structural elements.

Without live load reduction, the total floor load (in psf) obtained using the IBC load combinations (Section 2.1) is

$$w_{\text{TL}} = D + L = 40 + 50 = 90 \text{ psf}$$

Wall Dead Load

$\frac{1}{2}$ in. plywood (exterior face of wall)	= 1.6 psf
Stud wall framing (assume 2 × 6 studs at 2 ft o.c.)	= 0.9 psf
Wood shingles	= 3.0 psf
$\frac{5}{8}$ gypsum wallboard (interior face of wall)	= 3.2 psf
1 in. fiberglass insulation	= 1.1 psf
Total wall dead load D_{roof}	≈ 10 psf of vertical surface area

9.3 SEISMIC LATERAL LOADS

The lateral seismic loads on the building will now be calculated. The solution presented here will be in accordance with ASCE 7-05 Section 12.14, which is a simplified analysis procedure. This procedure is permitted for wood-framed structures that are three stories and less in height.

Given:

Floor dead load = 40 psf (partitions included)

Roof dead load ≈ 22.0 psf

Wall dead load = 10.0 psf (exterior walls)

Snow load $P_s = 35$ psf

Site class = D (ASCE 7 Section 20.1)

importance $I_e = 1.0$ (ASCE 7 Table 11.5-1)

$S_S = 0.25\%$ (ASCE Figure 22.1)

$S_1 = 0.07\%$ (ASCE 7 Figure 22.1)

$R = 6.5$ (ASCE 7 Table 12.14-1)

$F = 1.1$ for two-story building (ASCE 7 Section 12.14.8.1)

Calculate W for each level (see Table 9.1). W is the seismic weight tributary to each level; it includes the weight of the floor structure, the weight of the walls, and the weight of a portion of the snow load. Where the flat roof snow load P_f is greater than 30 psf, 20% of the flat roof snow load is included in the tributary seismic weight at the roof level (ASCE 7 12.7.2).

TABLE 9.1 Self-Weight of Each Level

Level	Area (ft²)	Tributary Height (ft)	Weight Floor	Weight Exterior Walls	W_{total}
Roof	30 ft × 48 ft = **1440 ft²**	(12 ft/2) + (5 ft/2) = **8.5 ft (short)** (12 ft/2) = 6 ft **(long)**	(1440 ft²)[22 psf + (0.2 × 35 psf)] = **41.8 kips**	[(8.5 ft)(10 psf)(2)(30 ft)] + [(6 ft)(10 psf)(2)(48 ft)] = **10.9 kips**	41.8 kips + 10.9 kips = **52.7 kips**
Second floor	30 ft × 48 ft = **1440 ft²**	(12 ft/2) + (12 ft/2) = **12 ft**	(1440 ft²)(40 psf) = **57.6 kips**	(12 ft)(10 psf)(2)(30 ft + 48 ft) = **18.8 kips**	57.6 kips + 18.8 kips = **76.4 kips**
				Total weight = 52.7 kips + 76.4 kips = **129.1 kips**	

Seismic variables:

$$F_a = 1.6 \text{ (ASCE 7 Table 11.4-1)}$$

$$F_v = 2.4 \text{ (ASCE 7 Table 11.4-2)}$$

$$S_{MS} = F_a S_S = (1.6)(0.25) = 0.40 \text{ (ASCE 7 Equation 11.4-1)}$$

$$S_{M1} = F_v S_1 = (2.4)(0.07) = 0.168 \text{ (ASCE 7 Equation 11.4-2)}$$

$$S_{DS} = \tfrac{2}{3} S_{MS} = (\tfrac{2}{3})(0.40) = \mathbf{0.267} \text{ (ASCE 7 Equation 11.4-3)}$$

$$S_{D1} = \tfrac{2}{3} S_{M1} = (\tfrac{2}{3})(0.168) = \mathbf{0.112} \text{ (ASCE 7 Equation 11.4-4)}$$

Base shear:

$$V = \frac{F S_{DS} W}{R} \text{ (ASCE 7 Equation 12.14-11)}$$

$$= \frac{(1.1)(0.267)(129.1)}{6.5} = \mathbf{5.84 \text{ kips}}$$

Force at each level:

$$F_x = \frac{F S_{DS} W_x}{R} \text{ (ASCE 7 Equations 12.14-11 and 12.14-12)}$$

$$F_R = \frac{(1.1)(0.267)(52.7)}{6.5} = \mathbf{2.38 \text{ kips}}$$

$$F_2 = \frac{(1.1)(0.267)(76.4)}{6.5} = \mathbf{3.46 \text{ kips}}$$

These values will be used later for the analysis and design of the diaphragm and shear walls. The seismic loads are summarized in Figure 9.2.

9.4 WIND LOADS

The wind loads on the building will now be calculated. For simplicity, the overhangs are neglected for this example.

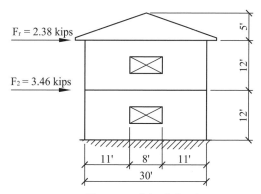

FIGURE 9.2 Summary of seismic loads.

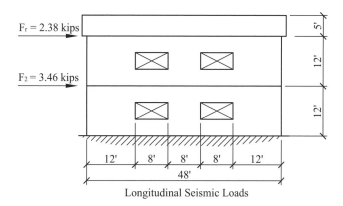

Given:

$$V = 90 \text{ mph}$$

$$I_w = 1.0$$

Mean roof height = 12 ft + 12 ft + (5 ft/2) = 26.5 ft

Exposure = C (assumed)

$$\theta = \tan^{-1} \frac{4}{12} = 18.4°$$

From ASCE 7 Figure 6-2 and with reference to Figure 9.3, the following base wind pressures are obtained:

Wind pressures on MWFRS (Transverse):

    Zone A = 17.8 psf

    Zone B = −4.7 psf (use 0 psf per ASCE 7 Figure 6.2)

    Zone C = 11.9 psf

    Zone D = −2.6 psf (use 0 psf per ASCE 7 Figure 6.2)

    Zone E = −15.4 psf

    Zone F = −10.7 psf

    Zone G = −10.7 psf

    Zone H = −8.1 psf

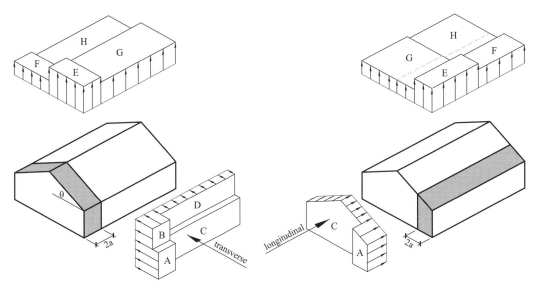

FIGURE 9.3 Wind loads on MWFRS. (Adapted from Ref. 5, Figure 1609.6.2.1.)

Wind pressures on MWFRS (Longitudinal):

$$\text{Zone A} = 12.8 \text{ psf}$$
$$\text{Zone C} = 8.5 \text{ psf}$$
$$\text{Zone E} = -15.4 \text{ psf}$$
$$\text{Zone F} = -8.8 \text{ psf}$$
$$\text{Zone G} = -10.7 \text{ psf}$$
$$\text{Zone H} = -6.8 \text{ psf}$$

Note: Zones A and C are horizontal wind pressures on vertical walls, B and D are horizontal wind pressures on the sloped roof, and E through G are vertical wind pressures on the roof. For the MWFRS, the end zone width = 2a (per ASCE 7 Figure 6-2), where

$$a \leq 0.1 \times \text{least horizontal dimension of building}$$
$$\leq 0.4 \times \text{mean roof height of the building and}$$
$$\geq 0.04 \times \text{least horizontal dimension of building}$$
$$\geq 3 \text{ ft}$$

Therefore, $a \leq (0.1)(30 \text{ ft}) = \mathbf{3 \text{ ft (governs)}}$
$$\leq (0.4)(26.5 \text{ ft}) = 10.6 \text{ ft}$$
$$\geq (0.04)(30 \text{ ft}) = 1.2 \text{ ft}$$
$$\geq 3 \text{ ft}$$
$$a = 3 \text{ ft}$$

Average horizontal wind pressures:
Transverse:

$$q_{\text{avg}} = \frac{(17.8 \text{ psf})(2)(3 \text{ ft}) + [(11.9 \text{ psf})(48 \text{ ft} - (2)(3 \text{ ft})]}{48 \text{ ft}} = \mathbf{12.64 \text{ psf}}$$

Longitudinal:

$$q_{\text{avg}} = \frac{(12.8 \text{ psf})(2)(3 \text{ ft}) + [(8.5 \text{ psf})(30 \text{ ft} - (2)(3 \text{ ft})]}{30 \text{ ft}} = \mathbf{9.36 \text{ psf}}$$

The average horizontal pressures must now be modified for height and exposure. The horizontal forces on the MWFRS are given in Table 9.2.

The vertical pressures (i.e., zones E through G) must now be modified for height and exposure ($h_{\text{avg}} = 26.5$ ft). The average uplift pressures on the roof are calculated by multiplying the adjusted wind pressures in each area by the corresponding roof area and summing up all these forces and dividing by the total roof area (see Figure 9.3). The vertical forces on the MWFRS are given in Table 9.3.

These uplift forces will need a load path from the roof to the foundation. The truss hold-down anchors were designed earlier for higher wind forces using components and cladding. The uplift forces shown in Table 9.3 can conservatively be added to the tension forces in the shear wall chords, which will be designed in a later section. Based on the shear wall layout, there are a total of ten shear walls (each 8 ft long) with hold-down anchors and chords at each end of the wall yielding a total of twenty shearwall chords and hold-down anchors. Therefore, the uplift load due to wind on each anchor and shearwall chord is:

$$U = \frac{14{,}200 \text{ lb}}{20 \text{ chords}} = 710 \text{ lb}$$

All of the adjusted wind pressures on the MWFRS are summarized in Figure 9.4.

9.5 COMPONENTS AND CLADDING WIND PRESSURES

The tabulated components and cladding (C&C) wind pressures on projected vertical and horizontal surface areas of the building are obtained below for each zone from ASCE 7 Figure 6-3,

TABLE 9.2 Transverse and Longitudinal Winds, MWFRS

Level	Elevation	Exposure/Height Coefficient at Mean Roof Height, λ (ASCE 7 Figure 6-2 by interpolation)	Average Horizontal Wind Pressure, q_{avg}	Design Horizontal Wind Pressure, $\lambda I_w q_{\text{avg}}$	Total Unfactored Wind Load on the Building at Each Level
Transverse Wind					
Roof	24 ft	1.37	12.64 psf	(1.37)(1.0)(12.64) = **17.3 psf**	(17.3 psf) (12 ft/2)(48 ft) = **5.0 kips**
Second floor	12 ft	1.37	12.64 psf	(1.37)(1.0)(12.64) = **17.3 psf**	(17.3 psf)(12 ft/2 + 12 ft/2)(48 ft) = **10.0 kips**
					Base shear = 5.0 + 10.0 = **15.0 kips**
Longitudinal Wind					
Roof	26.5 ft	1.37	9.36 psf	(1.37)(1.0)(9.36) = **12.9 psf**	(12.9 psf)(12 ft/2 + 5 ft/2)(30 ft) = **3.3 kips**
Seocnd floor	12 ft	1.37	9.36 psf	(1.37)(1.0)(9.36) = **12.9 psf**	(12.9 psf)(12 ft/2 + 12 ft/2)(30 ft) = **4.7 kips**
					Base shear = 3.3 + 4.7 = **8.0 kips**

Note: mean roof height = 26.5 ft

288 | STRUCTURAL WOOD DESIGN

TABLE 9.3 Average Uplift Pressures, MWFRS

Direction	Exposure / Height Coefficient at Mean Roof Height, λ (ASCE 7 Figure 6-2)	Total Vertical Wind Force	Average Uplift Pressure, q_u
Transverse	1.37	(15.4 psf + 10.7 psf)(2)(3 ft)(15 ft) + (10.7 psf + 8.1 psf)[48 ft − (2)(3 ft)](15 ft) = **14.2 kips**	$\dfrac{14{,}200}{(30\text{ ft})(48\text{ ft})}$ = **9.9 psf**
Longitudinal	1.37	(15.4 psf + 8.8 psf)(2)(3 ft)(24 ft) + (10.7 psf + 6.8 psf)[30 ft − (2)(3 ft)](24 ft) = **13.6 kips**	$\dfrac{13{,}600}{(30\text{ ft})(48\text{ ft})}$ = **9.5 psf**

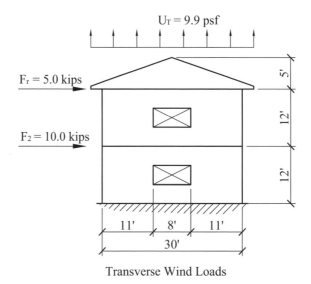

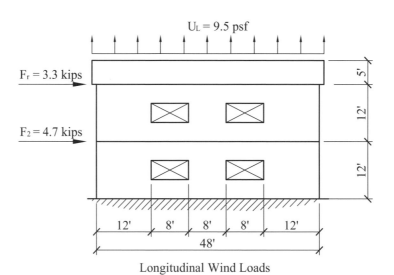

FIGURE 9.4 Adjusted wind loads on MWFRS.

based on a 90-mph wind speed, a roof slope of 18.43°, and the effective wind area A_e (see Tables 9.4 to 9.6).

Typical exterior wall stud:

>Floor-to-floor height = 12 ft
>
>Spacing of stud = 2 ft
>
>Effective width per ASCE 7 Section 6-2 = $\frac{1}{3}$span = $(\frac{1}{3})$(12 ft) = 4 ft
>
>Effective wind area A_e = (4 ft)(12 ft) = 48 ft^2

Typical roof truss:

Main span = 30 ft

Spacing of trusses = 2 ft

Effective width of truss per ASCE 7 Section 6-2 = $\frac{1}{3}$span = $(\frac{1}{3})$(30 ft) = 10 ft

Effective wind area A_e = (10 ft)(30 ft) = 300 ft^2

Roof overhang length = 2 ft

Effective width of roof overhang per ASCE 7 Section 6-2 = $\frac{1}{3}$span = $(\frac{1}{3})$(2 ft) = 0.67 ft

Effective wind area A_e = (0.67 ft)(2 ft) = 1.34 ft^2

The components and cladding (C&C) wind pressures below are obtained from ASCE 7 Table 6-3, and will be used later for the design of the exterior wall stud and the design of the roof truss hold-down strap anchors. The roof has a pitch of 4-in-12 which yields a roof slope of 18.4°.

For C&C, the end zone width = a per ASCE 7 Figure 6-3, where

>$a \leq 0.1 \times$ least horizontal dimension of building = (0.1)(30 ft) = 3 ft
>
>$\leq 0.4 \times$ mean roof height of the building = (0.4)(26.5 ft) = 10.6 ft
>
>≥ 3 ft

Therefore, the end zone width a = 3 ft.

>Exposure classification = C (assumed)
>
>Mean roof height = 26.5 ft

Linearly interpolating from ASCE 7 Figure 6-3, the height/exposure adjustment coefficient λ is calculated as shown in Table 9.5.

The mean roof height has been calculated previously to be 26.5 ft. The corresponding height/exposure adjustment factor, λ is calculated by linear interpolation of ASCE 7 Table 6-3 to be 1.37.

>Importance factor I_w = 1.0

The *design* C&C wind pressures are obtained by multiplying the *tabulated* C&C wind pressures above by the height/exposure adjustment factor and the importance factor.

Since the stud spacing is 2 ft and the edge zone width is 3 ft, the *ground-floor* and second floor studs located 2 ft from any corner of the building will be subjected to the maximum *horizontal* design wind suction of 22.6 psf and a pressure of 17.8 psf.

From the applicable gable roof plan (7° < θ ≤ 45°) in ASCE 7 Figure 6-3, the total wind uplift pressures on the roof are as shown in Figure 9.5. These values are used to calculate the

TABLE 9.4 Horizontal and Vertical Wind Pressures, Components and Cladding (psf)

Zone	Positive Pressure	Negative Pressure or Suction
Horizontal Pressures on Wall[a]		
Wall interior zone 4	13.0	−14.3
Wall end zone 5	13.0	−16.5
Vertical Pressures on Roof[b]		
Roof interior zone 1	5.9	−12.1
Roof end zone 2	5.9	−17.0
Roof corner zone 3	5.9	−26.9
Vertical Pressures on a 2-ft Roof Overhang[c]		
Roof end zone 2		−27.2
Roof corner zone 3		−45.7

[a] Horizontal wind pressures for longitudinal as well as transverse wind. The effective wind area A_e calculated ≈ 50 ft².
[b] Positive pressure indicates downward wind loads and negative pressure indicates upward wind load that causes uplift. The effective wind area A_e calculated is 300 ft². (Use the tabulated pressures for A_e = 100 ft².)
[c] The effective wind area calculated for the roof truss overhang is less than 10 ft². From ASCE 7 Figure 6-3, the roof overhang is in zone 2 for the interior zone roof trusses and in zone 3 for the end zone roof trusses (see Figure 9.5).

TABLE 9.5 Design Horizontal and Vertical Wind Pressures, $\lambda I_w P_{net30}$, Components and Cladding (psf)

Zone	Positive Pressure	Negative Pressure or Suction
Horizontal Pressures on Ground Floor and Second Floor Walls[a]		
Wall interior zone 4	(1.37)(1.0)(13.0) = 17.8	(1.37)(1.0)(−14.3) = **−19.6**
Wall end zone 5	(1.37)(1.0)(13.0) = 17.8	(1.37)(1.0)(−16.5) = **−22.6**

Vertical Pressures on Roof[b]

	Positive Pressure	Negative Pressure	Net Uplift Pressure (psf)
Roof interior zone 1	(1.37)(1.0)(5.9) = 8.1 < 10 Use 10 psf minimum	(1.37)(1.0)(−12.1) = −16.6	0.6(18.5 psf) − 16.6 psf = −5.5 psf
Roof end zone 2	(1.37)(1.0)(5.9) = 8.1 < 10 Use 10 psf minimum	(1.37)(1.0)(−17.0) = **−23.3**	0.6(18.5 psf) − 23.3 psf = −12.2 psf
Roof corner zone 3	(1.37)(1.0)(5.9) = 8.1 < 10 Use 10 psf minimum	(1.37)(1.0)(−26.9) = −36.9	0.6(18.5 psf) − 36.9 psf = −25.8 psf
Vertical Pressures on 2-ft Roof Overhang[b]			
Roof end zone 2^{ov}		(1.37)(1.0)(−27.2) = −37.3	0.6(18.5 psf) − 37.3 psf = −26.2 psf
Roof corner zone 3^{ov}		(1.37)(1.0)(−45.7) = **−62.6**	0.6(18.5 psf) − 62.6 psf = −51.5 psf

[a] Horizontal wind pressures for longitudinal as well as transverse wind.
[b] Positive pressure indicates downward wind loads, and negative pressure indicates upward wind load that causes uplift. Net uplift wind pressures on the roof are calculated above using the dead load of 18.5 psf which is the total roof dead load excluding the weight of the reroofing as calculated in Section 9.2

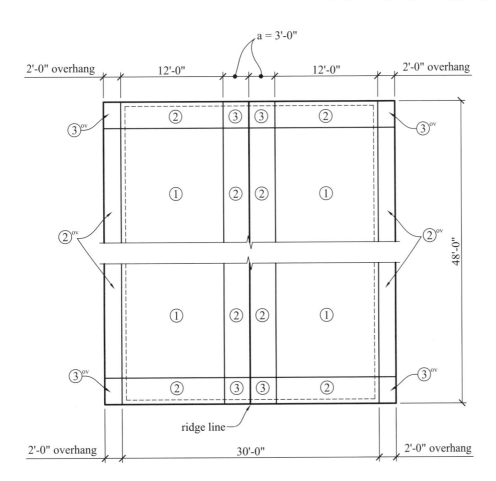

FIGURE 9.5 Uplift wind pressures on roof, components and cladding.

Zone 3^{ov} = −62.6 psf (use overhang wind pressure)
Zone 2^{ov} = −37.3 psf (use overhang wind pressure)
Zone 1 = −16.6 psf
Zone 2 = −23.3 psf
Zone 3 = −36.9 psf

net uplift wind pressures and the uplift load on a typical interior roof truss (see Figure 9.10). The net uplift load at the truss support is usually resisted with hurricane hold-down clips.

9.6 ROOF FRAMING DESIGN

The roof framing for this building consists of trusses spaced at 2 ft o.c. The truss configuration is shown in Figure 9.6. We will assume that the ends of the truss members are connected by a

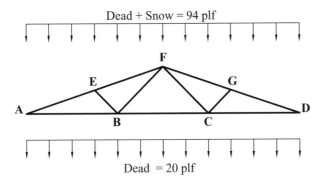

FIGURE 9.6 Roof truss profile and loading.

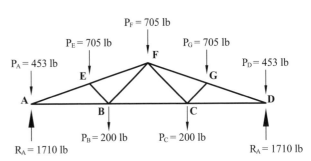

FIGURE 9.7 Free-body diagram of roof truss.

toothed plate or nailed plywood gusset plate; therefore, there is no reduction in the gross area. The span of the roof truss center to center of the exterior wall supports is 30 ft less the stud wall thickness. If we assume 2 × 6 studs, the actual center-to-center span of the roof truss will be 29.54 ft (i.e., 30 ft − 5.5 in./12). However, for simplicity, we will assume a span of 30 ft in the analysis and design of the roof truss, and this will not significantly affect or alter the truss design because of the very small difference (less than 2%) between the actual span and the 30-ft span used in our design.

Analysis of a Roof Truss

The total roof load of 57 psf was calculated in section 9.2 without consideration of the downward wind load acting on the roof. We will now determine if this downward wind load in combination with other loads is critical for this building.

The calculated roof dead and snow loads from Section 9.2 and the calculated downward wind load from Section 9.5 are as follows:

$$D = 22 \text{ psf}$$

$$S = 35 \text{ psf}$$

$$W = 10 \text{ psf (downwards)}$$

Using the normalized load method presented in Chapter 3 and using the applicable load combinations from Section 2.1 yields

$$D + S = 22 + 35 = 57 \text{ psf}; (D + S)/C_D = 57/1.15 = 49.6 \text{ psf (governs)}$$

$$D + W = 22 + 10 = 32 \text{ psf}, (D + W)/C_D = 32/1.6 = 20 \text{ psf}$$

$$D + 0.75W + 0.75S = 22 + 0.75(10 + 35) = 56 \text{ psf},$$

$$(D + 0.75W + 0.75S)/C_D = 56/1.6 = 35 \text{ psf}$$

Therefore, the $D + S$ combination is more critical than the wind load combinations as originally assumed. We will now proceed to design the roof truss members for the total load of 57 psf with a C_D of 1.15.

The concentrated gravity loads at the top and bottom chord joints of the roof truss (Figure 9.7) are calculated as follows, using the uniformly distributed total loads of 94 lb/ft and 20 lb/ft obtained in Section 9.2 for the truss top and bottom chords, respectively.

Top chord—roof dead load + snow load:

$$P_A \text{ (top)} = (7.5 \text{ ft}/2)(94 \text{ lb/ft}) = 353 \text{ lb}$$

$$P_E = (7.5 \text{ ft}/2 + 7.5 \text{ ft}/2)(94 \text{ lb/ft}) = 705 \text{ lb}$$

$$P_F = (7.5 \text{ ft}/2 + 7.5 \text{ ft}/2)(94 \text{ lb/ft}) = 705 \text{ lb}$$

$$P_G = (7.5 \text{ ft}/2 + 7.5 \text{ ft}/2)(94 \text{ lb/ft}) = 705 \text{ lb}$$

$$P_D \text{ (top)} = P_A \text{ (top)} = 353 \text{ lb}$$

Bottom chord—dead load only, due to ceiling (there is no live load on the bottom chord of this truss):

$$P_A \text{ (bottom)} = (10 \text{ ft}/2)(20 \text{ lb/ft}) = 100 \text{ lb}$$

$$P_C = (10 \text{ ft}/2 + 10 \text{ ft}/2)(20 \text{ lb/ft}) = 200 \text{ lb}$$

$$P_C = (10 \text{ ft}/2 + 10 \text{ ft}/2)(20 \text{ lb/ft}) = 200 \text{ lb}$$

$$P_D \text{ (bottom)} = (10 \text{ ft}/2)(20 \text{ lb/ft}) = 100 \text{ lb}$$

The truss is then analyzed using the method of joints *or* computer structural analysis software to obtain the member forces as follows (C = compression and T = tension).

Top chord members:

$$AE = GD = 3980 \text{ lb (C)}$$
$$EF = FG = 3420 \text{ lb (C)}$$

Bottom chord members:

$$AB = CD = 3770 \text{ lb (T)}$$
$$BC = 2520 \text{ lb (T)}$$

Web members:

$$EB = GC = 750 \text{ lb (C)}$$
$$FC = BF = 1030 \text{ lb (T)}$$

In addition to the axial forces on the top chord and bottom chord members, there are bending loads of 94 and 20 lb/ft on the top and bottom chords, respectively. These loads will be considered in the design of the top and bottom chord members of the roof truss.

Design of Truss Web Tension Members

1. The maximum tension force in the web tension members from the structural analysis above is 1030 lb.
2. Assume member size: Try a 2 × 4, which is dimension lumber. Use NDS-S Table 4A. From NDS-S Table 1B, the gross area A_g for a 2 × 4 = 5.25 in². The wood species (spruce-pine-fir) to be used is specified in the design brief; we assume a No. 3 stress grade.
3. From NDS-S Table 4A (SPF No. 3), we obtain the tabulated design tension stress and the adjustment factors as follows: Design tension stress F_t = 250 psi. Adjustment or C factors:

C_D = 1.15 (the C_D value for the shortest-duration load in the load combination is used, i.e., snow load; see Section 3.1)

C_t = 1.0 (normal temperature conditions)

C_M = 1.0 (dry service, since the truss members are protected from weather)

$C_F(F_t)$ = 1.5

Using the NDS applicability table (Tables 3.1), the allowable tension stress is given as

$$F'_t = F_t C_D C_M C_t C_F C_i$$
$$= (250)(1.15)(1.0)(1.0)(1.5)(1.0) = \mathbf{431.3 \text{ psi}}$$

Since toothed plate or nailed plywood gusset plate connections are assumed, the net area will be equal to the gross area since there are no bolt holes that will lead to a reduction in the gross area.

$$A_n = A_g = 5.25 \text{ in}^2$$

4. Applied tension stress

$$f_t = \frac{T}{A_n} = \frac{1030 \text{ lb}}{5.25 \text{ in}^2} = 197 \text{ psi} < F'_t = \mathbf{431.3 \text{ psi}} \quad \text{OK}$$

A 2 × 4 SPF No. 3 is adequate for the truss web tension member.

Design of Truss Web Compression Members

Since a 2 × 4 member was selected for the truss web tension member, we assume a 2 × 4 trial size for the web compression member. As determined for the web tension member, the 2 × 4 is dimension lumber, for which NDS-S Table 4A is applicable. From the table the stress adjustment or C factors are

$C_M = 1.0$ (normal moisture conditions)

$C_t = 1.0$ (normal temperature conditions)

$C_D = 1.15$ (the C_D value for the shortest-duration load in the load combination is used, i.e., snow load; see Section 3.1)

$C_F(F_c) = 1.15$

$C_i = 1.0$ (no incisions are specified in the wood for pressure treatment)

From NDS-S Table 4C, the tabulated design stresses are obtained as follows:

$F_c = 650$ psi (compression stress parallel to the grain)

$E = 1.2 \times 10^6$ psi (reference modulus of elasticity)

$E_{min} = 0.44 \times 10^6$ psi (buckling modulus of elasticity)

The member dimensions are $d_x = 3.5$ in. and $d_y = 1.5$ in.
The member length $= \sqrt{(2.5 \text{ ft})^2 + (2.5 \text{ ft})^2} = 3.54$ ft.
The cross-sectional area $A_g = 5.25$ in^2.

The effective length factor K_e is assumed to be 1.0 (see Section 5.4); therefore, the slenderness ratios about the x–x and y–y axes are given as

$$\frac{l_e}{d_x} = \frac{(1.0)(3.54 \text{ ft} \times 12)}{3.5 \text{ in.}} = 12.1 < 50 \quad \textbf{OK}$$

$$\frac{l_e}{d_y} = \frac{(1.0)(3.54 \text{ ft} \times 12)}{1.5 \text{ in.}} = \textbf{28.3 (governs)} < 50 \quad \textbf{OK}$$

Using the applicability table (Table 3.1), the allowable modulus of elasticity for buckling calculations is given as

$$E'_{min} = E_{min} C_M C_t C_i$$

$$= (0.44 \times 10^6)(1.0)(1.0)(1.0) = 0.44 \times 10^6 \text{ psi}$$

$c = 0.8$ (sawn lumber)

The Euler critical buckling stress is calculated as

$$F_{cE} = \frac{0.822 E'_{min}}{(l_e/d)^2} = \frac{(0.822)(0.44 \times 10^6)}{(28.3)^2} = 452 \text{ psi}$$

$$F_c^* = F_c C_D C_M C_t C_F C_i$$

$$= (650)(1.15)(1.0)(1.0)(1.15)(1.0) = 860 \text{ psi}$$

$$\frac{F_{cE}}{F_c^*} = \frac{452 \text{ psi}}{860 \text{ psi}} = 0.526$$

From equation (5.21), the column stability factor is calculated as

$$C_P = \frac{1 + 0.526}{(2)(0.8)} - \sqrt{\left[\frac{1 + 0.526}{(2)(0.8)}\right]^2 - \frac{0.526}{0.8}} = 0.452$$

The allowable compression stress

$$F'_c = F^*_c C_P = (860)(0.452) = \mathbf{389 \text{ psi}}$$

The applied compression stress

$$f_t = \frac{F}{A_g} = \frac{750 \text{ lb}}{5.25 \text{ in}^2} = 143 \text{ psi} < F'_c = \mathbf{389 \text{ psi}}$$

A 2 × 4 SPF No. 3 is adequate for the truss web compression member.

Design of Truss Bottom Chord Members

The bottom chord members are subjected to combined axial tension plus bending caused by the uniformly distributed load of 20 lb/ft from the insulation, ceiling, and mechanical and electrical equipment loads. The most critical bottom chord member is member AB or CD, and the forces in this member include an axial tension force of 3770 lb (caused by dead load plus snow load) combined with a uniformly distributed dead load of 20 lb/ft, as shown in Figure 9.8.

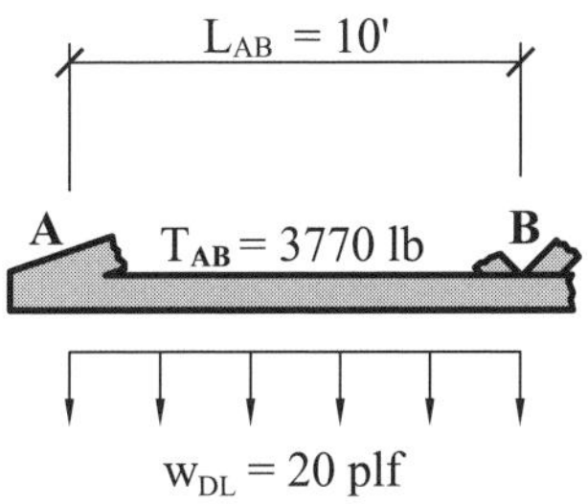

FIGURE 9.8 Free-body diagram of member AB.

Selection of a Trial Member Size

Assume 2 × 6 SPF No. 2 sawn lumber. The gross cross-sectional area and the section modulus can be obtained from NDS-S Table 1B as follows:

$$A_g = 8.25 \text{ in}^2$$

$$S_{xx} = 7.563 \text{ in}^3$$

The length of this member = 10 ft. Since the trial member is dimension lumber, NDS-S Table 4A is applicable. From the table we obtain the tabulated design stresses and stress adjustment factors.

F_b = 875 psi (tabulated bending stress)

F_t = 450 psi (tabulated tension stress)

$C_F(F_b)$ = 1.3 (size adjustment factor for bending stress)

$C_F(F_t)$ = 1.3 (size adjustment factor for tension stress)

C_D = 1.15 (for design check with snow load)

C_D = 0.9 (for design check with dead load only)

C_L = 1.0 (assuming lateral buckling is prevented by the ceiling and bridging)

C_r = 1.15 (all three repetitive member requirements are met; see Section 3.1)

C_i = 1.0 (there are no incisions specified in the wood for pressure treatment)

Design Check 1:

The applied tension force T = 3770 lb (caused by dead load + snow load).

The net area $A_n = A_g$ (toothed plated or nailed plywood plate connection assumed) = 8.25 in².

From equation (5.4), the applied tension stress at the supports,

$$f_t = \frac{T}{A_n} = \frac{3770 \text{ lb}}{8.25 \text{ in}^2} = 457 \text{ psi}$$

Using the NDS applicability table (Table 3.1), the allowable tension stress is given as

$$F'_t = F_t C_D C_M C_t C_F C_i$$

$$= (450)(1.15)(1.0)(1.0)(1.3)(1.0) = \mathbf{673 \text{ psi}} > f_t \quad \mathbf{OK}$$

Design Check 2:
From equation (5.5), the applied tension stress at the midspan

$$f_t = \frac{T}{A_g} = \frac{3770 \text{ lb}}{8.25 \text{ in}^2} = 457 \text{ psi} < F'_t = 673 \text{ psi} \quad \mathbf{OK}$$

Design Check 3:
The maximum moment which for this member occurs at its midlength is given as

$$M_{max} = \frac{(20 \text{ lb/ft})(10 \text{ ft})^2}{8} = 250 \text{ ft-lb} = 3000 \text{ in.-lb}$$

This moment is caused by the ceiling *dead* load only (since no ceiling live load is specified), therefore, the controlling load duration factor C_D for this case is 0.9. Using the NDS applicability table (Table 3.1), the allowable bending stress is given as

$$F'_b = F_b C_D C_M C_t C_L C_F C_i C_r C_{fu}$$

$$= (875)(0.9)(1.0)(1.0)(1.0)(1.3)(1.0)(1.15)(1.0)$$

$$= \mathbf{1177 \text{ psi}}$$

From equation (5.6) the applied bending stress (i.e., tension stress due to bending) is,

$$f_{bt} = \frac{M_{max}}{S_x} = \frac{3000 \text{ in.-lb}}{7.563 \text{ in}^3} = 397 \text{ psi} < F'_b$$

Design Check 4:
For this case, the applicable load is dead load plus snow load; therefore, the controlling load duration factor C_D is 1.15 (see Chapter 3). Using the NDS applicability table (Table 3.1), the allowable axial tension stress and the allowable bending stress are calculated as follows:

$$F'_t = F_t C_D C_M C_t C_F C_i$$

$$= (450)(1.15)(1.0)(1.0)(1.3)(1.0) = \mathbf{673 \text{ psi}}$$

$$F^*_b = F_b C_D C_M C_t C_F C_i C_r C_{fu}$$

$$= (875)(1.15)(1.0)(1.0)(1.3)(1.0)(1.15)(1.0) = 1504 \text{ psi}$$

The applied axial tension stress f_t at the midspan is 457 psi, as calculated for design check 2, while the applied tension stress due to bending f_{bt} is 397 psi, as calculated in design check 3. From equation (5.7) the combined axial tension plus bending interaction equation for the stresses in the tension fiber of the member is given as

$$\frac{f_t}{F'_t} + \frac{f_{bt}}{F^*_b} \leq 1.0$$

Substitution into the interaction equation yields

$$\frac{457 \text{ psi}}{673 \text{ psi}} + \frac{397 \text{ psi}}{1504 \text{ psi}} = \mathbf{0.94 < 1.0} \quad \mathbf{OK}$$

Design Check 5:
The applied compression stress due to bending is

$$f_{bc} = \frac{M_{max}}{S_x} = \frac{3000 \text{ in.-lb}}{7.563 \text{ in}^3} = 397 \text{ psi} \ (= f_{bt} \text{ since this is a rectangular cross section})$$

It should be noted that f_{bc} and f_{bt} are equal for this problem because the wood member is rectangular in cross section. The load duration factor for this case $C_D = 1.15$ (combined dead + snow loads). Using the NDS applicability table (Table 3.1), the allowable compression stress due to bending is

$$F'_b = F_b C_D C_M C_t C_L C_F C_i C_r$$
$$= (875)(1.15)(1.0)(1.0)(1.0)(1.3)(1.0)(1.15) = 1504 \text{ psi}$$

From equation (5.10) the interaction equation for this design check is given as

$$\frac{f_{bc} - f_t}{F'_b} \leq 1.0$$

Substituting in the interaction equation yields

$$\frac{397 \text{ psi} - 457 \text{ psi}}{1504 \text{ psi}} = -0.04 < 1.0 \quad \text{OK}$$

From all of the steps above, we find that all the five design checks are satisfied; therefore, use a 2 × 6 SPF No. 2 for the bottom chord.

Design of Truss Top Chord Members

The top chord members are subjected to combined axial compression plus bending caused by the uniformly distributed dead load and snow load of 94 lb/ft. The most critical top chord member is member *AE* or *GD*, and the forces in this member include an axial compression force of 3980 lb combined with a uniformly load of 94 lb/ft, as shown in Figure 9.9.

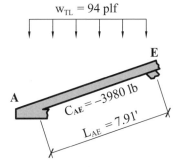

FIGURE 9.9 Load on truss top chord.

Selection of a Trial Member Size
Assume 2 × 6 SPF sawn lumber. The design aids in Appendix B are utilized to design this member. The length of this member $= \sqrt{(7.5 \text{ ft})^2 + (2.5 \text{ ft})^2} = 7.91$ ft. It should be noted that for each design check, the controlling load duration factor value will correspond to the C_D value of the shortest-duration load in the load combination for that design case. Thus, the C_D value for the various design cases may vary. The reader will find that design check 4 will govern, and therefore the design aids in Appendix B can conservatively be used to select an appropriate member.

Design Check 1: Compression on Net Area
This condition occurs at the *ends* (i.e., *supports*) of the member. For this load case, the axial load *P* is caused by dead load plus snow load (*D* + *S*); therefore, the load duration factor C_D from Chapter 3 is 1.15. The worst-case axial load for this case is $P = 3980$ lb.

Design Check 2: Compression on Gross Area
This condition occurs at the *midspan* of the member. The axial load *P* for this load case is also caused by dead load plus snow load (*D* + *S*); therefore, the load duration factor C_D from Chapter 3 is 1.15. The worst-case axial load for this case is $P = 3980$ lb.

Design Check 3: Bending Only
This condition occurs at the point of maximum moment (i.e., at the *midspan* of the member), and the moment is caused by the uniformly distributed gravity load on the top chord (member

AE), and since this load is dead load plus snow load, the controlling load duration factor (see Chapter 3) is 1.15.

The horizontal span of the sloped member, $AE = 7.5$ ft.

The maximum moment in member AD

$$M_{max} = \frac{w_{D+S}L^2}{8}$$

$$= \frac{(94 \text{ lb/ft})(7.5 \text{ ft})^2}{8} = 661 \text{ ft-lb}$$

Design Check 4: Bending plus Axial Compression Force

This condition occurs at the *midspan* of the member. For this load case, the loads causing the combined stresses are dead load plus snow load ($D + S$); therefore, the controlling load duration factor C_D is 1.15. It should be noted that because the load duration factor C_D for this combined load case is the same as for load cases 2 and 3, the parameters to be used in the combined stress interaction equation can be obtained from design checks 2 and 3. The reader is cautioned to be aware that the C_D value for all four design checks are not necessarily always equal and has to be determined for each case.

The loads for this case are

$$P = 3980 \text{ lb}$$

$$M = 661 \text{ ft-lb}$$

By inspection, the loads from this case will control. From Figure B.43, the axial load capacity at a moment of 661 ft-lb is approximately 4000 lb. Note that the load duration factor used in Figure B.43 is $C_D = 1.0$, so the use of this design aid is somewhat conservative, but appropriate. Therefore, use a 2 × 6 SPF Select Structural for the top chord.

The truss member sizes and stress grades are summarized in Table 9.7.

Gable and hip roofs must not only be designed for the balanced snow loads discussed in Chapter 2, and used to design the roof truss in this section, but the unbalanced snow loads prescribed in the ASCE 7 load standard must also be considered in design, because often times, these unbalanced loads might lead to higher stresses in the truss members compared to the stresses due to the balanced snow load. In our building design case study, the applied stresses due to unbalanced snow load are smaller that those due to the balanced snow load except in the truss web members where there was approximately a 20% increase in stress due to unbalanced snow loads. However, since the applied stress in the actual web members selected were only about 50% of the allowable stress, the roof truss members in this building design case study would still be adequate to support the unbalanced snow loads. The reader should refer to Sections 7.6 though 7.9 of ASCE 7 for a full discussion of unbalanced snow load, snow drift loads and sliding snow loads.

TABLE 9.7 Summary of Truss Member Sizes and Stress Grades

Truss Member	Member Size and Stress Grade
Top chord	2 × 6 SPF Select Structural
Bottom chord	2 × 6 SPF No. 2[a]
Web members	2 × 4 SPF No. 3

[a]For simplicity and to avoid construction errors, it may be prudent to use the same size and stress grade for the top and bottom chords. In that case, 2 × 6 SPF Select Structural would be used for the bottom chord as well as the top chord.

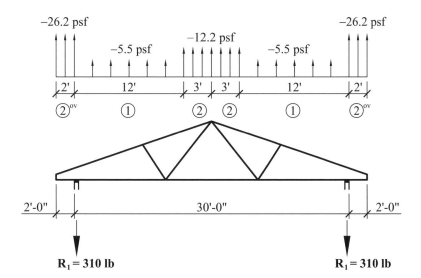

FIGURE 9.10a Net wind uplift pressures on a typical interior roof truss.

$R_1 = [(26.2 \text{ psf})(2') + (5.5 \text{ psf})(12') + (12.2 \text{ psf})(3')] (2' \text{ trib. width}) = 310 \text{ lb}$

(a)

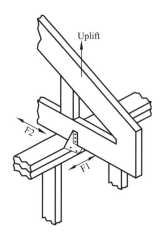

Model #	Fasteners		DF-L or S.P. Allowable Loads (lb)			S.P.F. or H.F. Allowable Loads (lb)		
	To truss	To plates	Uplift (160%)	Lateral (160%)		Uplift (160%)	Lateral (160%)	
				F1	F2		F1	F2
HT1	(6)-8d	(4)-8d	585	450	160	390	405	120
xx	xx	xx	xx	xx	xx	xx	xx	xx

(b)

FIGURE 9.10b Roof truss uplift connector.

Net Uplift Load on a Roof Truss

The net uplift load on a typical interior zone roof truss is calculated using the design uplift wind pressures calculated in Table 9.5. This uplift load will be used to determine the size of the roof truss hold-down straps or hurricane clips. Using Table 9.5, Figures 9.5 and 9.10a, the total net uplift load at the typical interior roof truss supports is 310 lb. The roof truss hold-down strap will be designed to resist this net uplift force of 310 lb. A hurricane tie down anchor HT-1 is adequate to resist this force and has an allowable uplift resistance of 390 lb with a load duration factor of 1.6 as indicated in the generic connector selection table shown in Table 9.10b.

9.7 SECOND-FLOOR FRAMING DESIGN

In this section the typical sawn-lumber floor joist and typical glulam girder are designed (Figure 9.11). It is assumed that the floor joist will be simply supported and supported off the face of

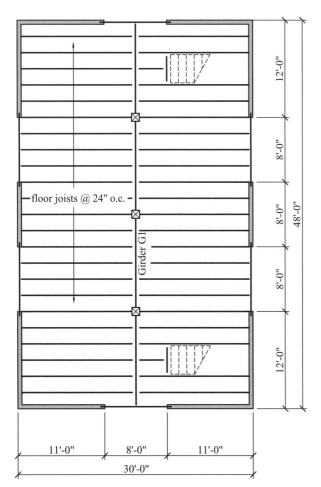

FIGURE 9.11 Second-floor framing plan.

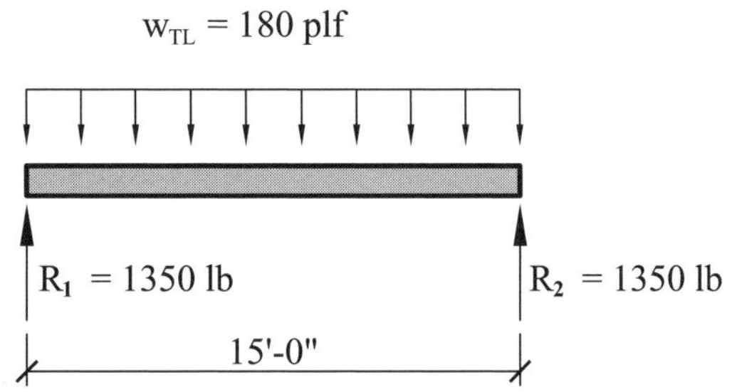

FIGURE 9.12 Free-body diagram of a typical floor joist.

the glulam girder with joist hangers. To illustrate the use of the design aids presented in Appendix B, the floor joist will be designed using the design aids.

Design of a Typical Floor Joist

Joist span = 15 ft (the actual center-to-center span of the joist will be slightly less than this). From Section 9.2 the dead load was calculated as 40 psf and the floor live load is 50 psf. The total floor load (in psf) using the IBC load combinations (See Section 2.1) is

$$w_{TL} = D + L = 40 + 50 = 90 \text{ psf}$$

Tributary area of typical floor joist = (2 ft)(15 ft) = 30 ft^2 < 150 ft^2

From Section 2.4,

$$A_T = \text{tributary area} = 30 \text{ ft}^2 \quad \text{and} \quad K_{LL} = 2 \text{ (interior beam)}$$

$$K_{LL}A_T = (2)(30 \text{ ft}^2) = 60 \text{ ft}^2 < 400 \text{ ft}^2$$

Therefore, live load reduction is *not* permitted.

The tributary width (TW) of a typical floor joist = 2 ft. The total uniform load on the joist that will be used to design the joist for bending, shear, and bearing is

$$w_{TL} = (D + L)(TW) = (40 \text{ psf} + 50 \text{ psf})(2\text{ft}) = 180 \text{ lb/ft}$$

Using the free-body diagram of the typical floor joist (Figure 9.12), the load effects are calculated as follows:

$$\text{maximum shear } V_{max} = \frac{w_{TL}L}{2} = \frac{(180 \text{ lb/ft})(15 \text{ ft})}{2} = 1350 \text{ lb}$$

$$\text{maximum reaction } R_{max} = 1350 \text{ lb}$$

$$\text{maximum moment } M_{max} = \frac{w_{TL}L^2}{8} = \frac{(180 \text{ lb/ft})(15 \text{ ft})^2}{8} = 5063 \text{ ft-lb} = 60{,}756 \text{ in.-lb}$$

Using Figure B.8, the allowable uniform load for a 2 × 14 SPF No. 1/No. 2 is approximately 120 plf. Therefore, use two 2 × 14's at 24 in. o.c. since the total load carrying capacity is (2)(120 plf) = 240 plf, which is greater than the applied load of 180 plf.

The required adjustment factors for the floor joists are:

load duration $C_D = 1.0$ (dead load plus live load)

moisture factor $C_M = 1.0$ (dry service conditions assumed)

temperature factor $C_t = 1.0$ (normal temperature conditions apply)

bearing factor $C_b = 1.0$

Check of Bearing Stress (Compression Stress Perpendicular to the Grain)

Maximum reaction at the support $R_1 = 1350$ lb.
Thickness of two 2 × 14 sawn-lumber joists $b = (2) \times (1.5 \text{ in}) = 3$ in.
The allowable bearing stress or compression stress parallel to grain is

$$F'_{c\perp} = F_{c\perp} C_M C_t C_b = (425)(1)(1)(1) = 425 \text{ psi}$$

From equation (4.9) the minimum required bearing length l_b is

$$l_{b,\text{req'd}} \geq \frac{R_1}{bF_{c\perp}} = \frac{1350 \text{ lb}}{(3.0 \text{ in.})(425 \text{ psi})} = 1.1 \text{ in.}$$

The floor joists will be connected to the floor girder using face-mounted joist hangers with the top of the joists at the same level as the top of the girders (see Figure 4.6b). The reader should select from one of the proprietary wood connector catalogs a joist hanger that meets the following requirements:

- The bearing length provided by the joist hanger should be at least $1\frac{1}{8}$ in.
- The joist hanger should have enough width to accommodate the two 2 × 14 floor joists.
- The joist hanger must have a capacity of at least 1350 lb using a load duration factor of 100% (i.e., $C_D = 1$ for dead load + floor live load).

Use two 2 × 14's at 24 in. o.c. SPF No. 1/No. 2 for the floor joists.

Design of a Glulam Floor Girder

This glulam girder could be designed as continuous over several supports, which will result in a smaller member compared to simply supported girders. However, hauling a 48-ft-long girder to the building site and maneuvering a beam that long on site may present problems for the contractor. Alternatively, the glulam girder could be delivered to the site in two 24-ft-long pieces, thus indicating two two-span continuous girders. However, to give the contractor wide flexibility in choosing the best way to install this girder, we design the girders as four simply supported members with 12-ft spans. This assumption will result in a larger member than if a single 48-ft-long girder continuous over several supports was used. It is assumed that the floor joists are supported on the face of the girder with face-mounted joist hangers; thus, the top of the glulam girder will be at the same elevation as the top of the floor joists. Assume that the girder self-weight = 15 lb/ft (this will be checked later). As calculated in Section 9.2, the dead and live loads are as follows:

$$\text{dead load } D = 40 \text{ psf}$$

$$\text{floor live load } L = 50 \text{ psf}$$

$$\text{tributary width (TW) of girder} = \frac{15 \text{ ft}}{2} + \frac{15 \text{ ft}}{2} = 15 \text{ ft}$$

$$\text{tributary area (TA) of girder G1} = (15 \text{ ft})(12 \text{ ft}) = 180 \text{ ft}^2$$

From Section 2.4,

$$A_T = \text{tributary area} = 180 \text{ ft}^2 \quad \text{and} \quad K_{LL} = 2 \text{ (interior girder)}$$

$$K_{LL}A_T = (2)(180 \text{ ft}^2) = 360 \text{ ft}^2 < 400 \text{ ft}^2$$

Therefore, live-load reduction is *not* permitted.

The total uniform load on the girder that will be used to design for bending, shear, and bearing is obtained below using the dead load, the live load, and the assumed self-weight of the girder, which will be checked later.

$$w_{TL} = (D + L)(\text{tributary width}) + \text{girder self-weight}$$

$$= (40 \text{ psf} + 50 \text{ psf})(15 \text{ ft}) + 15 \text{ lb/ft} = \textbf{1365 lb/ft}$$

Using the free-body diagram of the girder (Figure 9.13), the load effects are calculated as follows:

$$\text{maximum shear } V_{max} = \frac{w_{TL}L}{2} = \frac{(1365 \text{ lb/ft})(12 \text{ ft})}{2} = 8190 \text{ lb}$$

$$\text{maximum reaction } R_{max} = 8190 \text{ lb}$$

$$\text{maximum moment } M_{max} = \frac{w_{TL}L^2}{8} = \frac{(1365 \text{ lb/ft})(12 \text{ ft})^2}{8} = 24{,}570 \text{ ft-lb} = 294{,}840 \text{ in.-lb}$$

The following loads will be used for calculating the joist deflections:

$$\text{uniform dead load } w_{DL} = (40 \text{ psf})(15 \text{ ft}) + 15 \text{ lb/ft} = 615 \text{ lb/ft} = 51.3 \text{ lb/in.}$$

$$\text{uniform live load } w_{LL} = (50 \text{ psf})(15 \text{ ft}) = 750 \text{ lb/ft} = 62.5 \text{ lb/in.}$$

Note that if a four-span continuous girder had been assumed for this design, the maximum moment would have been 21,052 ft-lb [i.e., (0.1071)(1365 lb/ft)(12 ft)(12 ft)], a 17% reduction. The maximum shear would have been 9943 lb [i.e., (0.607)(1365 lb/ft)(12 ft)]. With a two-span continuous girder, the maximum moment would have been 24,570 ft-lb [i.e., (0.125)(1365 lb/ft)(12 ft)(12 ft)], which is the same maximum moment calculated for a simply supported girder. The maximum shear would have been 10,238 lb [i.e., (0.625)(1365 lb/ft)(12 ft)].

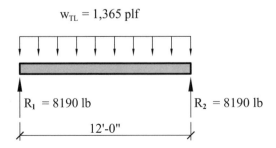

FIGURE 9.13 Free-body diagram of girder G1.

We will now proceed with the design following the steps described in Section 4.1.

Check of Bending Stress (Girders)

1. Summary of load effects (the self-weight of the girder was assumed, but this will be checked later in the design process):

$$\text{maximum shear } V_{max} = 8190 \text{ lb}$$

$$\text{maximum reaction } R_{max} = 8190 \text{ lb}$$

$$\text{maximum moment } M_{max} = 294{,}840 \text{ in.-lb}$$

$$\text{uniform dead load } w_{DL} = 51.3 \text{ lb/in.}$$

$$\text{uniform live load } w_{LL} = 62.5 \text{ lb/in.}$$

2. For a glulam used primarily in bending, use NDS-S Table 5A–Expanded.

3. Using NDS-S Table 5A–Expanded, assume a 24F–E11 HF/HF glulam, therefore, $F_{bx}^+ = 2400$ psi. Assume initially that $F_{bx}'^+ = F_{bx}^+ = 2400$ psi. From equation (4.1) the required approximate section modulus of the member is given as

$$S_{xx,\text{ req'd}} \geq \frac{M_{max}}{F_{b,\text{ NDS-S}}} = \frac{294{,}840 \text{ in.-lb}}{2400 \text{ psi}} = 122.9 \text{ in}^3$$

4. From NDS-S Table 1C (for western species glulam), the possible trial sizes that satisfy the section modulus requirement of step 3 are

SIZE (in.)	b (in.)	d (in.)	S_{xx} (in^3)	A (in^2)	I_x (in^4)
$2\frac{1}{2} \times 18$	2.5	18	135	45	1215
$3\frac{1}{8} \times 16\frac{1}{2}$	3.125	16.5	141.8	51.56	1170
$3\frac{1}{2} \times 15$	3.5	15	131.3	52.5	984.4

Although the $2\frac{1}{2} \times 18$ in. member has the least area, it is prudent in this case to limit the depth of the girder because we will be using the bottom edge of the girder to provide lateral support to the columns since these girders are supported on top of the columns. To ensure this lateral support, the difference between the joist depth and the girder depth should be kept to a minimum.

Try $3\frac{1}{8} \times 16\frac{1}{2}$ in.

$$b = 3.125 \text{ in.} \quad \text{and} \quad d = 16.5 \text{ in.}$$

$$S_{xx} \text{ provided} = 141.8 \text{ in}^3 > 122.9 \text{ in}^3 \quad \text{OK}$$

$$A \text{ provided} = 51.56 \text{ in}^2$$

$$I_{xx} \text{ provided} = 1170 \text{ in}^4$$

5. The NDS-S Table 5A–Expanded tabulated stresses are

$$F_{bx}^+ = 2400 \text{ psi}$$

$$F_{vxx} = 215 \text{ psi}$$

$$F_{c\perp xx,\text{ tension lam}} = 500 \text{ psi}$$

$$E_x = 1.8 \times 10^6 \text{ psi}$$

$$E_y = 1.5 \times 10^6 \text{ psi}$$

$$E_{y,min} = 0.78 \times 10^6 \text{ psi}$$

($E_{y,\min}$, not $E_{x,\min}$, is required for lateral buckling of the girder about the weak axis.) The adjustment or C factors are given in Table 9.8.

From the adjustment factor applicability table for glulams (Table 3.2), we obtain the allowable bending stress of the glulam girder with C_V and C_L equal to 1.0 as

$$F_{bx}^{*+} = F_{bx}^{+}C_D C_M C_t C_{fu} C_c = (2400)(1.0)(1.0)(1.0)(1.0)(1.0) = \mathbf{2400\ psi}$$

The allowable pure bending modulus of elasticity E_x' and the bending stability modulus of elasticity $E_{y,\min}'$ are calculated as

$$E_x' = E_x C_M C_t = (1.8 \times 10^6)(1.0)(1.0) = 1.8 \times 10^6 \text{ psi}$$

$$E_{y,\min}' = E_{y,\min} C_M C_t = (0.78 \times 10^6)(1.0)(1.0) = 0.78 \times 10^6 \text{ psi}$$

Calculating the Beam Stability Factor C_L

From equation (3.2) the beam stability factor is now calculated: The unsupported length of the compression edge of the beam *or* distance between points of lateral support preventing rotation and/or lateral displacement of the compression edge of the beam is

$$l_u = 2 \text{ ft} = 24 \text{ in. (i.e., the distance between lateral supports provided by the joists)}$$

$$\frac{l_u}{d} = \frac{24 \text{ in.}}{16.5 \text{ in.}} = 1.45$$

We have previously assumed a uniformly loaded girder; therefore, using the l_u/d value, the effective length of the beam of the beam is obtained from Table 3.9 (or NDS Code Table 3.3.3) as

$$l_e = 2.06 l_u = (2.06)(24 \text{ in.}) = 50 \text{ in.}$$

$$R_B = \sqrt{\frac{l_e d}{b^2}} = \sqrt{\frac{(50)(16.5 \text{ in.})}{(3.125 \text{ in.})^2}} = 9.2 \leq 50 \qquad \text{OK}$$

$$F_{bE} = \frac{1.20 E_{\min}'}{R_B^2} = \frac{(1.20)(0.78 \times 10^6)}{(9.2)^2} = 11{,}058 \text{ psi}$$

$$\frac{F_{bE}}{F_b^*} = \frac{11{,}058 \text{ psi}}{2400 \text{ psi}} = 4.61$$

From equation (3.2), the beam stability factor is calculated as

TABLE 9.8 Stress Adjustment or C Factors for Glulam Floor Girders

Adjustment Factor	Symbol	Value	Rationale for the Value Chosen
Beam stability factor	C_L	0.986	See calculation
Volume factor	C_V	1.0	See calculation
Curvature factor	C_c	1.0	Glulam girder is straight
Flat use factor	C_{fu}	1.0	Glulam is bending about its strong x–x axis
Moisture or wet service factor	C_M	1.0	Equilibrium moisture content is < 16%
Load duration factor	C_D	1.0	The largest C_D value in the load combination of dead load plus snow load (i.e., $D + L$)
Temperature factor	C_t	1.0	Insulated building; therefore, normal temperature conditions apply
Bearing stress factor	C_b	1.0	$C_b = \dfrac{l_b + 0.375}{l_b}$ for $l_b \leq 6$ in. $= 1.0$ for $l_b > 6$ in. $= 1.0$ for bearings at the ends of a member (see Section 3.1)

$$C_L = \frac{1 + F_{bE}/F_b^*}{1.9} - \sqrt{\left(\frac{1 + F_{bE}/F_b^*}{1.9}\right)^2 - \frac{F_{bE}/F_b^*}{0.95}}$$

$$= \frac{1 + 4.61}{1.9} - \sqrt{\left[\frac{(1 + 4.61)}{1.9}\right]^2 - \frac{4.61}{0.95}} = 0.986$$

Calculating the Volume Factor C_V

L = length of beam in feet between points of zero moment = 12 ft
(conservatively, assume that L = span of beam)

d = depth of beam = 16.5 in.

b = width of beam, in. ($\leq$ 10.75 in.) = 3.125 in.

x = 10 (for western species glulam)

From equation (3.3),

$$C_V = \left(\frac{21}{L}\right)^{1/x}\left(\frac{12}{d}\right)^{1/x}\left(\frac{5.125}{b}\right)^{1/x} = \left(\frac{1291.5}{bdL}\right)^{1/x}$$

$$= \left[\frac{1291.5}{(3.125 \text{ in.})(16.5 \text{ in.})(12 \text{ ft})}\right]^{1/10} = 1.08 \leq 1.0$$

Therefore, C_V is 1.0.

The smaller of C_V and C_L will govern and is used in the allowable bending stress calculation. Since C_L = 0.986 < C_V = 1.0, use C_L = 0.0.986.

Using Table 3.2 (i.e., adjustment factor applicability table for glulam), we obtain the allowable bending stress as

$$F_b' = F_b C_D C_M C_t C_F C_r(C_L \text{ or } C_V) = F_b^*(C_L \text{ or } C_V) = (2400)(0.986) = \textbf{2366.4 psi}$$

6. Using equation (4.10), the actual applied bending stress is

$$f_b = \frac{M_{max}}{S_{xx}} = \frac{294{,}840 \text{ lb-in.}}{141.8 \text{ in}^3} = 2079.3 \text{ psi} < F_b' = 2366.4 \text{ psi} \quad \textbf{OK}$$

Check of Shear Stress

V_{max} = 8190 lb. The beam cross-sectional area A = 51.56 in^2. The applied shear stress in the beam at the centerline of the girder support is

$$f_v = \frac{1.5V}{A} = \frac{(1.5)(8190 \text{ lb})}{51.56 \text{ in}^2} = 238.3 \text{ psi}$$

Using the adjustment factor applicability table for glulam (Table 3.2), we obtain the allowable shear stress as

$$F_v' = F_v C_D C_M C_t = (215)(1)(1)(1) = 215 \text{ psi} < f_v \quad \text{N.G.}$$

Recalculate the shear stress at a distance d from the face of the column support, assuming an 8 × 8 column at this stage (Figure 9.14). Therefore,

$$V' = \frac{(6 \text{ ft}) - [(7.5 \text{ in.}/2)/12] - (16.5 \text{ in.}/12)}{6 \text{ ft}}(8190 \text{ lb}) = 5887 \text{ lb}$$

Recalculating the shear stress at a distance of d from the face of support yields

$$f_v' = \frac{1.5V'}{A} = \frac{(1.5)(5887 \text{ lb})}{51.56 \text{ in}^2} = 171.3 \text{ psi} < F_v' \quad \textbf{OK}$$

FIGURE 9.14 Shear force diagram of a glulam girder.

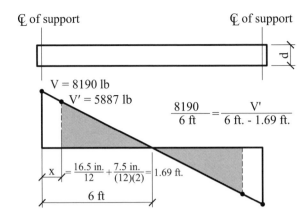

Check of Deflection (see Table 9.9)

The allowable pure bending modulus of elasticity for strong (i.e., x–x)-axis bending of the girder was calculated previously:

$$E'_x = E_x C_M C_t = (1.8 \times 10^6)(1.0)(1.0) = 1.8 \times 10^6 \text{ psi}$$

The moment of inertia about the strong axis $I_{xx} = 1170$ in^4.
The uniform dead load is $w_{DL} = 51.3$ lb/in.
The uniform live load is $w_{LL} = 62.5$ lb/in.

The dead-load deflection is

$$\Delta_{DL} = \frac{5wL^4}{384EI} = \frac{(5)(51.3 \text{ lb/in.})(12 \text{ ft} \times 12)^4}{(384)(1.8 \times 10^6 \text{ psi})(1170 \text{ in}^4)} = 0.14 \text{ in.}$$

The live-load deflection is

$$\Delta_{LL} = \frac{5wL^4}{384EI} = \frac{(5)(62.5 \text{ lb/in.})(12 \text{ ft} \times 12)^4}{(384)(1.8 \times 10^6 \text{ psi})(1170 \text{ in}^4)} = 0.17 \text{ in.} < \frac{L}{360} = 0.4 \text{ in.} \quad \textbf{OK}$$

Since seasoned wood in dry service conditions is assumed to be used in this building, the creep factor $k = 0.5$. The total incremental dead plus floor live-load deflection is

$$\Delta_{TL} = k\Delta_{DL} + \Delta_{LL}$$

$$= (0.5)(0.14 \text{ in.}) + 0.17 \text{ in.} = 0.24 \text{ in.} < \frac{L}{240} = 0.6 \text{ in.} \quad \textbf{OK}$$

Check of Bearing Stress (Compression Stress Perpendicular to the Grain)

Maximum reaction at the support $R_1 = 8190$ lb.
Width of $3\frac{1}{8} \times 16\frac{1}{2}$ in. glulam girder, $b = 3.125$ in.

TABLE 9.9 Floor Girder Deflection Limits

Deflection	Deflection Limit
Live-load deflection Δ_{LL}	$\dfrac{L}{360} = \dfrac{(12 \text{ ft})(12)}{360} = 0.4 \text{ in.}$
Incremental long-term deflection due to dead load plus live load (including creep effects), $\Delta_{TL} = k\Delta_{DL} + \Delta_{LL}$	$\dfrac{L}{240} = \dfrac{(12 \text{ ft})(12)}{240} = 0.6 \text{ in.}$

The allowable bearing stress or compression stress parallel to the grain is

$$F'_{c\perp xx(T)} = F_{c\perp xx(T)} C_M C_t C_b = (500)(1)(1)(1) = 500 \text{ psi}$$

From equation (4.9) the minimum required bearing length L_b is

$$l_{b,\text{req'd}} \geq \frac{R_1}{bF_{c\perp}} = \frac{8190 \text{ lb}}{(3.125 \text{ in.})(500 \text{ psi})} = 5.24 \text{ in., say } 5.25 \text{ in.}$$

The girders will be connected to the column using a U-shaped column cap detail as shown in Fig. 4.6d, and the minimum length of the column cap required is

$$(5.25 \text{ in.})(2) + \tfrac{1}{2}\text{-in. clearance between ends of girders} = 11.0 \text{ in.}$$

Check of Assumed Self-Weight of Girder

The density of Hem Fir is 27 lb/ft³ and is used to calculate the actual weight of the girder selected.

- Assumed girder self-weight = 15 lb/ft.
- Actual weight of the $3\tfrac{1}{8} \times 16\tfrac{1}{2}$ in. glulam girder selected (from NDS-S Table 1C) is

$$\frac{(51.56 \text{ in}^2)(27 \text{ lb/ft}^3)}{(12 \text{ in.})(12 \text{ in.})} = 9.7 \text{ lb/ft, which is less than the}$$

$$\text{assumed girder selfweight of 15 lb/ft} \quad \textbf{OK}$$

Use a $3\tfrac{1}{8} \times 16\tfrac{1}{2}$ in. 24F–E11 HF/HF glulam girder.

Design of Header Beams

With reference to Figure 9.15, there are several header beams with different loading and span conditions. However, they are all very similar, and therefore only two header beams will be considered (H-1 and H-2 at the second floor).

Design of Header Beam H-1

Dead load on header:

(40 psf)(15 ft/2)	= 300 lb/ft	(second-floor dead load)
(10 psf)(6 ft height above opening)	= 60 lb/ft	(weight of wall above opening)
Self-weight (assume three 2 × 12's)	= 12 lb/ft	
w_{DL}	**= 372 lb/ft**	

Live load on header:

$$w_{\text{LL}} = (50 \text{ psf})\left(\frac{15 \text{ ft}}{2}\right) = \textbf{375 lb/ft} \quad \text{(second-floor live load)}$$

Total load on header:

$$w_{\text{TL}} = 372 + 375 = 747 \text{ lb/ft}$$

Adjusting the loads for seasoned lumber (recall that $k = 0.5$ for seasoned lumber),

$$kw_{\text{DL}} + w_{\text{LL}} = (0.5)(372) + 375 = 561 \text{ lb/ft}$$

(Use this value for calculating deflections only.)

FIGURE 9.15 Header beam layout.

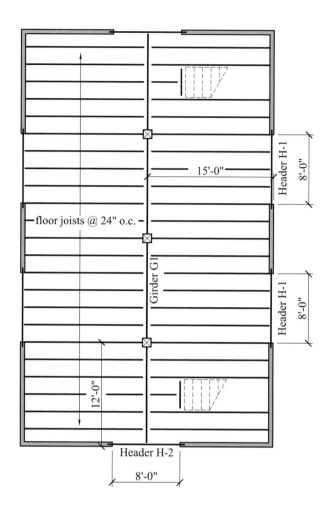

$$\text{maximum moment } M = \frac{w_{TL}L^2}{8} = \frac{(747)(8)^2}{8} = 5976 \text{ ft-lb}$$

$$\text{maximum shear and reaction } V_{TL} = \frac{w_{TL}L}{2} = \frac{(747)(8)}{2} = 2988 \text{ lb}$$

Bending:

$$F'_b = F_b C_D C_M C_t C_L C_F C_{fu} C_i C_r$$

$$C_D(\text{DL} + \text{LL}) = 1.0$$

$$C_M = 1.0$$

$$C_t = 1.0$$

$$C_F(F_b) = 1.0$$

$$C_{fu} = 1.0$$

$$C_i = 1.0$$

$$C_r = 1.0$$

From NDS Table 4A (spruce-pine-fir, No. 1/No. 2)

$$F_b = 875 \text{ psi}$$
$$F_v = 135 \text{ psi}$$
$$F_{c\perp} = 425 \text{ psi}$$
$$E = 1.4 \times 10^6 \text{ psi}$$
$$E_{min} = 0.51 \times 10^6 \text{ psi}$$

Since the beam is unbraced laterally for its full length, the beam stability factor C_L will have to be calculated. See Chapter 3 for the C_L equations. For three 2 × 12's:

$$d = 11.25 \text{ in.}$$
$$b = (3)(1.5 \text{ in.}) = 4.5 \text{ in.}$$
$$A = (3)(16.88 \text{ in}^2) = 50.63 \text{ in}^2$$
$$S_x = (3)(31.64 \text{ in}^3) = 94.92 \text{ in}^3$$
$$I_x = (3)(178 \text{ in}^4) = 534 \text{ in}^4$$
$$l_u = 8 \text{ ft} = 96 \text{ in. (span of header)}$$
$$\frac{l_u}{d} = \frac{96 \text{ in.}}{11.25 \text{ in.}} = 8.53$$
$$l_e = 1.63 l_u + 3d \quad \text{(Table 3.9 or NDS Table 3.3.3)}$$
$$= (1.63)(96 \text{ in.}) + (3)(11.25 \text{ in.}) = 191 \text{ in.}$$
$$R_B = \sqrt{\frac{l_e d}{b^2}} \leq 50 \quad \text{(NDS Equation 3.3-5)}$$
$$= \sqrt{\frac{(191)(11.25)}{(4.5)^2}} = \mathbf{10.3} < 50 \quad \mathbf{OK}$$
$$E'_{min} = E_{min} C_M C_t C_i = (0.51 \times 10^6)(1.0)(1.0)(1.0) = 0.51 \times 10^6$$
$$F_{bE} = \frac{1.20 E_{min}}{R_B^2} = \frac{(1.20)(0.51 \times 10^6)}{(10.3)^2} = 5768 \text{ psi}$$
$$F_b^* = F_b C_D C_M C_t C_F C_{fu} C_i C_r = (875)(1.0)(1.0)(1.0)(1.0)(1.0)(1.0)(1.0) = 875 \text{ psi}$$
$$C_L = \frac{1 + F_{bE}/F_b^*}{1.9} - \sqrt{\left(\frac{1 + F_{bE}/F_b^*}{1.9}\right)^2 - \frac{F_{bE}/F_b^*}{0.95}} \quad \text{(NDS Equation 3.3-6)}$$
$$= \frac{1 + 5768/875}{1.9} - \sqrt{\left(\frac{1 + 5768/875}{1.9}\right)^2 - \frac{5768/875}{0.95}} = 0.991$$

The allowable bending stress

$$F'_b = F_b^* C_L = (875)(0.991) = 867 \text{ psi}$$

The bending stress applied

$$f_b = \frac{M}{S_x} = \frac{(5976)(12)}{94.92} = \mathbf{756 \text{ psi}} < F'_b \quad \mathbf{OK}$$

Shear. The allowable shear stress

$$F'_v = F_v C_D C_M C_t C_i = (135)(1.0)(1.0)(1.0)(1.0) = 135 \text{ psi}$$

The shear stress applied

$$f_v = \frac{1.5V}{A} = \frac{2988}{50.63} = \mathbf{88.6 \text{ psi}} < F'_v \quad \mathbf{OK}$$

Bearing. The allowable bearing stress perpendicular to the grain is

$$F'_{c\perp} = F_{c\perp} C_M C_t C_i = (425)(1.0)(1.0)(1.0) = 425 \text{ psi}$$

The required bearing length

$$l_{b,\text{req'd}} = \frac{R}{F_{c\perp} b}$$

$$= \frac{2988}{(425)(4.5)} = 1.6 \text{ in.}$$

The headers will bear directly on two jack studs (see Figure 9.16).
Deflection. The total load deflection (including creep effects) is

$$E' = E C_M C_t C_i = (1.4 \times 10^6)(1.0)(1.0)(1.0) = 1.4 \times 10^6$$

$$\Delta_{kDL+LL} = \frac{5wL^4}{385EI} = \frac{(5)(561/12)(96)^4}{(384)(1.4 \times 10^6)(534)} = \mathbf{0.07 \text{ in.}} < \frac{L}{240} = \frac{96}{240} = 0.4 \text{ in.} \quad \mathbf{OK}$$

The live-load deflection is

$$\Delta_{LL} = \frac{5wL^4}{385EI} = \frac{(5)(375/12)(96)^4}{(384)(1.4 \times 10^6)(534)} = \mathbf{0.046 \text{ in.}} < \frac{L}{360} = \frac{96}{360} = 0.27 \text{ in.} \quad \mathbf{OK}$$

Use three 2 × 12's for header H-1, bear on two 2 × 6 jack studs.

Design of Header Beam H-2

Header beam H-2 has a $16\frac{1}{2}$-in.-deep girder framing into it. For practical framing considerations, the header should be at least as deep as the girder. For this example, a preengineered header beam will be selected with the following assumed properties:

$$F_b = 2600 \text{ psi}$$
$$F_v = 285 \text{ psi}$$
$$E = 1900 \text{ ksi}$$
$$F_{c\perp} = 750 \text{ psi}$$

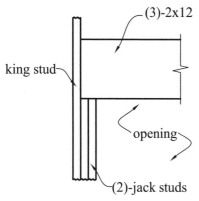

FIGURE 9.16 Header (H-1) bearing.

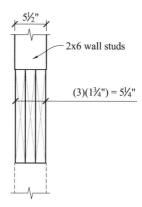

FIGURE 9.17 Header (H-2) section.

For this example, three $1\frac{3}{4} \times 16$ in. LVL headers will be assumed. The design will show that one header would be adequate, but it is practical to use at least three members in a 2×6 stud wall so that a "void" is not created above the opening (Figure 9.17).

Section properties:

$$d = 16 \text{ in.}$$

$$b = (3)(1.75 \text{ in.}) = 5.25 \text{ in.}$$

$$A = (3)(1.75)(16) = 84 \text{ in}^2$$

$$S_x = (3)\frac{bd^2}{6} = (3)\frac{(1.75)(16)^2}{6} = 224 \text{ in}^3$$

$$I_x = (3)\frac{bd^3}{12} = (3)\frac{(1.75)(16)^3}{12} = 1792 \text{ in}^4$$

Loads on header:

$$P_D = (40 \text{ psf})\left(\frac{12 \text{ ft}}{2}\right)(15 \text{ ft}) + 15 \text{ plf}\left(\frac{12 \text{ ft}}{2}\right) = 3690 \text{ lb}$$

$$P_L = (50 \text{ psf})\left(\frac{12 \text{ ft}}{2}\right)(15 \text{ ft}) = 4500 \text{ lb}$$

$$P_T = 3690 + 4500 = \textbf{8190 lb}$$

$$\text{Maximum moment } M = \frac{PL}{4} = \frac{(8190)(8)}{4} = 16{,}380 \text{ ft-lb}$$

$$\text{Maximum shear and reaction } V_{TL}\frac{P}{2} = \frac{8190}{2} = 4095 \text{ lb}$$

The bending stress applied is

$$f_b = \frac{M}{S_x} = \frac{(16{,}380)(12)}{224} = \textbf{878 psi} < 2600 \text{ psi} \qquad \textbf{OK}$$

The shear stress applied is

$$f_v = \frac{1.5V}{A} = \frac{(1.5)(4095)}{84} = \textbf{74 psi} < F'_v = 285 \text{ psi} \qquad \textbf{OK}$$

The required bearing length is

$$l_{b,\text{req'd}} = \frac{R}{F_{c\perp}b}$$

$$= \frac{4095}{(750)(5.25)} = 1.04 \text{ in.}$$

The headers will bear directly on at least one jack stud.

Deflection. Since the dead load is more than one-half of the live load, the $L/240$ criteria will control. Therefore,

$$\Delta_{\text{DL+LL}} = \frac{PL^3}{48EI} = \frac{(8190)(96)^3}{(48)(1.9 \times 10^6)(1792)} = \textbf{0.05 in.} < \frac{L}{240} = \frac{96}{240} = 0.4 \text{ in.} \qquad \textbf{OK}$$

Use three $1\frac{3}{4} \times 16$ in. LVLs for header H-2, bear on one 2×6 jack stud.

9.8 DESIGN OF A TYPICAL GROUND-FLOOR COLUMN

It is assumed that the ground-floor column in this building will be a wood column instead of the steel pipe column that is frequently used in residential structures. However, the wood column base should be protected from moisture at the slab-on-grade/footing level using a column base detail similar to that shown in Figure 9.18.

Determining the Column Axial Load

We recall that this is a two-story building, but the roof tributary area for this column is zero since the column supports only the second floor.

$$\text{Tributary area of typical interior column} = (15 \text{ ft})(12 \text{ ft}) = 180 \text{ ft}^2$$

From Chapter 2,

$$A_T = \text{tributary area} = 180 \text{ ft}^2 \text{ (only the second floor live load is reducible for this column)}$$
$$K_{LL} = 4 \text{ (interior column)}$$

$K_{LL} A_T = (4)(180 \text{ ft}^2) = 720 \text{ ft}^2 > 400 \text{ ft}^2$.

Floor live load $L = 50$ psf < 100 psf.

Floor occupancy is not assembly occupancy.

Since all the conditions above are satisfied, live-load reduction is permitted.

The governing IBC load combination from Chapter 2 for calculating the column axial loads is $D + 0.75L + 0.75(L_r \text{ or } S \text{ or } R)$. The reduced or design floor live loads for the second and third floors are calculated using Table 9.10

Using the loads calculated in Section 9.2, the column axial loads are calculated using Table 9.11. Therefore, the ground-floor column will be designed for a cumulative maximum axial compression load $P = 14.5$ kips.

It should be noted that the column load calculated above assumes simply supported glulam girders. However, if the contractor chooses to use two 24-ft-long pieces for the girder, thus indicating two two-span continuous girders, the interior column load would be amplified due to the girder continuity effect. For a two-span continuous girder, this will result in a 25% increase in the second-floor load on the column. Therefore, the revised axial load on the column will be

$$P = (1.25)(14.5) = 18.1 \text{ kips}$$

Using Figure B.17, the allowable axial load at an unbraced length of 10.5 ft is approximately 25.0 kips with a load duration factor of 1.0. Therefore, use a 8 × 8 SPF No. 2 Column. The reader should note that the load capacity of a 6 × 6 column is approximately 11 kips and would not be adequate.

9.9 DESIGN OF A TYPICAL EXTERIOR WALL STUD

The typical exterior wall stud is subjected to combined concentric axial compression load plus bending due to the wind loads acting perpendicular to the face of the exterior wall. The ground-

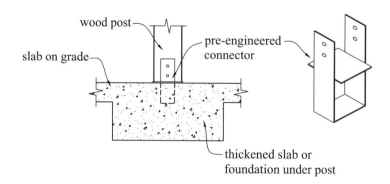

FIGURE 9.18 Column base support detail.

TABLE 9.10 Reduced or Design Floor Live-Load Calculation Table

Member	Levels Supported	Summation of Floor Tributary Area, A_T	K_{LL}	Unreduced Floor Live load, L_0	Live-Load Reduction Factor, $0.25 + 15/\sqrt{K_{LL}A_T}$	Design Floor Live Load, L
Second-floor column (i.e., column below roof)	Roof only	Floor live-load reduction *not* applicable to roofs!	—	—	—	35 psf (snow load)
Ground-floor column (i.e., column below the second floor)	One floor + roof	One floor × 180 ft² = **180 ft²**	4	50 psf	$0.25 + 15/\sqrt{(4)(180)}$ = 0.809	(0.809)(50) = **40.5 psf** ≥ $0.50L_0$ = 25 psf

Note: $K_{LL}A_T = 720$ ft² > 400 ft²; therefore, live load reduction allowed

floor exterior wall stud, which is the more critically loaded than the second-floor stud, is designed in this section. The size obtained is also used for the second-floor stud.

Determining the Gravity and Lateral Loads Acting on a Typical Wall Stud

Floor-to-floor height = 12 ft

Tributary width of wall stud = 2 ft

Tributary area per stud at the *roof* level = (30 ft/2 + 2 ft truss overhang) × (2 ft) = 34 ft²

Tributary area per stud at the *second-floor* level = (15 ft/2) × (2 ft) = 15 ft²

From Chapter 2,

A_T = floor tributary area = 15 ft² (only the second-floor live load is reducible for this stud)

K_{LL} = 2 (exterior column)

$K_{LL}A_T = (2)(15 \text{ ft}^2) = 30 \text{ ft}^2 < 400 \text{ ft}^2$

Therefore, floor live-load reduction is *not* permitted for this stud.

From Section 9.2, the gravity loads have been calculated as follows:

roof dead load $D \approx 22$ psf

roof snow load $S = 35$ psf

roof live load $L_r = 20$ psf < S; therefore, the snow load controls

second-floor dead load $D = 40$ psf

second-floor live load $L = 50$ psf

second-floor wall self-weight = 10 psf

ground-floor wall self-weight = 10 psf

The lateral or horizontal wind pressure W acting perpendicular to the face of the exterior wall represents the components and cladding wall pressures calculated in Section 9.5 as $W = 22.6$ psf.

Gravity loads. The total *dead load* on the ground-floor studs is

TABLE 9.11 Column Load Summation Table (With Floor Live-Load Reduction)

Level	Tributary Area, TA (ft²)	Dead Load, D (psf)	Live Load, L_0 (S or L_r or R on the Roof) (psf)	Design Live Load Roof: S or L_r or R Floor: L (psf)	Unfactored Total Load at Each Level, w_{s1} Roof: D Floor: D + L (psf)	Unfactored Total Load at Each Level, w_{s2} Roof: D + 0.75S Floor: D + 0.75L (psf)	Unfactored Column Axial Load at Each Level, $P = (TA)(w_{s1})$ or $(TA)(w_{s2})$ (kips)	Cumulative Unfactored Axial Load, ΣP_{D+L} (kips)	Cumulative Unfactored Axial Load, $\Sigma P_{D+0.75L+0.75S}$ (kips)	Maximum Cumulative Unfactored Axial Load, ΣP (kips)
Roof	0	22	35	35	22	48.3	0	0	0	0
Second floor	180	40	50	40.5	80.5	70.4	14.5 or 12.7	14.5	12.7	**14.5**

$$P_D = P_{D(\text{roof})} + P_{D(2\text{nd floor})} + P_{D(\text{wall})}$$

$$= (22 \text{ psf})(34 \text{ ft}^2) + (40 \text{ psf})(15 \text{ ft}^2)$$

$$+ (10 \text{ psf})(2\text{-ft stud spacing})(12\text{-ft wall height})$$

$$= \mathbf{1588 \text{ lb}}$$

The total *snow load* on the ground-floor studs is

$$P_S = (35 \text{ psf})(34 \text{ ft}^2) = \mathbf{1190 \text{ lb}}$$

The total *floor live load* on the ground-floor studs is

$$P_L = (50 \text{ psf})(15 \text{ ft}^2) = \mathbf{750 \text{ lb}}$$

Lateral loads. The wind load acts perpendicular to the face of the stud wall, causing bending of the stud about the x–x (strong) axis (see Figure 9.19). The lateral wind load $w_{\text{wind}} = (22.6 \text{ psf})(2\text{-ft tributary width}) = 45.2$ plf. The maximum moment in the exterior wall stud due to wind load is calculated using the 12-ft floor-to-floor height (conservative for the ground-floor studs on the east and west face of the building) as

$$M_w = \frac{wL^2}{8}$$

$$= \frac{(45.2 \text{ lb/ft})(12)^2}{8} = \mathbf{814 \text{ ft-lb}}$$

The most critical axial load combination and the most critical combined axial and bending load combination are determined using the normalized load method outlined in Chapter 3. The axial load P in the wall stud will be caused by the gravity loads D, L, and S, while the bending load and moment will be caused by the lateral wind load W. The applicable nonzero loads and load effect in the load combinations are

$$D = 1588 \text{ lb}$$
$$L = 750 \text{ lb}$$
$$S = 1190 \text{ lb}$$
$$W = 814 \text{ ft-lb}$$

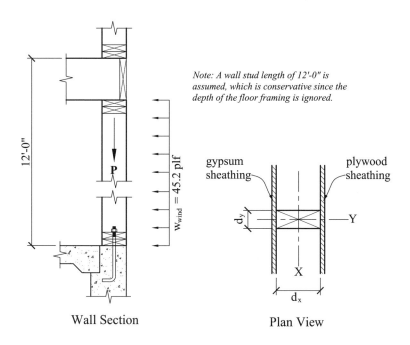

FIGURE 9.19 Stud wall section and plan view.

316 | STRUCTURAL WOOD DESIGN

We can separate the load cases in Table 9.12 into two types of loads: pure axial load cases and combined load cases. Since not all the loads on this wall stud are pure axial loads only or bending loads only, the normalized load method discussed in Chapter 3 can then only be used to determine the most critical pure axial load case, not the most critical combined load case. The most critical combined axial plus bending load case would have to be determined by carrying out the design (or analysis) for all the combined load cases, with some of the load cases eliminated by inspection.

Pure axial load case. For the pure axial load cases, the load combination with the highest normalized load P/C_D from the table above is

$$D + 0.75(L + S) \quad P = 3043 \text{ lb with } C_D = 1.15$$

Combined load case. For the combined load cases, the load combinations with the highest normalized load P/C_D and normalized moment M/C_D from Table 9.12 are

$$D + 0.75(L + S + W) \quad \boldsymbol{P = 3043 \text{ lb}} \quad \boldsymbol{M = 611 \text{ ft-lb}} \quad \boldsymbol{C_D = 1.6}$$

or

$$0.6D + W \quad \boldsymbol{P = 953 \text{ lb}} \quad \boldsymbol{M = 814 \text{ ft-lb}} \quad \boldsymbol{C_D = 1.6}$$

By inspection, it would appear as if the load combination $D + 0.75(L + S + W)$ will control for the combined load case, but this should be verified through analysis, and therefore, both of these combined load cases will have to be investigated. The figures in Appendix B will be utilized. A 2 × 6 wall stud will be assumed and SPF lumber will be used.

For the pure axial load case, the applied axial load is $P = 3043$ lb. Using Figure B.44, the allowable axial load capacity is **4400 lb** at an unbraced length of 12 ft with no applied moment. Note that the load duration factor for this design aid is 1.0, so the use of this design aid is conservative since the actual load duration is 1.15. Note also that the lumber here is 2 × 6, SPF No. 1/No. 2.

For the first combined load case, the applied axial load is $P = 3043$ lb and the applied moment is $M = 611$ ft-lb. Using Figure B.27 ($C_D = 1.6$), the allowable axial load capacity is approximately **2900 lb** at an unbraced length of 12 ft with an applied moment of **611 ft-lb** which is very close to the applied axial load of 3043 lb. Therefore, the stud is deemed to be adequate since the length of the stud is actually less than 12 ft. Note here that the lumber assumed is 2 × 6, SPF Select Structural. The reader should verify that the use of SPF No. 1/No. 2 is not adequate.

For the second combined load case, the applied axial load is $P = 953$ lb and the applied moment is $M = 814$ ft-lb. Using Figure B.27 ($C_D = 1.6$), the allowable axial load capacity is approximately **2400 lb** at an unbraced length of 12 ft with an applied moment of **814 ft-lb.** Note that the lumber assumed is 2 × 6, SPF Select Structural.

Therefore, use 2 × 6 at 24 in. o.c. SPF Select Structural wall studs.

TABLE 9.12 Applicable and Governing Load Combinations

Load Combination	Axial Load, P (lb)	Moment, M (ft-lb)	C_D	Normalized Load and Moment	
				P/C_D	M/C_D
D	1588	0	0.9	1764	0
$D + L$	1588 + 750 = 2338	0	1.0	2388	0
$D + S$	1588 + 1190 = 2778	0	1.15	2416	0
$D + 0.75L + 0.75S$	1588 + (0.75)(750 + 1190) = **3043**	0	1.15	2646	0
$D + 0.75W + 0.75L + 0.75S$	1588 + (0.75)(750 + 1190) = **3043**	(0.75)(814) = **611**	1.6	1902	382
$0.6D + W$	(0.6)(1588) = **953**	**814**	1.6	596	509

9.10 DESIGN OF ROOF AND FLOOR SHEATHING

Gravity Loads

Roof Sheathing

From the floor load calculations in Section 9.2, the total loads acting on the roof sheathing are as follows:

$$\begin{aligned}
\text{Asphalt shingles} &= 2.5 \text{ psf} \\
\text{Reroofing} &= 2.5 \text{ psf} \\
\underline{\text{Plywood sheathing}} &= \underline{1.2 \text{ psf}} \\
\text{Dead load on sheathing, } D_{sheathing} &= 6.2 \text{ psf} \\
\text{Snowload, S} &= 35 \text{ psf}
\end{aligned}$$

The total gravity load to roof sheathing = 6.2 + 35 = **41.2 psf.** Assuming a roof truss spacing of 24 in., the following is selected from IBC Table 2304.7(3):

$\frac{7}{16}$-in. CD–X. (minimum thickness)

Strength axis perpendicular to the supports.

Span rating = 24/16.

Edge support not required (may be required for diaphragm strength).

Total load capacity = 50 psf > 41.2 psf, so OK.

Live load capacity = 40 psf > 35 psf, so OK.

Floor Sheathing

From the floor load calculations in Section 9.2, the total loads acting on the floor sheathing alone are as follows:

$$\begin{aligned}
\text{Floor covering} &= 4.0 \text{ psf} \\
\text{Plywood sheathing} &= 3.2 \text{ psf} \\
\underline{\text{Partitions}} &= \underline{20.0 \text{ psf}} \\
DL_{sheathing} &= \mathbf{27.2 \text{ psf}} \\
LL &= \mathbf{50 \text{ psf}}
\end{aligned}$$

The total gravity load to floor sheathing = 27.2 + 50 = **77.2 psf.** Assuming that the floor framing is spaced at 24 in. o.c., the following is selected from IBC Table 2304.7(3) or 2304.7(4):

$\frac{3}{4}$-in. Structural I sheathing. (minimum thickness)

Note that Table 2304.7(4) indicates that a plywood panel with veneers made from Group 1 wood species (i.e., Structural I) must be used if $\frac{3}{4}$ in. plywood thickness is specied. On the other hand, Table 2304.7(3) does not indicate the specie of wood used for the veneer.

Strength axis perpendicular to the supports.

Span rating = 24 in.

Tongue-and-grooved joints.

Total load capacity = 100 psf > 77.2 psf, so OK.

Lateral Loads

Based on the lateral wind and seismic loads calculated in Sections 9.4 and 9.5, the loads to the floor and roof diaphragms (Figure 9.20) are as summarized in Table 9.13. By inspection, wind loads will govern the diaphragm design in the transverse direction, and seismic loads will govern the longitudinal direction (recall that the allowable shear values can be increased by 40% for

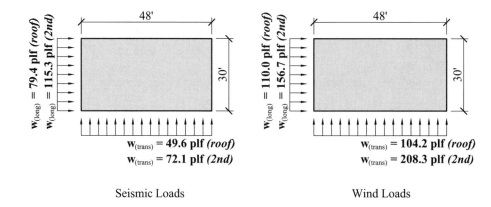

FIGURE 9.20 Lateral loads on the roof and floor diaphragms.

wind loads when using IBC Table 2306.3.1). From the gravity load design in Section 9.5, the minimum thickness for roof sheathing is $\frac{7}{16}$ in. and $\frac{3}{4}$ in. for the floor sheathing. From Table 6.3, the minimum required fastening is 8d nails spaced 6 in. o.c. at the supported edges and 12 in. o.c. on the intermediate members for both the roof and floor sheathing. These minimum parameters will be considered in the design of the diaphragm fasteners.

Transverse Direction (Wind Loads Govern)

The applied diaphragm unit shears (see Table 9.13) are:

$$v_{dr} = 83.3 \text{ plf (roof level)}$$

$$v_{d2} = 166.7 \text{ plf (second floor level)}$$

For the *roof sheathing,* using load case 1 from IBC Table 2306.3.1, the capacity of a $\frac{3}{8}$-in. CD–X Structural I panel with 8d nails into 2× framing members is 240 plf for an unblocked diaphragm. This value can be increased by 40% for wind loads but must also be decreased since IBC Table 2306.3.1 is valid only for framing of Douglas fir-larch or southern pine. The adjusted allowable shear is calculated as follows

$$\text{SGAF} = 1 - (0.5 - G) \leq 1.0 \qquad G = 0.42 \text{ (spruce-pine-fir lumber is specified)}$$

$$= 1 - (0.5 - 0.42) = 0.92$$

$$v_{\text{allowable}} = (1.4)(0.92)(240 \text{ plf}) = \mathbf{309 \text{ plf}} > v_{dr} = 83.3 \text{ plf}$$

Therefore, the required fastening is 8d nails at 6 in. o.c. (EN) and 12 in. o.c. (FN).

TABLE 9.13 Seismic and Wind Loads at Each Level

Direction	Level	F (lb)	w	V	v_d
			Seismic Loads		
Transverse	Roof	2380[a]	2380 lb / (48 ft) = 49.6 plf	(2380 lb)/2 = 1190 lb	1190 lb / (30 ft) = **39.7 plf**
	Second floor	3460	3460 lb / (48 ft) = 72.1 plf	(3460 lb)/2 = 1730 lb	1730 lb / (30 ft) = **57.7 plf**
Longitudinal	Roof	2380	2380 lb / (30 ft) = 79.4 plf	(2380 lb)/2 = 1190 lb	1190 lb / (48 ft) = **24.8 plf**
	Second floor	3460	3460 lb / (30 ft) = 115.3 plf	(3460 lb)/2 = 1730 lb	1730 lb / (48 ft) = **36.1 plf**
			Wind Loads		
Transverse	Roof	5000[b]	5000 lb / (48 ft) = 104.2 plf	(5000 lb)/2 = 2500 lb	2500 lb / (30 ft) = **83.3 plf**
	Second floor	10,000	10,000 lb / (48 ft) = 208.3 plf	(10,000 lb)/2 = 5000 lb	5000 lb / (30 ft) = **166.7 plf**
Longitudinal	Roof	3300	3300 lb / (30 ft) = 110 plf	(3300 lb)/2 = 1650 lb	1650 lb / (48 ft) = **34.4 plf**
	Second floor	4700	4700 lb / (30 ft) = 156.7 plf	(4700 lb)/2 = 2350 lb	2350 lb / (48 ft) = **49.0 plf**

[a] See Section 9.3.
[b] See Table 9.2.

For the *floor sheathing*, using load case 1 from IBC Table 2306.3.1, the capacity of a $\frac{15}{32}$-in. CD–X Structural I panel with 10d nails into 2× framing members is 285 plf for an unblocked diaphragm. Adjusting this value for wind loads and spruce-pine-fir lumber yields

$$v_{\text{allowable}} = (1.4)(0.92)(285 \text{ plf}) = \mathbf{367 \text{ plf}} > v_{dr} = 166.7 \text{ plf}$$

Note here that a comparison was made with $\frac{15}{32}$-in. panels, which is conservative when a $\frac{3}{4}$-in. panel is used. Also, the minimum fastener penetration is $1\frac{1}{2}$ in. (from IBC Table 2306.3.1). Since the length of an 10d nail is 3 in. (from NDS Table L4), the actual penetration is greater than the required penetration (3 in. $- \frac{3}{4}$ in. $= 2\frac{1}{4}$ in. $> 1\frac{1}{2}$ in.). The required fastening is 10d nails at 6 in. on centers edge nailing (EN) and 12 in. on centers fielding nailing (FN).

Longitudinal Direction (Seismic Loads Govern)

The applied diaphragm unit shears (See Table 9.13) are:

$$v_{dr} = 24.8 \text{ plf (roof level)}$$
$$v_{d2} = 36.1 \text{ plf (second floor level)}$$

By inspection, the diaphragm shear in the transverse direction is much greater than the diaphragm shear in the longitudinal direction. Since the diaphragm design did not require blocking in the transverse direction, it would be economical to select a diaphragm design that did not include blocking for the longitudinal direction.

For the *roof sheathing*, using load case 3 from IBC Table 2306.3.1, the capacity of a $\frac{3}{8}$-in. panel with 8d nails into 2× framing members is 180 plf for an unblocked diaphragm. Adjusting this value for spruce-pine-fir lumber yields

$$v_{\text{allowable}} = (0.92)(180 \text{ plf}) = \mathbf{165 \text{ plf}} > v_{dr} = 24.8 \text{ plf}$$

Therefore, the specified fastening used for the transverse direction is also adequate for the longitudinal direction.

For the *floor sheathing*, using load case 3 from IBC Table 2306.3.1, the capacity of a $\frac{15}{32}$-in. panel with 10d nails into 2× framing members is 215 plf for an unblocked diaphragm. Adjusting this value for spruce-pine-fir lumber yields

$$v_{\text{allowable}} = (0.92)(215 \text{ plf}) = \mathbf{197 \text{ plf}} > v_{dr} = 36.1 \text{ plf}$$

Therefore, the specified fastening used for the transverse direction is also adequate for the longitudinal direction.

The attachment of the horizontal diaphragm to the shear walls requires consideration. A variety of diaphragm-to-shear wall connections are possible (see Figure 9.21 for typical details). The most efficient connection would be one where the horizontal wood panels meet the vertical panels (Figure 9.21), where the boundary nails in the horizontal diaphragm transfer loads directly to the boundary nails in the vertical diaphragm. Other connection types often require more analysis and fasteners.

9.11 DESIGN OF WALL SHEATHING FOR LATERAL LOADS

From IBC Table 2308.9.3(3), a $\frac{7}{16}$-in. panel with a minimum of four plies and a span rating of wall-24 is adequate to resist a wind pressure perpendicular to the face of the wall sheathing. This panel can be oriented either parallel or perpendicular to the wall studs and allows the builder some latitude in selecting the most economical layout. Table 6-1 of the *Wood Structural Panels Supplement* indicates that the minimum fastening is 6d nails spaced 6 in. o.c. at the supported edges and 12 in. o.c. in the field of the plywood panel. These minimum parameters will be considered in the design of the shear wall fasteners.

The aspect ratio of the available shear walls must be calculated. The shortest available wall length is 8 ft and the total height of the walls is 24 ft. Therefore,

$$\frac{h}{w} = \frac{24 \text{ ft}}{8 \text{ ft}} = 3.0 < 3.5 \qquad \text{OK for wind loads}$$

FIGURE 9.21
Diaphragm-to-shearwall connections.

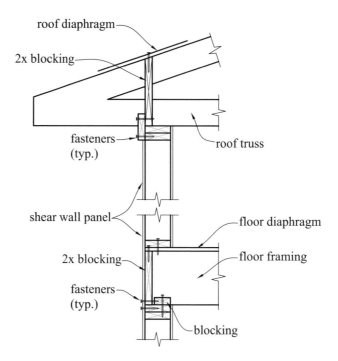

For seismic loads, the maximum aspect ratio of 3.5 is permitted if the allowable unit shears from IBC Table 2306.4.1 are multiplied by an adjustment factor $2w/h$, where w is the length of the shear wall and h is the height of the wall. Therefore, the following adjustment factor will be applied to the tabulated allowable unit shears for seismic loads:

$$\frac{2w}{h} = \begin{cases} \dfrac{(2)(11 \text{ ft})}{24 \text{ ft}} = \mathbf{0.92} & \text{(aspect ratio adjustment factor for the 11 ft long wall (SW1)} \\ & \text{in the transverse direction)} \\ \dfrac{(2)(8 \text{ ft})}{24} = \mathbf{0.67} & \text{(aspect ratio adjustment factor for the 8 ft long wall (SW3)} \\ & \text{in the longitudinal direction)} \end{cases}$$

The unit shear due to lateral loads will now be calculated. Note that the base shear V to each line of shear walls is cumulative at the ground level. With reference to Figure 9.22, the unit shear in the shear walls is as summarized in Table 9.14. By inspection, wind loads will govern the shear wall design in the transverse direction, and seismic loads will govern in the longitudinal direction (recall that the allowable shear values can be increased by 40% for wind loads when using IBC Table 2306.4.1.

Transverse Direction (*Wind Governs*)

The shear wall unit shears (See Table 9.14) are:

$$v_{wr} \text{ (roof level)} = 113.7 \text{ plf}$$

$$v_{w2} \text{ (second-floor level)} = 340.9 \text{ plf}$$

From IBC Table 2306.4.1, the capacity of a $\frac{15}{32}$-in. panel with 8d nails into 2× framing members is 280 plf. This value can be increased by 40% for wind loads but must be decreased since IBC Table 2306.3.1 is valid only for framing of Douglas fir-larch or southern pine. The species adjustment factor to be applied to the allowable shear is

$$\text{SGAF} = 1 - (0.5 - G) \leq 1.0 \qquad G = 0.42 \text{ (spruce-pine-fir lumber is specified)}$$

$$= 1 - (0.5 - 0.42) = 0.92$$

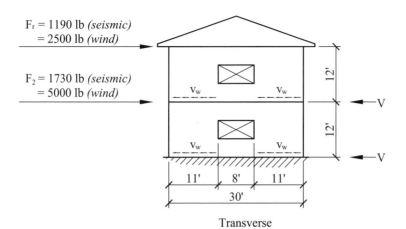

FIGURE 9.22 Lateral loads on the shear walls.

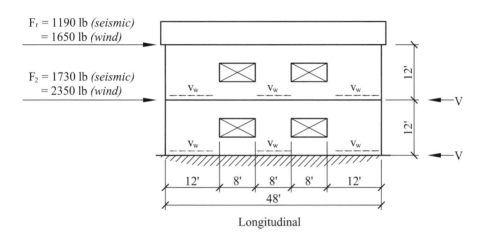

TABLE 9.14 Summary of Shear Wall Unit Shears

Direction	Load	Level	V (lb) (from Section 9.22)	$v_w = \dfrac{V}{\Sigma L_w}$	v_w (for Drag Struts)
Transverse	Seismic	Roof	1190	(1190 lb) / (11 ft + 11 ft) = **54.1 plf**	54.1 plf
		Second floor	1730	(1190 + 1730 lb) / (11 ft + 11 ft) = **132.8 plf**	(1730 lb) / (11 ft + 11 ft) = **78.7 plf**
	Wind	Roof	2500	(2500 lb) / (11 ft + 11 ft) = **113.7 plf**	113.7 plf
		Second floor	5000	(2500 + 5000 lb) / (11 ft + 11 ft) = **340.9 plf**	(5000 lb) / (11 ft + 11 ft) = **227.3 plf**
Longitudinal	Seismic	Roof	1190	(1190 lb) / (12 ft + 8 ft + 12 ft) = **37.2 plf**	37.2 plf
		Second floor	1730	(1190 + 1730 lb) / (12 ft + 8 ft + 12 ft) = **91.3 plf**	(1730 lb) / (12 ft + 8 ft + 12 ft) = **54.1 plf**
	Wind	Roof	1650	(1650 lb) / (12 ft + 8 ft + 12 ft) = **51.6 plf**	51.6 plf
		Second floor	2350	(1650 lb + 2350 lb) / (12 ft + 8 ft + 12 ft) = **125 plf**	(2350 lb) / (12 ft + 8 ft + 12 ft) = **73.4 plf**

The adjusted allowable unit shear is

$$v_{\text{allowable (wind)}} = (1.4)(0.92)(280 \text{ plf}) = \mathbf{361 \text{ plf}} > v_{wr} = 113.7 \text{ plf}$$

$$> v_{w2} = 340.9 \text{ plf} \quad \text{OK}$$

Therefore, the required fastening for the shear walls between the second floor and the roof and between the ground floor and the second floor is 8d nails at 6 in. on centers edge nailing (EN) and 12 in. on centers field nailing (FN).

Longitudinal Direction (Seismic Governs)
The shear wall unit shears are:

$$V_{Er} \text{ (roof level)} = 37.2 \text{ plf}$$

$$V_{E2} \text{ (second-floor level)} = 91.3 \text{ plf}$$

The $\frac{15}{32}$-in. panel with 8d nails will be the minimum required for these walls; therefore, the base capacity of the shear walls remains at 280 plf. Adjusting this value for spruce-pine-fir lumber and applying the aspect ratio adjustment factor 0.67 for 8-ft long wall which is applicable for seismic loads only yields

$$v_{\text{allowable (seismic)}} = (0.92)(0.67)(280 \text{ plf}) = \mathbf{171 \text{ plf}} > v_{Er} = 37.2 \text{ plf}$$

$$> v_{E2} = 91.3 \text{ plf}$$

Therefore, the fastening specified for the shear wall in the transverse direction is adequate for the shear walls in the longitudinal direction. Note that the 40% increase in allowable unit shears applies only to wind design but not to seismic design. It should be noted that though the unit shears due to longitudinal wind were higher than those due to seismic loads (see Table 9.14), the seismic loads still govern. Compare the seismic unit shears to the corresponding wind load values and including the 40% increase in capacity for wind loads yields,

$$v_{\text{allowable (wind)}} = (0.92)(280 \text{ plf})(1.40) = 361 \text{ plf}$$

$$\gg v_{wr} = 51.6 \text{ plf}$$

$$v_{w2} = 125 \text{ plf}$$

9.12 OVERTURNING ANALYSIS OF SHEAR WALLS: SHEAR WALL CHORD FORCES

The shear walls (Figure 9.23) will now be analyzed for overturning. Through this analysis, the maximum compression and tension forces in the shear wall chords will be calculated. The first step will be to calculate the lateral and gravity loads to each shear wall. The lateral load to each shear wall is summarized in Tables 9.15 and 9.16.

The gravity loads to each shear wall will now be calculated. Recall the following from Section 9.2:

$$\text{roof dead load} = 22 \text{ psf}$$

$$\text{roof snow load} = 35 \text{ psf}$$

$$\text{floor dead load} = 40 \text{ psf}$$

$$\text{floor live load} = 50 \text{ psf}$$

$$\text{wall dead load} = 10 \text{ psf}$$

The calculation of gravity loads on shear walls is covered in Chapter 7. The summary of the loads on walls SW1, SW2, and SW3 are shown in Figures 9.24, 9.25, and 9.26.

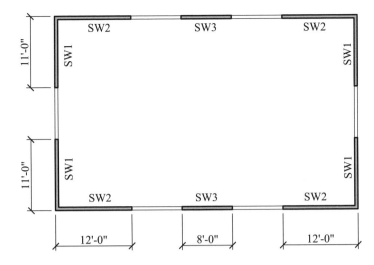

FIGURE 9.23 Shear wall layout.

TABLE 9.15 Summary of Lateral Forces on a Building

Direction	Load	Level	F (lb) (from Figures 9.2 and 9.4)
Transverse	Seismic	Roof	2380
		Second floor	3460
	Wind	Roof	5000
		Second floor	10,000
Longitudinal	Seismic	Roof	2380
		Second floor	3460
	Wind	Roof	3300
		Second floor	4700

TABLE 9.16 Lateral Forces on Each Shear Wall

Wall	Load	Level	F (lb) (from Figures 9.2 and 9.4)	$F_w = \dfrac{F}{\Sigma L_w}(L_{wn})$
SW1	Seismic	Roof	2380	$\dfrac{(2380 \text{ lb})}{(4)(11 \text{ ft})}(11 \text{ ft}) = \mathbf{595\ lb}$
		Second floor	3460	$\dfrac{(3460 \text{ lb})}{(4)(11 \text{ ft})}(11 \text{ ft}) = \mathbf{865\ lb}$
	Wind	Roof	5000	$\dfrac{(5000 \text{ lb})}{(4)(11 \text{ ft})}(11 \text{ ft}) = \mathbf{1250\ lb}$
		Second floor	10,000	$\dfrac{(10{,}000 \text{ lb})}{(4)(11 \text{ ft})}(11 \text{ ft}) = \mathbf{2500\ lb}$
SW2	Seismic	Roof	2380	$\dfrac{(2380 \text{ lb})}{(4)(12 \text{ ft}) + (2)(8 \text{ ft})}(12 \text{ ft}) = \mathbf{447\ lb}$
		Second floor	3460	$\dfrac{(3460 \text{ lb})}{(4)(12 \text{ ft}) + (2)(8 \text{ ft})}(12 \text{ ft}) = \mathbf{649\ lb}$
	Wind	Roof	3300	$\dfrac{(3300 \text{ lb})}{(4)(12 \text{ ft}) + (2)(8 \text{ ft})}(12 \text{ ft}) = \mathbf{619\ lb}$
		Second floor	4700	$\dfrac{(4700 \text{ lb})}{(4)(12 \text{ ft}) + (2)(8 \text{ ft})}(12 \text{ ft}) = \mathbf{882\ lb}$
SW3	Seismic	Roof	2380	$\dfrac{(2380 \text{ lb})}{(4)(12 \text{ ft}) + (2)(8 \text{ ft})}(8 \text{ ft}) = \mathbf{298\ lb}$
		Second floor	3460	$\dfrac{(3460 \text{ lb})}{(4)(12 \text{ ft}) + (2)(8 \text{ ft})}(8 \text{ ft}) = \mathbf{433\ lb}$
	Wind	Roof	3300	$\dfrac{(3300 \text{ lb})}{(4)(12 \text{ ft}) + (2)(8 \text{ ft})}(8 \text{ ft}) = \mathbf{413\ lb}$
		Second floor	4700	$\dfrac{(4700 \text{ lb})}{(4)(12 \text{ ft}) + (2)(8 \text{ ft})}(8 \text{ ft}) = \mathbf{588\ lb}$

FIGURE 9.24 Free-body diagram of shear wall SW1.

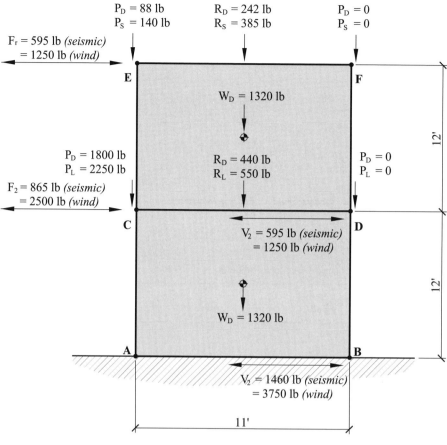

FIGURE 9.25 Free-body diagram of shear wall SW2.

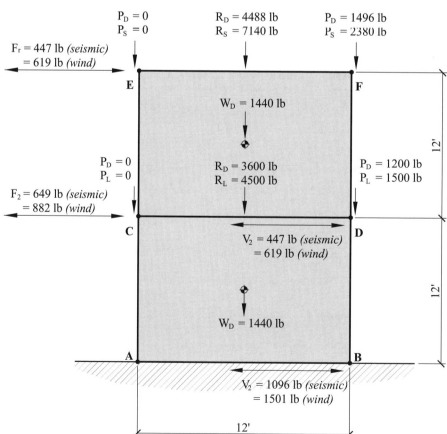

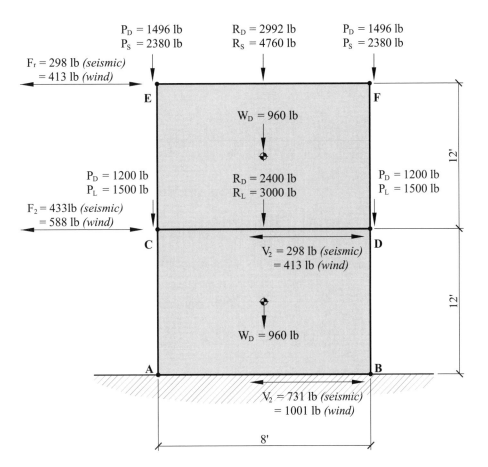

FIGURE 9.26 Free-body diagram of shear wall SW3.

Maximum Force in the Tension Chord

The maximum force that can occur in the tension chord force will now be calculated for each shear wall using the applicable IBC load combinations from Chapter 2. The reader should refer to Section 7.2 where the equations for the forces in the tension and compression chords of the shear wall are derived. It will be recalled that because the lateral wind or seismic loads can act in any direction, the summation of moments in the shear wall is taken about a point on the left or right end chord, whichever produces the greater chord forces.

SW1 wind loads (refer to Figure 9.24 for loads):

$$OM_2 = (1250)(12 \text{ ft}) = 15{,}000 \text{ ft-lb}$$

$$RM_{D2} = (242 + 1{,}320)\left(\frac{11 \text{ ft}}{2}\right) = 8591 \text{ ft-lb}$$

$$T_2 = \frac{OM_2 - 0.6 RM_{D2}}{w} \text{ [from equation (7.3)]}$$

$$= \frac{15{,}000 - (0.6)(8591)}{11 \text{ ft}} = \textbf{895 lb} \text{ (net uplift)}$$

$$OM_1 = (1250)(12 \text{ ft} + 12 \text{ ft})] + [(2500)(12 \text{ ft}) = 60{,}000 \text{ ft-lb}$$

$$RM_{D1} = \left[(242 + 440 + 1320 + 1320)\left(\frac{11 \text{ ft}}{2}\right)\right] = 18{,}271 \text{ ft-lb}$$

$$T_1 = \frac{OM_1 - 0.6RM_{D1}}{w} \text{ [from equation (7.4)]}$$

$$= \frac{60{,}000 - (0.6)(18{,}271)}{11 \text{ ft}} = \textbf{4458 lb} \text{ (net uplift)}$$

SW1 seismic loads (refer to Figure 9.24 for loads):
$\rho = 1.0$ and $S_{D5} = 0.267$

$$OM_2 = (595)(12 \text{ ft}) = 7140 \text{ ft-lb}$$

$$T_2 = \frac{0.7OM_2 - 0.56RM_{D2}}{w} \text{ [from equation (7.5)]}$$

$$= \frac{(0.7)(7140) - (0.56)(8591)}{11 \text{ ft}} = \textbf{17 lb} \text{ (net uplift)}$$

$$OM_1 = [(595)(12 \text{ ft} + 12 \text{ ft})] + [(865)(12 \text{ ft})] = 24{,}660 \text{ ft-lb}$$

$$T_1 = \frac{0.7OM_1 - 0.56RM_{D1}}{w} \text{ [from equation (7.6)]}$$

$$= \frac{(0.7)(24{,}660) - (0.56)(18{,}271)}{11 \text{ ft}} = \textbf{639 lb}$$

SW2 wind loads (refer to Figure 9.25 for loads):

$$OM_2 = (619)(12 \text{ ft}) = 7428 \text{ ft-lb}$$

$$RM_{D2} = (4488 + 1440)\left(\frac{12 \text{ ft}}{2}\right) = 35{,}568 \text{ ft-lb}$$

$$T_2 = \frac{OM_2 - 0.6RM_{D2}}{w} \text{ [from equation (7.3)]}$$

$$= \frac{7428 - (0.6)(35{,}568)}{12 \text{ ft}} = \textbf{-1159 lb} \text{ (minus sign indicates no net uplift)}$$

$$OM_1 = (619)(12 \text{ ft} + 12 \text{ ft}) + (882)(12 \text{ ft}) = 25{,}440 \text{ ft-lb}$$

$$RM_{D1} = \left[(4488 + 3600 + 1440 + 1440)\left(\frac{12 \text{ ft}}{2}\right)\right] = 65{,}808 \text{ ft-lb}$$

$$T_1 = \frac{OM_1 - 0.6RM_{D1}}{w} \text{ [from equation (7.4)]}$$

$$= \frac{25{,}440 - (0.6)(65{,}808)}{12 \text{ ft}} = \textbf{-1170 lb} \text{ (minus sign indicates no net uplift)}$$

SW2 seismic loads (refer to Figure 9.25 for loads):
$\rho = 1.0$ and $S_{D5} = 0.267$

$$OM_2 = (447)(12 \text{ ft}) = 5364 \text{ ft-lb}$$

$$T_2 = \frac{0.7OM_2 - 0.56RM_{D2}}{w} \text{ [from equation (7.5)]}$$

$$= \frac{(0.7)(5364) - (0.56)(35{,}568)}{12 \text{ ft}} = \textbf{-1346 lb} \text{ (minus sign indicates no net uplift)}$$

$OM_1 = (447)(12 \text{ ft} + 12 \text{ ft})] + (649)(12 \text{ ft}) = 18{,}516 \text{ ft-lb}$

$T_1 = \dfrac{0.7 OM_1 - 0.56 RM_{D1}}{w}$ [from equation (7.6)]

$= \dfrac{(0.7)(18{,}516) - (0.56)(65{,}808)}{12 \text{ ft}} = \mathbf{-1991 \text{ lb}}$ (minus sign indicates no net uplift)

SW3 wind loads (refer to Figure 9.26 for loads):

$OM_2 = (413)(12 \text{ ft}) = 4956 \text{ ft-lb}$

$RM_{D2} = [(1496)(8 \text{ ft})] + \left[(2992 + 960)\left(\dfrac{8 \text{ ft}}{2}\right)\right] = 27{,}776 \text{ ft-lb}$

$T_2 = \dfrac{OM_2 - 0.6 RM_{D2}}{w}$ [from equation (7.3)]

$= \dfrac{4956 - (0.6)(27{,}776)}{8 \text{ ft}} = \mathbf{-1463 \text{ lb}}$ (minus sign indicates no net uplift)

$OM_1 = (413)(12 \text{ ft} + 12 \text{ ft}) + (588)(12 \text{ ft}) = 16{,}968 \text{ ft-lb}$

$RM_{D1} = [(1496 + 1200)(8 \text{ ft})] + \left[(2992 + 2400 + 960 + 960)\left(\dfrac{8 \text{ ft}}{2}\right)\right] = 50{,}816 \text{ ft-lb}$

$T_1 = \dfrac{OM_1 - 0.6 RM_{D1}}{w}$ [from equation (7.4)]

$= \dfrac{(16{,}968) - (0.6)(50{,}816)}{8 \text{ ft}} = \mathbf{-1690 \text{ lb}}$ (minus sign indicates no net uplift)

SW3 seismic loads (refer to Figure 9.26 for loads):

$\rho = 1.0$ and $S_{D5} = 0.267$

$OM_2 = (298)(12 \text{ ft}) = 3576 \text{ ft-lb}$

$T_2 = \dfrac{0.7 OM_2 - 0.56 RM_{D2}}{w}$ [from equation (7.5)]

$= \dfrac{(0.7)(3576) - (0.56)(27{,}776)}{8 \text{ ft}} = \mathbf{-1631 \text{ lb}}$ (minus sign indicates no net uplift)

$OM_1 = (298)(12 \text{ ft} + 12 \text{ ft}) + (433)(12 \text{ ft}) = 12{,}348 \text{ ft-lb}$

$T_1 = \dfrac{0.7 OM_1 - 0.56 RM_{D1}}{w}$ [from equation (7.6)]

$= \dfrac{(0.7)(12{,}348) - (0.56)(50{,}816)}{8 \text{ ft}} = \mathbf{-2476 \text{ lb}}$ (minus sign indicates no net uplift)

Maximum Force in Compression Chord

The maximum compression chord force will now be calculated for each shear wall using the applicable load combinations from Chapter 2. Recall that in calculating the resisting moments (RM_{D1}, RM_{D2}, RM_{T1}, RM_{T2}) in equations (7.7) to (7.10), it should be noted that the full P loads or header reactions (see Figure 7.8) and the tributary R and W (i.e., distributed) loads should be used since the compression chords only support distributed gravity loads that are tributary to it in addition to the concentrated header reactions or P loads (see Section 7.2 for further discussion).

SW1 wind loads (refer to Figure 9.24 for loads):

$OM_2 = 15{,}000$ ft-lb (from tension chord force calculations)

$$RM_{T2} = (385)\left(\frac{2 \text{ ft}}{2}\right) + 140(11 \text{ ft}) = 1925 \text{ ft-lb}$$

$$RM_{D2} = (242 + 1320)\left(\frac{2 \text{ ft}}{2}\right) + 88(11 \text{ ft}) = 2530 \text{ ft-lb}$$

$$C_2 = \frac{0.75 OM_2 + RM_{D2} + 0.75 RM_{T2}}{w} \text{ [from equation (7.7)]}$$

$$= \frac{(0.75)(15{,}000) + (2530) + (0.75)(1925)}{11 \text{ ft}} = \mathbf{1384 \text{ lb}}$$

$OM_1 = 60{,}000$ ft-lb (from tension chord force calculations)

$$RM_{T1} = [(2{,}250 + 140)(11 \text{ ft})] + \left[(385 + 550)\left(\frac{2 \text{ ft}}{2}\right)\right] = 27{,}225 \text{ ft-lb}$$

$$RM_{D1} = [(1800 + 88)(11 \text{ ft})] + \left[(242 + 440 + 1320 + 1320)\left(\frac{2 \text{ ft}}{2}\right)\right] = 24{,}090 \text{ ft-lb}$$

$$C_1 = \frac{0.75 OM_1 + RM_{D1} + 0.75 RM_{T1}}{w} \text{ [from equation (7.8)]}$$

$$= \frac{(0.75)(60{,}000) + (24{,}090) + (0.75)(27{,}225)}{11 \text{ ft}} = \mathbf{8137 \text{ lb}}$$

SW1 seismic loads (refer to Figure 9.24 for loads):

$\rho = 1.0$ and $S_{DS} = 0.267$

$OM_2 = 7140$ ft-lb (from tension chord force calculations)

$RM_{T2} = 1925$ ft-lb (from wind loads)

$RM_{D2} = 2530$ ft-lb (from wind loads)

$$C_2 = \frac{0.525 OM_2 + 1.03 RM_{D2} + 0.75 RM_{T2}}{w} \text{ [from equation (7.9)]}$$

$$= \frac{(0.525)(7140) + (1.03)(2530) + (0.75)(1925)}{8 \text{ ft}} = \mathbf{975 \text{ lb}}$$

$OM_1 = 24{,}660$ ft-lb (from tension chord force calculations)

$RM_{T1} = 27{,}225$ ft-lb (from wind loads)

$RM_{D1} = 24{,}090$ ft-lb (from wind loads)

$$C_1 = \frac{0.525 OM_1 + 1.03 RM_{D1} + 0.75 RM_{T1}}{w} \text{ [from equation (7.10)]}$$

$$= \frac{(0.525)(24{,}660) + (1.03)(24{,}090) + (0.75)(27{,}225)}{8 \text{ ft}} = \mathbf{7272 \text{ lb}}$$

SW2 wind loads (refer to Figure 9.25 for loads):

$$OM_2 = 7428 \text{ ft-lb (from tension chord force calculations)}$$

$$RM_{T2} = (2380)(12 \text{ ft}) + \left[(7140)\left(\frac{2 \text{ ft}}{2}\right)\right] = 35{,}700 \text{ ft-lb}$$

$$RM_{D2} = (1496)(12 \text{ ft}) + \left[(4488 + 1440)\left(\frac{2 \text{ ft}}{2}\right)\right] = 23{,}880 \text{ ft-lb}$$

$$C_2 = \frac{0.75 OM_2 + RM_{D2} + 0.75 RM_{T2}}{w} \text{ [from equation (7.7)]}$$

$$= \frac{(0.75)(7428) + (23{,}880) + (0.75)(35{,}700)}{12 \text{ ft}} = \mathbf{4686 \text{ lb}}$$

$$OM_1 = 25{,}440 \text{ ft-lb (from tension chord force calculations)}$$

$$RM_{T1} = [(2380 + 1500)(12 \text{ ft})] + \left[(7140 + 4500)\left(\frac{2 \text{ ft}}{2}\right)\right] = 58{,}200 \text{ ft-lb}$$

$$RM_{D1} = [(1496 + 1200)(12 \text{ ft})] + \left[(4488 + 3600 + 1440 + 1440)\left(\frac{2 \text{ ft}}{2}\right)\right]$$

$$= 43{,}320 \text{ ft-lb}$$

$$C_1 = \frac{(0.75 OM_1 + RM_{D1} + 0.75 RM_{T1})}{w} \text{ [from equation (7.8)]}$$

$$= \frac{(0.75)(25{,}440) + (43{,}320) + (0.75)(58{,}200)}{12 \text{ ft}} = \mathbf{8838 \text{ lb}}$$

SW2 seismic loads (refer to Figure 9.25 for loads):

$$\rho = 1.0 \text{ and } S_{DS} = 0.267$$

$$OM_2 = 5364 \text{ ft-lb (from tension chord force calculations)}$$

$$RM_{T2} = 35{,}700 \text{ ft-lb (from wind loads)}$$

$$RM_{D2} = 23{,}880 \text{ ft-lb (from wind loads)}$$

$$C_2 = \frac{0.525 OM_2 + 1.03 RM_{D2} + 0.75 RM_{T2}}{w} \text{ [from equation (7.9)]}$$

$$= \frac{(0.525)(5364) + (1.03)(23{,}880) + (0.75)(35{,}700)}{12} = \mathbf{4516 \text{ lb}}$$

$$OM_1 = 18{,}516 \text{ ft-lb (from tension chord force calculations)}$$

$$RM_{T1} = 58{,}200 \text{ ft-lb (from wind loads)}$$

$$RM_{D1} = 43{,}320 \text{ ft-lb (from wind loads)}$$

$$C_1 = \frac{0.525 OM_1 + 1.03 RM_{D1} + 0.75 RM_{T1}}{w} \text{ [from equation (7.10)]}$$

$$= \frac{(0.525)(18{,}516) + (1.03)(43{,}320) + (0.75)(58{,}200)}{12 \text{ ft}} = \mathbf{8166 \text{ lb}}$$

SW3 wind loads (refer to Figure 9.26 for loads):

$OM_2 = 4956$ ft-lb (from tension chord force calculations)

$$RM_{T2} = (2380)(8 \text{ ft}) + \left[(4760)\left(\frac{2 \text{ ft}}{2}\right)\right] = 23{,}800 \text{ ft-lb}$$

$$RM_{D2} = (1496)(8 \text{ ft}) + \left[(2992 + 960)\left(\frac{2 \text{ ft}}{2}\right)\right] = 15{,}920 \text{ ft-lb}$$

$$C_2 = \frac{0.75OM_2 + RM_{D2} + 0.75RM_{T2}}{w} \text{ [from equation (7.7)]}$$

$$= \frac{(0.75)(4956) + (15{,}920) + (0.75)(23{,}800)}{8 \text{ ft}} = \mathbf{4686 \text{ lb}}$$

$OM_1 = 16{,}968$ ft-lb (from tension chord force calculations)

$$RM_{T1} = [(2380 + 1500)(8 \text{ ft})] + \left[(4760 + 3000)\left(\frac{2 \text{ ft}}{2}\right)\right] = 38{,}800 \text{ ft-lb}$$

$$RM_{D1} = [(1496 + 1200)(8 \text{ ft})] + \left[(2992 + 2400 + 960 + 960)\left(\frac{2 \text{ ft}}{2}\right)\right] = 28{,}880 \text{ ft-lb}$$

$$C_1 = \frac{(0.75OM_1 + RM_{D1} + 0.75RM_{T1})}{w} \text{ [from equation (7.8)]}$$

$$= \frac{(0.75)(16{,}968) + (28{,}880) + (0.75)(38{,}800)}{8 \text{ ft}} = \mathbf{8838 \text{ lb}}$$

SW3 seismic loads (refer to Figure 9.26 for loads):

$\rho = 1.0$ and $S_{D5} = 0.267$

$OM_2 = 3576$ ft-lb (from tension chord force calculations)

$RM_{T2} = 23{,}800$ ft-lb (from wind loads)

$RM_{D2} = 15{,}920$ ft-lb (from wind loads)

$$C_2 = \frac{0.525OM_2 + 1.03RM_{D2} + 0.75RM_{T2}}{w} \text{ [from equation (7.9)]}$$

$$= \frac{(0.525)(3576) + (1.03)(15{,}920) + (0.75)(23{,}800)}{8 \text{ ft}} = \mathbf{4516 \text{ lb}}$$

$OM_1 = 12{,}348$ ft-lb (from tension chord force calculations)

$RM_{T1} = 38{,}800$ ft-lb (from wind loads)

$RM_{D1} = 28{,}880$ ft-lb (from wind loads)

$$C_1 = \frac{0.525OM_1 + 1.03RM_{D1} + 0.75RM_{T1}}{w} \text{ [from equation (7.10)]}$$

$$= \frac{(0.525)(12{,}348) + (1.03)(28{,}880) + (0.75)(38{,}800)}{8 \text{ ft}} = \mathbf{8166 \text{ lb}}$$

The shear wall design forces are summarized in Table 9.17.

TABLE 9.17 Summary of Shear Wall Chord Design Forces[a] (lb)

		Level 2			Ground Level		
Load	Wall	V_2	T_2	C_2	V_1	T_1	C_1
Seismic	SW1	595	17	975	1460	639	7272
	SW2	447	−1346	4516	1096	−1991	8166
	SW3	298	−1631	4516	731	−2476	8166
Wind	SW1	1250	**895**	1384	**3750**	**4458**	8137
	SW2	619	−1159	4686	1501	−1170	8838
	SW3	413	−1463	4686	1001	−1690	**8838**

[a] A negative value for T_1 and T_2 forces actually indicates net a compressive force; hold-down anchors would theoretically, not be required. Bold values indicate controlling values that will be used for design. Note that the value of the wind uplift force of 710 lb in the shearwall chords calculated in Section 9.4 must be added to the net tension forces due to overturning from wind obtain the net uplift force on the MWFRS (see Section 9.4). For wall SW1 at the ground floor level, the total net uplift force, T_1 will become 5168 lb (i.e., 4458 lb + 710 lb).

9.13 FORCES IN HORIZONTAL DIAPHRAGM CHORDS, DRAG STRUTS, AND LAP SPLICES

With reference to Figure 9.20, the diaphragm chord forces are summarized in Table 9.18. With reference to Figures 9.27 and 9.28, the drag strut forces are as summarized in Table 9.19.

Design of Chords, Struts, and Splices

The maximum force in the chord and drag struts is a diaphragm chord force of 2000 lb (see Table 9.18). It is common practice to use the maximum force in the design of all chords and lap splices.

Design for Tension

The maximum applied tension force, $T_{applied} = 2000$ lb. For stress grade, assume spruce-pine-fir Stud grade. Since a 2 × 6 is dimension lumber, use NDS-S Table 4A. From the table we obtain the design stress values:

TABLE 9.18 Seismic and Wind Loads on Horizontal Diaphragm Chords

Direction	Level	w (plf) (from Figure 9.20)	L (ft)	b (ft)	$M = \dfrac{wL^2}{8}$ (ft-lb)	$F_{dc} = \dfrac{M}{b}$ (lb)
		Seismic Loads				
Transverse	Roof	49.6	48	30	$(49.6)(48^2)/8 = $ **14,285**	$14{,}285/30 = $ **477**
	Second floor	72.1	48	30	$(72.1)(48^2)/8 = $ **20,765**	$20{,}765/30 = $ **693**
Longitudinal	Roof	79.4	30	48	$(79.4)(30^2)/8 = $ **8933**	$8933/48 = $ **187**
	Second floor	115.3	30	48	$(115.3)(30^2)/8 = $ **12,972**	$12{,}972/48 = $ **271**
		Wind Loads				
Transverse	Roof	104.2	48	30	$(104.2)(48^2)/8 = $ **30,010**	$30{,}010/30 = $ **1001**
	Second floor	208.3	48	30	$(208.3)(48^2)/8 = $ **59,990**	$59{,}990/30 = $ **2000**
Longitudinal	Roof	110	30	48	$(110)(30^2)/8 = $ **12,375**	$12{,}375/48 = $ **258**
	Second floor	156.7	30	48	$(156.7)(30^2)/8 = $ **17,629**	$17{,}629/48 = $ **367**

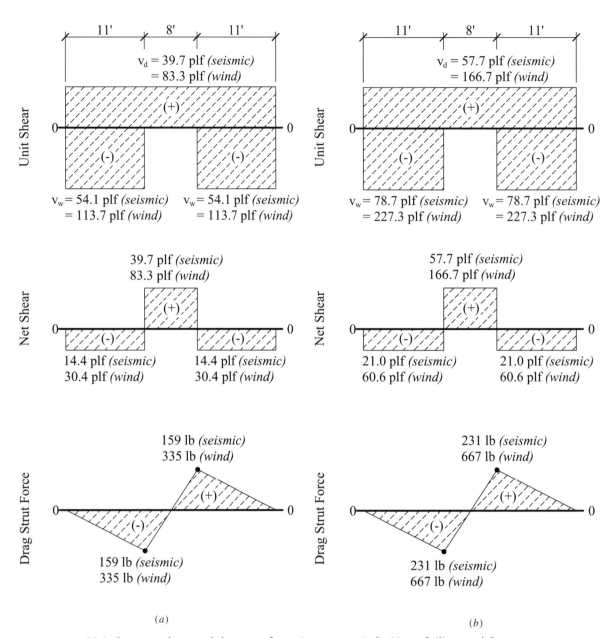

FIGURE 9.27 Unit shear, net shear, and drag strut forces (transverse wind): (a) roof; (b) second floor.

$$F_t = 350 \text{ psi}$$

$$C_D(\text{wind}) = 1.6$$

$$C_M = 1.0$$

$$C_t = 1.0$$

$$C_F(F_t) = 1.0$$

$$A_g = 8.25 \text{ in}^2 \text{ (only one 2} \times \text{6 is effective in tension)}$$

The allowable tension stress is

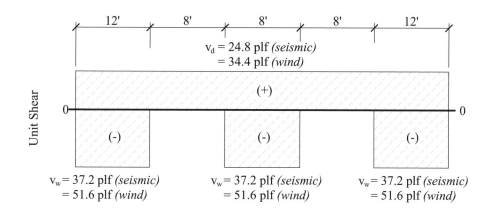
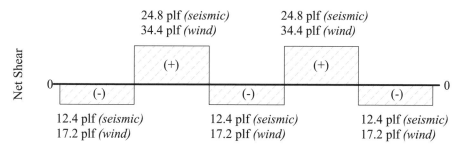
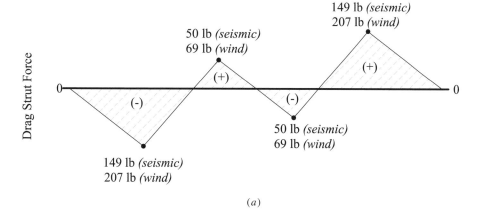

FIGURE 9.28
Unit shear, net shear, and drag strut forces (longitudinal wind): (*a*) roof; (*b*) second floor.

(*a*)

$$F'_t = F_t C_D C_M C_t C_F C_i$$
$$= (350)(1.6)(1)(1)(1)(1) = 560 \text{ psi}$$

The allowable tension force in the top plates is

$$T_{\text{allowable}} = F'_t A_g = (560 \text{ psi})(8.25 \text{ in}^2) = \mathbf{4620 \text{ lb}} > T_{\text{applied}} = 2000 \text{ lb} \quad \mathbf{OK}$$

Design for Compression

The top plate is fully braced about both axes of bending. It is braced by the stud wall for y–y or weak-axis bending, and it is braced by the roof or floor diaphragm for x–x or strong-axis bending, Therefore, the column stability factor $C_p = 1.0$. The maximum applied compression force $P_{\text{max}} = 2000$ lb. For stress grade, assume spruce-pine-fir Stud grade. Since a 2 × 6 is dimension lumber, use NDS-S Table 4A for design values. From the table the following design stress values are obtained:

FIGURE 9.28 (Continued)

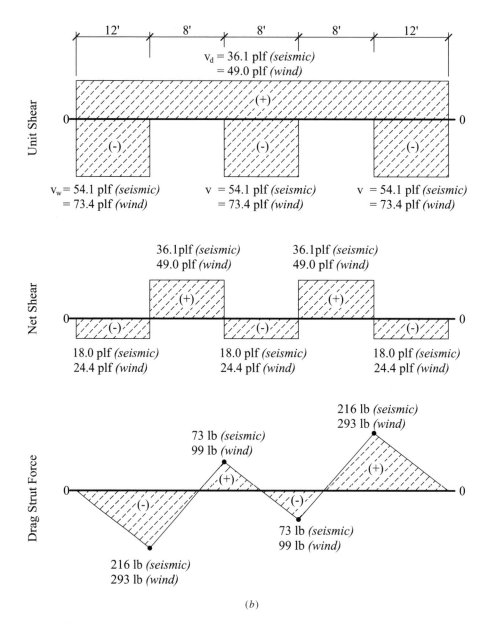

(b)

TABLE 9.19 Seismic and Wind Loads on Drag Struts

Direction	Level	v_d (plf) (from Table 9.13)	v_w (plf) (from Table 9.14)	Net Shear, (plf) $V_w - V_d$	F_{ds} (lb)
		Seismic Loads			
Transverse	Roof	39.7	54.1	14.4	**159**
	Second floor	57.7	78.7	21.0	**231**
Longitudinal	Roof	24.8	37.2	12.4	**149**
	Second floor	36.1	54.1	18.0	**216**
		Wind Loads			
Transverse	Roof	83.3	113.7	30.4	**335**
	Second floor	166.7	227.3	60.6	**667**
Longitudinal	Roof	34.4	51.6	17.2	**207**
	Second floor	49.0	73.4	24.4	**293**

$$F_c = 725 \text{ psi}$$

$$C_D(\text{wind}) = 1.6$$

$$C_M = 1.0$$

$$C_t = 1.0$$

$$C_F(F_c) = 1.0$$

$$C_i = 1.0 \text{ (assumed)}$$

$$C_P = 1.0$$

$$A_g = 16.5 \text{ in}^2 \text{ (for two 2} \times \text{6's)}$$

The allowable compression stress is

$$F'_c = F_c C_D C_M C_t C_F C_i C_P$$
$$= (725)(1.6)(1.0)(1.0)(1.0)(1.0)(1.0) = 1160 \text{ psi}$$

The allowable compression force in the top plates is

$$P_{\text{allowable}} = F'_c A_g = (1160 \text{ psi})(16.5 \text{ in}^2) = \textbf{19,140 lb} > P_{\text{applied}} = 2000 \text{ lb} \quad \textbf{OK}$$

Lap Splice

The IBC recommended minimum length of a top plate lap splice is 4 ft. If the length of a lap splice length is *more than twice* the Code minimum length of 4 ft (i.e. 8 ft or more), the nailed or bolted connection each side of the splice can be designed for *one-half* of the full chord force for double top plates and *one-third* of the full chord force for triple top plates. For lap splice lengths less than 8 ft, the splice is designed for the full chord or drag strut force. The IBC requires a minimum of 8–16d nails each side of a splice connection.

We will design both nailed and bolted lap splices.

(a) Using a nailed connection and assuming a lap splice length of 6 ft, we will design the connection for the full chord force. The maximum applied tension or compression load on the top plate splice $T = 2000$ lb. Try 16d common nails.

$$L = 3\tfrac{1}{2} \text{ (length of nail)}$$

$$D = 0.162 \text{ in. (diameter of nail)}$$

Minimum spacing (see Table 8.1):

$$\text{end distance} = 15D = 2.43 \text{ in.} \Rightarrow \textbf{2.5 in.}$$

$$\text{edge distance} = 2.5D = 0.41 \text{ in.} \Rightarrow \textbf{0.5 in.}$$

$$\text{center-to-center spacing} = 15D = 2.43 \text{ in.} \Rightarrow \textbf{2.5 in.}$$

$$\text{row spacing} = 5D = 0.81 \text{ in.} \Rightarrow \textbf{1 in.}$$

Allowable lateral resistance per nail:

$$Z' = Z C_D C_M C_t C_g C_\Delta C_{eg} C_{di} C_{tn}$$

$$C_D = 1.6 \text{ (wind)}$$

$$C_M = 1.0 \text{ (MC} \leq 19\%)$$

$$C_t = 1.0 \text{ (}T \leq 100°\text{F)}$$

$$C_g, C_\Delta, C_{eg}, C_{di}, C_{tn} = 1.0$$

The tabulated single shear value for 16d nail (spruce pine-fir framing), $Z = (120 \text{ lb})(p/10D)$.

$$Z = (120)[1.5 \text{ in.}/(10 \times 0.162)] = \mathbf{111 \text{ lb}} \text{ (see NDS Code Table 11N and footnote 3)}$$

The allowable shear per nail is

$$Z' = ZC_D C_M C_t C_g C_\Delta C_{eg} C_{di} C_{tn}$$
$$= (111)(1.6)(1)(1)(1)(1)(1)(1)(1) = 177 \text{ lb per nail}$$

$$\text{No. of nails req'd} = \frac{T}{Z'} = \frac{2000 \text{ lb}}{177 \text{ lb}} = 11.3 \Rightarrow \mathbf{12 \text{ nails}}$$

Using a 6 ft 0 in. lap splice length and two rows of nails (which gives 6 nails per row), and assuming a center to center spacing of 3 inches, we obtain

$$\text{minimum nail penetration } P \geq 6D = 6(0.162) = 0.972 \text{ in.} \quad \mathbf{OK}$$

Use twelve 16d nails in two rows *each side* of the splice (Figure 9.29).

(b) *Bolted Lap Splice*

Nailed lap splices are more commonly used for top plates, but as an alternate, we will design a bolted lap splice in this section. Assuming a *lap splice length of 6 ft*, which is less than twice the Code minimum lap splice length of 4 ft, therefore, the lap splice connection must be designed for the full chord force. The maximum applied tension or compression force on the top plate splice (see Tables 9.18 and 9.19) is

$$T = 2000 \text{ lb}$$

Assume $\frac{5}{8}$ in. diameter thru-bolts
The length of the fastener is given as

$$L = 1.5 \text{ in. (side member)} + 1.5 \text{ in. (main member)} = 3 \text{ in.}$$

The minimum bolt spacing to achieve a geometry factor C_Δ of 1.0 are as follow:

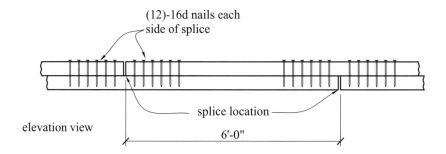

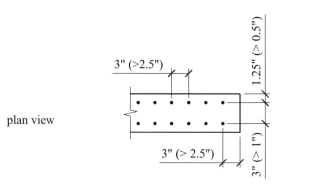

FIGURE 9.29 Nailed lap splice detail.

$$\text{end distance} = 7D = (7)(\tfrac{5}{8}\text{ in.}) = 4.38\text{ in.} \Rightarrow \textbf{4.5 in.}$$

edge distance: For $L/D = 3/0.625 = 4.8 < 6$,

$$= 1.5D = 0.94\text{ in.}$$

$$\text{center-to-center spacing} = 4D = 2.5\text{ in.}$$

$$\text{row spacing} = 1.5D = 0.94\text{ in.}$$

Allowable lateral load per bolt, $Z' = ZC_D C_M C_t C_g C_\Delta C_{eg} C_{di} C_{tn}$
where
$C_D = 1.6$ (wind)
$C_M = 1.0$ (MC $\leq$ 19%)
$C_t = 1.0$ (T $\leq$ 100° F)
$C_\Delta, C_{eg}, C_{di}, C_{tn} = 1.0$

C_g: $A_s = A_m = (1.5\text{ in.}) \times (5.5\text{ in.}) = 8.25\text{ in.}^2 \Rightarrow A_s/A_m = 1.0$

Assume number of fasteners in a row = 3 (this will be verified later)

Entering NDS Code Table 10.3.6A with A_s/A_m of 1.0, A_s of 8.25 in^2, and 3 bolts in a row, the group factor, C_g can be obtained by linear interpolation as

$$C_g = 0.97 + \left(\frac{0.99 - 0.97}{12\text{ in.}^2 - 5\text{ in.}^2}\right)(8.25\text{ in.}^2 - 5\text{ in.}^2)$$

$$= 0.98$$

Since the lateral load is *parallel* to the grain in both the *side* and *main* members,

$$\mathbf{Z} = \mathbf{Z}_\parallel = 510\text{ lb per bolt (See NDS Code Table 11A)}$$

$$\mathbf{Z'} = \mathbf{Z}_\parallel C_D C_M C_t C_g C_\Delta C_{eg} C_{di} C_{tn}$$

$$= 510 \times 1.6 \times 1 \times 1 \times 0.98 \times 1 \times 1 \times 1 \times 1 = 799\text{ lb per bolt}$$

Number of bolts required to resist the axial force in the top plate splice is,

$$= \mathbf{T/Z'} = 2000/799 = 2.5 < \mathbf{3\ bolts}$$

Since the number of bolts required above is *less than or equal to* the number of bolts initially assumed in calculating C_g, this implies that the C_g value obtained above is conservative, and the 3 bolts in a row initially assumed is adequate.

It should noted, however, that if the calculated number of bolts required had been *greater* than the number of bolts initially assumed, the C_g value would have had to be recalculated using the newly calculated required number of bolts; and the process is repeated until the number of bolts required in the current cycle is *less than or equal to* the number of bolts used in the previous cycle.

Therefore, **use 3 – $\tfrac{5}{8}$ in. diameter bolts** in a single row each side of the lap splice

Hold-Down Anchors

From Table 9.17 the maximum uplift at the second floor was 895 lb due to windloads. At the ground level, the maximum tension was 4458 lb also due to wind loads. Adding the 710 lb from the uplift on the MWFRS (see Section 9.4), the total net uplift forces become 1605 lb and 5168 lb at the second floor and ground level, respectively. Although both connectors are typically selected from a variety of preengineered products, the connector at the second floor will be designed.

Hold-Down Strap

Assuming an 18-gage ASTM A653 grade 33 strap and 10d common nails, determine the strap length and nail layout.

$$L = 3 \text{ in. (length of nail)}$$
$$D = 0.148 \text{ in. (diameter of nail)}$$

Minimum spacing (see Table 8.1):

$$\text{end distance} = 15D = 2.22 \text{ in.} \Rightarrow \textbf{2.5 in.}$$
$$\text{edge distance} = 2.5D = 0.37 \text{ in.} \Rightarrow \textbf{0.5 in.} \text{ (0.75 in. provided)}$$
$$\text{center-to-center spacing} = 15D = 2.22 \text{ in.} \Rightarrow \textbf{2.5 in.}$$

row spacing not applicable since there is only one row of nails

$$\text{minimum } p = 10D = 1.48 \text{ in. (approx. 3 in. provided)}$$

From NDS Table 11P, the nominal lateral design value is $Z = 99$ lb. The allowable shear per nail

$$Z' = ZC_D C_M C_t C_g C_\Delta C_{eg} C_{di} C_{tn}$$
$$= (99)(1.6)(1)(1)(1)(1)(1)(1)(1) = 158.4 \text{ lb per nail}$$

$$\text{No. of nails req'd} = \frac{T}{Z'} = \frac{1605 \text{ lb}}{158.4 \text{ lb}} = 10.2 \Rightarrow \textbf{6 nails each side}$$

(a steel strap will be provided on both sides of the studs to avoid eccentric loading).
The required length of strap

$$L_{min} = (2)[(2.5 \text{ in. end dist.}) + (5 \text{ spaces})(2.5 \text{ in.}) + (2.5 \text{ in. end dist.})]$$
$$+ (18 \text{ in. floor structure}) = \textbf{53 in.}$$

See Figure 9.30 for strap detail.

Hold-down Anchor
A generic anchor selection table is shown in Figure 9-31. Comparing the maximum tension load of 4088 lb., the hold-down anchor HD4 is selected. With Spruce-Pine-Fir lumber and a load duration factor of 1.6 (160%), the capacity of this connector is 5200 lb. Note that this hold-down connector will be required at each end of the shear wall since the lateral loads can act in either direction. The effect of eccentricity on this connection will be discussed in a later section when the chords are designed.

Sill Anchors

From Table 9.17 the maximum shear at the ground level is 3750 lb shear wall SW1. The anchor bolts will now be designed.

Assuming $\frac{1}{2}$-in. anchor bolts embedded 7 in. into concrete with a minimum 28-day strength of 2500 psi, the allowable shear in the anchor bolts parallel to the grain of the sill plate

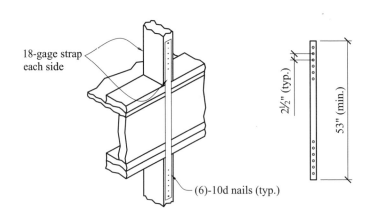

FIGURE 9.30 Hold-down strap detail.

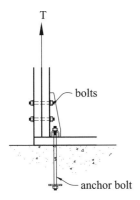

FIGURE 9.31 Hold-down anchor detail.

Model #	Fasteners		DF-L or S.P. Allowable Tension Load (lb) 160%		S.P.F. or H.F. Allowable Tension Load (lb) 160%	
	Anchor Bolts	To studs (bolts)	bolt length in wood		bolt length in wood	
			1½"	3"	1½"	3"
HD1	(1)-⅝"	(2)-½"	2,900	4,200	2,200	3,600
HD4	(1)-⅞"	(2)-½"	3,700	6,300	3,200	5,200
xx	xx	xx	xx	xx	xx	xx

$$Z' = Z C_D C_M C_t C_g C_\Delta C_{eg} C_{di} C_{tn}$$

$$C_D = 1.6 \text{ (wind or seismic)}$$

$$C_M = 1.0 \text{ (MC} \leq 19\%)$$

$$C_t = 1.0 \text{ (T} \leq 100° \text{F)}$$

$$C_g, C_\Delta, C_{eg}, C_{di}, C_{tn} = 1.0$$

Since the base shear is parallel to the grain of the sill plate; use

$$Z_\| = 590 \text{ lb (NDS Table 11E)}$$

Minimum spacing (see Table 8.3):

$$\text{end distance} = 7D = \textbf{3.5 in.}$$

$$\text{edge distance} = 1.5D = \textbf{0.75 in.}$$

$$\text{center-center-spacing} = 4D = \textbf{2.0 in.}$$

row spacing is not applicable here since there is only one row of bolts.

These distances will be maintained in the detailing; therefore, $C_\Delta = 1.0$.

The allowable shear per bolt

$$Z' = Z C_D C_M C_t C_g C_\Delta C_{eg} C_{di} C_{tn}$$

$$= (590)(1.6)(1)(1)(1)(1)(1)(1) = 944 \text{ lb per bolt}$$

$$\Rightarrow \text{number of sill anchor bolts required} = 3750 \text{ lb}/944 \text{ lb} = 3.97 \simeq 4 \text{ bolts}$$

Maximum bolt spacing = (11 ft wall length − 1 ft edge distance − 1 ft edge distance)/(4 bolts − 1) = 3 ft 0 in. This is less than the maximum spacing of 6 ft 0 in. permitted by the code.

Use ½-in.-diameter anchor bolts at 3 ft 0 in. o.c. See Figure 9.32.

9.14 SHEAR WALL CHORD DESIGN

The shear wall chords will now be designed. From Table 9.17 the maximum uplift at the ground level is 5168 lb and the maximum compression is 8838 lb. In addition to designing the shear wall chords for the maximum tension and compression loads, a connector needs to be provided to transfer the uplift load to the foundation. It is common practice to select a preengineered connector to resist this load (see Figure 9.33 various hold-down anchors).

Note that if the connector is one-sided such that the load path is eccentric to the shear wall chord, the chord must be designed to resist the tension load plus the bending moment induced by the eccentric load (see Figure 9.34) using the methods discussed in Chapter 5. If the hold-down connector requires bolts through the chords, the net area of the chord should be used in

FIGURE 9.32 Anchor bolt details.

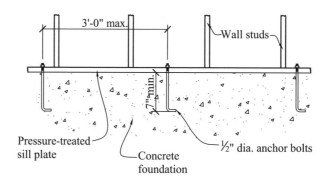

the design. For this example, a hold-down anchor with a single anchor bolt placed 1.25 in. from the face of the inner stud and two $\frac{1}{2}$-in. bolts through the chords was selected. Two 2 × 6 chords will be assumed. Note that multiple chords need to be nailed or bolted together in accordance with NDS Section 15.3.3 (nailed) or 15.3.4 (bolted).

The following data will be used for the design of the shear wall chords:

$$C_D \text{ (wind)} = 1.6$$

$$C_M = 1.0$$

$$C_t = 1.0$$

$$C_F(F_t, F_b) = 1.3$$

$$C_F(F_c) = 1.1$$

$$C_{fu} = 1.15 \text{ (for the tension case)}$$

$$C_i = 1.0$$

$$C_r = 1.0$$

From NDS Table 4A (spruce-pine-fir, select structural):

$$F_b = 1250 \text{ psi}$$

$$F_t = 700 \text{ psi}$$

$$F_c = 1400 \text{ psi}$$

$$E_{min} = 0.55 \times 10^6 \text{ psi}$$

For two 2 × 6's:

$$d = 5.5 \text{ in.}$$

$$b = (2)(1.5 \text{ in.}) = 3 \text{ in.}$$

$$A_g = (2)(8.25 \text{ in}^2) = 16.5 \text{ in}^2$$

$$A_n = A_g - A_{holes} = (16.5) - (2)(1.5 \text{ in.})(\tfrac{1}{2} \text{ in.} + \tfrac{1}{8} \text{ in.}) = 14.6 \text{ in}^2$$

$$S_y = (3)(31.64 \text{ in}^3) = 94.92 \text{ in}^3$$

$$L_u = 12 \text{ ft} = 144 \text{ in. (floor-to-floor height)}$$

Design for Compression
Check the slenderness ratio:

$$\frac{\ell_e}{d} = \frac{(1.0)(144 \text{ in.})}{5.5 \text{ in.}} = \mathbf{26.1 < 50} \qquad \text{OK (from Section 5.4, where } l_e = K_e l_u\text{)}$$

Column stability factor C_P:

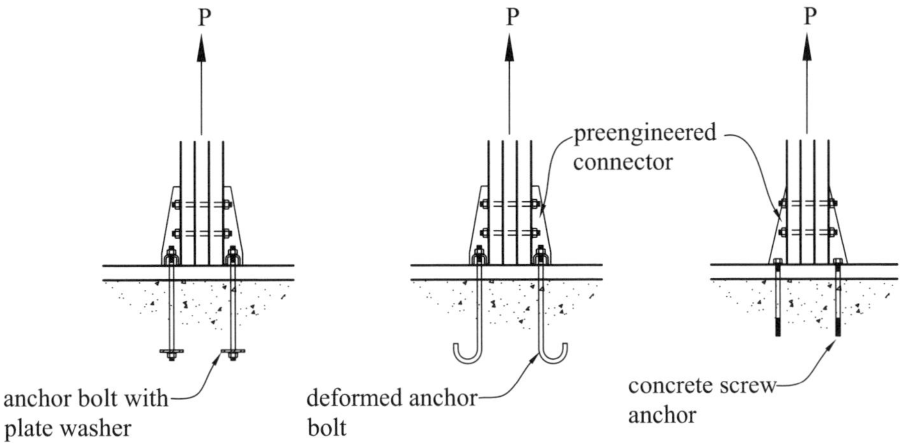

FIGURE 9.33 Hold-down anchor types.

$c = 0.8$ (sawn lumber)

$K_f = 1.0$ (Section 5.4)

$E'_{min} = E_{min} C_M C_t C_i = (0.55 \times 10^6)(1.0)(1.0)(1.0) = 0.55 \times 10^6$

$F_{cE} = \dfrac{0.822 E'_{min}}{\ell_e / d} = \dfrac{(0.822)(0.55 \times 10^6)}{(26.1)^2} = 659 \text{ psi}$

$F_c^* = F_c C_D C_M C_t C_F C_i$

$ = (1400)(1.6)(1.0)(1.0)(1.0)(1.1)(1.0) = 2464 \text{ psi}$

$C_P = K_f \left[\dfrac{1 + F_{cE}/F_c^*}{2c} - \sqrt{\left(\dfrac{1 + F_{cE}/F_c^*}{2c}\right)^2 - \dfrac{F_{cE}/F_c^*}{2c}} \right]$ [equation (5.21)]

$ = (1.0)\left[\dfrac{1 + 659/2464}{(2)(0.8)} - \sqrt{\left[\dfrac{1 + 659/2464}{(2)(0.8)}\right]^2 - \dfrac{659/2464}{0.8}} \right] = 0.250$

Check compression loads:

$F'_c = F_c^* C_P = (2464 \text{ psi})(0.250) = 616 \text{ psi}$

$P_{allow} = \begin{cases} F'_c A_g = (616 \text{ psi})(16.5 \text{ in}^2) = \mathbf{10{,}164 \text{ lb}} > \mathbf{8838 \text{ lb}} & \text{(OK at midheight)} \\ F_c^* A_n = (2464 \text{ psi})(14.6 \text{ in}^2) = \mathbf{35{,}974 \text{ lb}} > \mathbf{8838 \text{ lb}} & \text{(OK at base)} \end{cases}$

Design for Tension

The total tension force of 5168 lb (i.e., 4458 lb + 710 lb) is applied eccentrically to the shear wall chords. The applied bending moment is

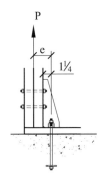

FIGURE 9.34 Eccentric hold-down anchor.

TABLE 9.20 Shear Wall and Hold-down Schedule

Mark	Height (ft)	Length (ft)	Unit Shear (lb/ft)	Shear Wall Chords	Compression Force (lb)	Tension, Hold-Down (lb)	Plywood thickness and Nailing	Sides	Level	Hold-Down Connector	Hold-down Anchor Bolt Embedment	Sill Anchors Bolts
SW1	12	11	114/341	2-2 × 6 2-2 × 6	1384/8137	895/4458 (1605/5168)[7]	$\frac{15}{32}$" Str.-1 C-DX, 8d nails @ 6" o.c. at edges, 12" o.c. Field	1 1	Second flr Ground floor	18 ga.-53" strap, (12)-10d nails HD4	7"	$\frac{1}{2}$" dia. @ 3'-0" o.c.
SW2	12	12	51.6/125	2-2 × 6 2-2 × 6	4686/8838	No Uplift	$\frac{15}{32}$" Str.-1 C-DX, 8d nails @ 6" o.c. at edges, 12" o.c. Field	1 1	Second flr Ground floor	18 ga.-53" strap, (12)-10d nails HD1	7"	$\frac{1}{2}$" dia. @ 3'-0" o.c.
SW3	12	8	51.6/125	2-2 × 6 2-2 × 6	4686/8838	No Uplift	$\frac{15}{32}$" Str.-1 C-DX, 8d nails @ 6" o.c. at edges, 12" o.c. Field	1 1	Second flr Ground floor	18 ga.-53" strap, (12)-10d nails HD1	7"	$\frac{1}{2}$" dia. @ 3'-0" o.c.

Notes:
1. All plywood panel edges shall be blocked with 2× wood blocking.
2. All plywood panels shall extend from bottom of sole or sill plate to top of the top plates.
3. Where oversized bolt holes in the sill plate are needed for ease of placement of the sill anchors, sleeve the bolt holes with metal sleeves and fill the gaps between the bolts and the inside face of the sleeve with expansive cement to ensure snug tight condition.
4. Plywood panels shall have a minimum span rating of $\frac{24}{16}$.
5. Locate edge nails at an edge distance of not less than $\frac{3}{8}$" from penal edges and at equal distance from the panel edges and the edge of the wood framing.
6. All shear wall anchor bolts shall have a minimum embedment of 7 inches into the concrete foundation wall.
7. 1605 lb and 5168 lb Uplift forces include 710 lb force due to wind uplift.

$$M = Te = (5168)(1.5 \text{ in.} + 1.25 \text{ in.}) = 14{,}212 \text{ in.-lb}$$

The allowable stresses are

$$F'_t = F_t C_D C_M C_t C_F C_i$$
$$= (700)(1.6)(1.0)(1.0)(1.0)(1.3)(1.0) = 1456 \text{ psi}$$

$$F'_b = F_b C_D C_M C_t C_F C_{fu} C_i C_r$$
$$= (1250)(1.6)(1.0)(1.0)(1.3)(1.15)(1.0)(1.0) = 2990 \text{ psi}$$

The actual stresses are

$$f_t = \frac{T}{A_n} = \frac{5168}{14.6} = 354 \text{ psi}$$

$$S_y = \frac{b_{\text{net}} d^2}{6}$$

The width of the chords is reduced by the presence of the bolts that connect the hold-down anchors to the chord member.

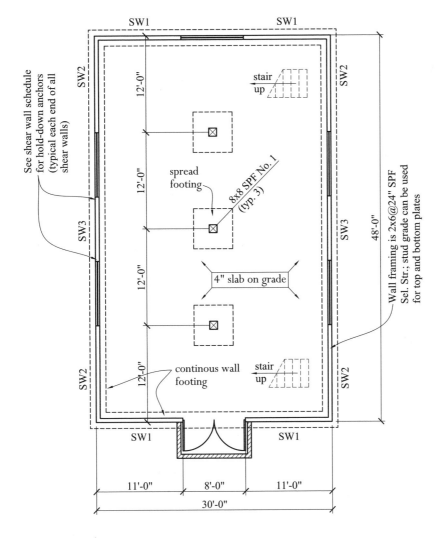

Fig. 9-35: Ground Floor Plan

FIGURE 9.35 Ground floor plan.

$$b_{net} = 5.5 \text{ in.} - (\tfrac{1}{2} \text{ in.} + \tfrac{1}{8} \text{ in.}) = 4.88 \text{ in. (effective width of chords)}$$

$$d = (2)(1.5 \text{ in.}) = 3 \text{ in.}$$

$$S_y = \frac{(4.88)(3 \text{ in.})^2}{6} = 7.31 \text{ in}^3$$

$$f_b = \frac{M}{S_y} = \frac{14{,}212}{7.31} = 1944 \text{ psi}$$

Check the combined stresses:

$$\frac{f_t}{F'_t} + \frac{f_b}{F'_b} \leq 1.0$$

$$\frac{354}{1456} + \frac{1944}{2990} = 0.89 \leq 1.0 \quad \textbf{OK for combined stresses}$$

Two 2 × 6 chords are structurally adequate for the worst-case loading conditions. The other chords could be designed, but it is common practice to provide the same chords and hold-down anchors at the ends of all shear walls for buildings of this size.

9.15 CONSTRUCTION DOCUMENTS

In a typical wood building project, the construction documents would include the structural, mechanical, electrical and architectural drawings as well as the construction specifications that

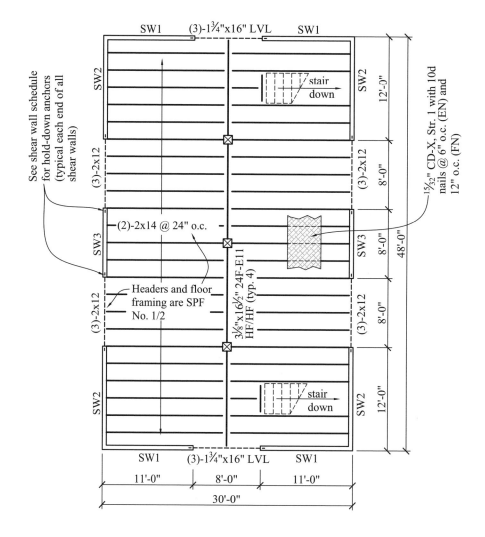

FIGURE 9.36 Second floor framing plan.

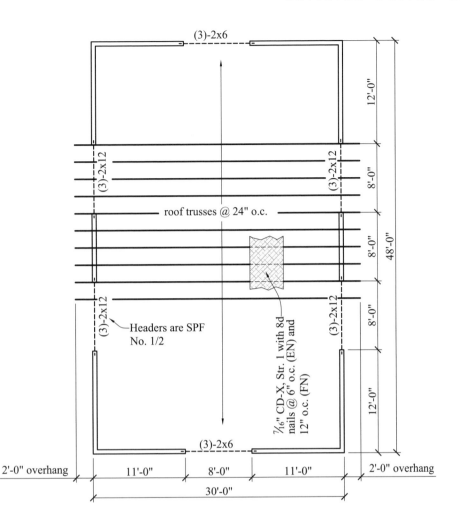

FIGURE 9.37 Roof framing plan.

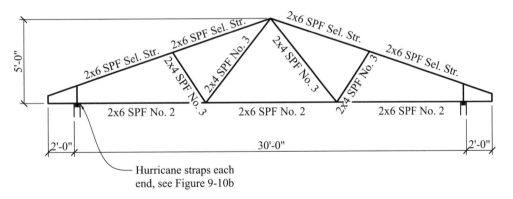

FIGURE 9.38 Roof truss members.

would enable the builder construct the building. For the building design case study, the drawings shown in Figures 9.35 through 9.38 is representative of what would typically be shown on a construction set of structural drawings.

REFERENCES

1. ICC (2006), *International Building Code,* International Code Council, Washington, DC.
2. ASCE (2005), *Minimum Design Loads for Buildings and Other Structures,* American Society of Civil Engineers, Reston, VA.

3. NAHB (2000), *Residential Structural Design Guide—2000,* National Association of Home Builders Research Center.
4. Cohen, Albert H. (2002), *Introduction to Structural Design: A Beginner's Guide to Gravity Loads and Residential Wood Structural Design,* AHC, Edmonds, WA.
5. ICC (2003), *International Building Code,* International Code Council, Washington, DC.

appendix A

WEIGHTS OF BUILDING MATERIALS

WOOD FRAMING

Weights shown are for Douglas fir-larch, $G = 0.50$, which is conservative for most structures.

Nominal Size	Weight (psf) for Spacing:			
	12 in. o.c	16 in. o.c.	19.2 in. o.c.	24 in. o.c.
2 × 4	1.2	0.9	0.8	0.6
2 × 6	1.8	1.4	1.2	0.9
2 × 8	2.4	1.8	1.5	1.2
2 × 10	3.1	2.3	1.9	1.5
2 × 12	3.7	2.8	2.3	1.9
2 × 14	4.4	3.3	2.7	2.2
2 × 16	5.0	3.8	3.1	2.5
3 × 4	1.9	1.5	1.2	1.0
3 × 6	3.0	2.3	1.9	1.5
3 × 8	4.0	3.0	2.5	2.0
3 × 10	5.1	3.8	3.2	2.6
3 × 12	6.1	4.6	3.9	3.1
3 × 14	7.2	5.4	4.5	3.6
3 × 16	8.3	6.2	5.2	4.2
4 × 4	2.7	2.0	1.7	1.4
4 × 6	4.2	3.2	2.7	2.1
4 × 8	5.5	4.2	3.5	2.8
4 × 10	7.1	5.3	4.4	3.6
4 × 12	8.6	6.4	5.4	4.3
4 × 14	10.1	7.6	6.3	5.1
4 × 16	11.6	8.7	7.3	5.8

PREENGINEERED FRAMING

Roof Trusses

Span (ft.)	Weight (psf of horizontal plan area)
30	5.0
35	5.5
40	6.0
50	6.5
60	7.5
80	8.5
100	9.5
120	10.5
150	11.5

Floor Joists (I-Joists and Open Web)

Depth (in.)	Weight (psf of horizontal plan area)
$9\frac{1}{2}$	2.5
$11\frac{7}{8}$	3.5
14	4.0
16	4.5
18	5.0
20	5.5
24	6.0
26	6.3
28	6.6
30	7.0

OTHER MATERIALS

	Weight (psf)
Sheathing	
Plywood (per $\frac{1}{8}$-in. thickness)	0.4
Gypsum sheathing (per 1-in. thickness)	5.0
Lumber sheathing (per 1-in. thickness)	2.5
Floors and floor finishes	
Asphalt mastic (per 1-in. thickness)	12.0
Ceramic or quarry tile ($\frac{3}{4}$-in. thickness)	10.0
Concrete fill (per 1-in. thickness)	12.0
Gypsum fill (per 1-in. thickness)	6.0
Hardwood flooring ($\frac{7}{8}$-in. thickness)	4.0
Linoleum or asphalt tile (per $\frac{1}{4}$-in. thickness)	1.0
Slate (per 1-in. thickness)	15.0
Softwood ($\frac{3}{4}$-in. thickness)	2.5
Terrazzo (per 1-in. thickness)	13.0
Vinyl tile ($\frac{1}{8}$-in. thickness)	1.4
Roofs	
Copper or tin	1.0
Felt	
3-ply ready roofing	1.5
3-ply with gravel	5.5
5-ply	2.5
5-ply with gravel	6.5
Insulation (per 1-in. thickness)	
Cellular glass	0.7
Expanded polystyrene	0.2
Fiberboard	1.5
Fibrous glass	1.1
Loose	0.5
Perlite	0.8
Rigid	1.5
Urethane foam with skin	0.5
Metal deck	3.0
Roll roofing	1.0
Shingles	
Asphalt ($\frac{1}{4}$-in.)	2.0
Clay tile	12.0
Clay tile with mortar	22.0
Slate ($\frac{1}{4}$-in.)	10.0
Wood	3.0
Ceilings	
Acoustical fiber tile	1.0
Channel-suspended system	
Steel	2.0
Wood furring	2.0
Plaster on wood lath	8.0
Walls and partitions	
Brick (4-in. nominal thickness)	40
Glass (per 1-in. thickness)	15
Glass block (4-in. thickness)	18
Glazed tile	18
Plaster	8.0
Stone (4-in. thickness)	55
Stucco ($\frac{7}{8}$-in. thickness)	10.0
Vinyl siding	1.0
Windows (glass, frame, sash)	8.0
Wood paneling	2.5

appendix *B*

DESIGN AIDS

NOTES ON DESIGN AIDS

Allowable Uniform Loads on Floor Joists

The allowable uniform load is calculated with respect to joist length for common 2× lumber sizes and wood species. The repetitive member factor applies and normal load duration was assumed. Joists were analyzed for bending and deflection (creep factor $k = 1.0$). Shear is typically not a controlling factor and was not used; consequently, the allowable uniform load usually does not exceed 300 plf in the charts. For practical reasons, joist lengths exceeding 30 ft were not considered. Furthermore, joists in the deflection-controlled region should be investigated for vibration.

Allowable Axial Loads on Columns

The allowable axial load is calculated with respect to the unbraced length for common square column sizes (4 × 4 through 10 × 10) using common wood species. Normal load duration was assumed. Values for southern pine in wet service conditions were included since southern pine is commonly used in exterior applications (such as wood-framed decks).

Combined Axial and Bending Loads on Wall Studs

A unity curve is plotted in accordance with NDS Equation 3.9.2 for wall studs supporting vertical gravity loads and lateral wind loads. The repetitive member factor applies and a load duration factor of $C_D = 1.6$ was assumed. The weak axis is assumed to be fully braced by wall sheathing.

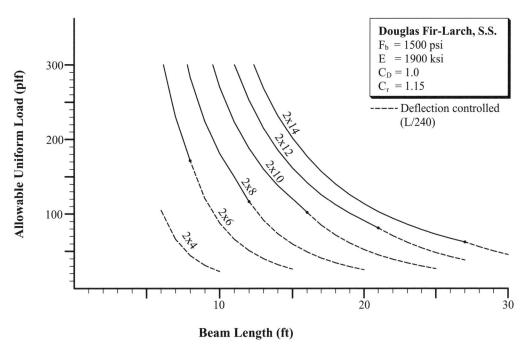

FIGURE B.1 Allowable uniform loads on floor joists (DF-L, Sel. Str.).

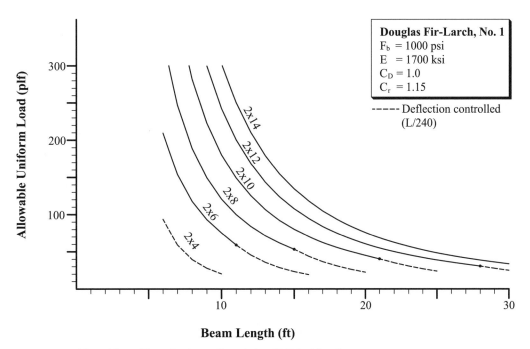

FIGURE B.2 Allowable uniform loads on floor joists (DF-L, No. 1).

352 | STRUCTURAL WOOD DESIGN

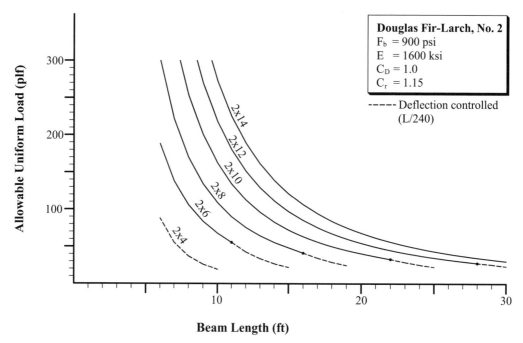

FIGURE B.3 Allowable uniform loads on floor joists (DF-L, No. 2).

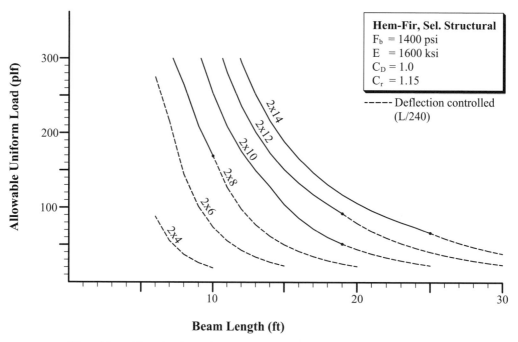

FIGURE B.4 Allowable uniform loads on floor joists (hem-fir, Sel. Str.).

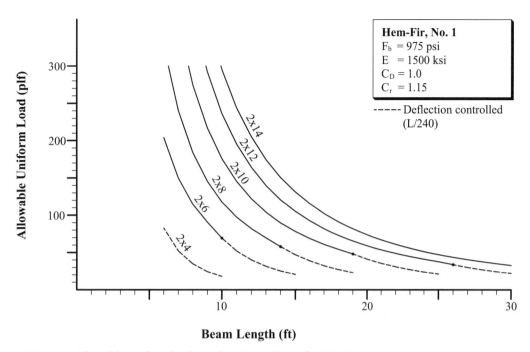

FIGURE B.5 Allowable uniform loads on floor joists (hem-fir, No. 1).

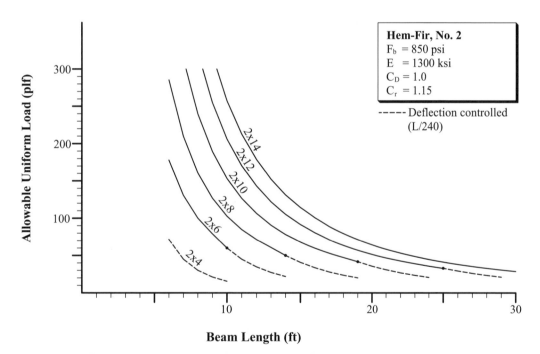

FIGURE B.6 Allowable uniform loads on floor joists (hem-fir, No. 2).

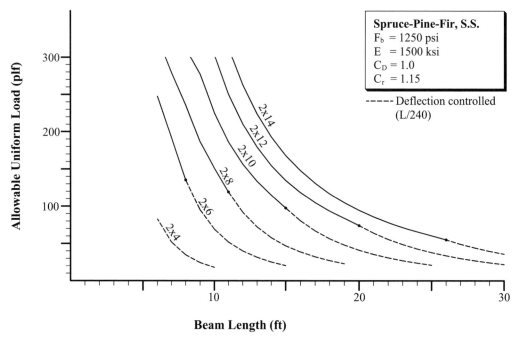

FIGURE B.7 Allowable uniform loads on floor joists (spruce-pine-fir, Sel. Str.).

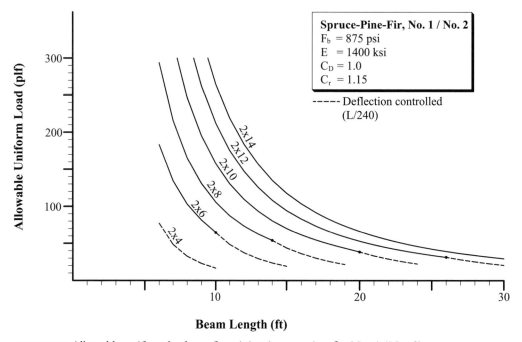

FIGURE B.8 Allowable uniform loads on floor joists (spruce-pine-fir, No. 1 / No. 2).

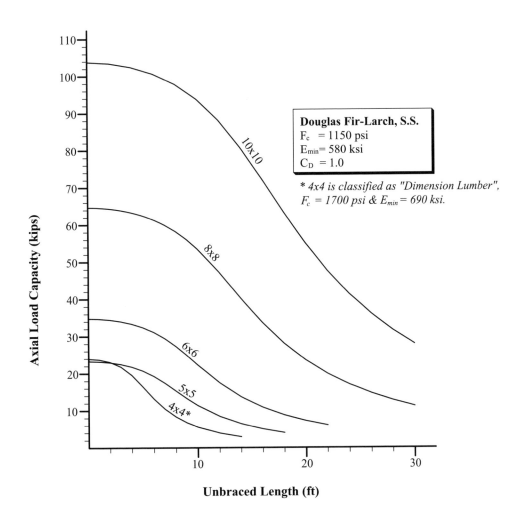

FIGURE B.9 Allowable axial loads on columns (DF-L, Sel. Str.).

FIGURE B.10
Allowable axial loads on columns (DF-L, No. 1).

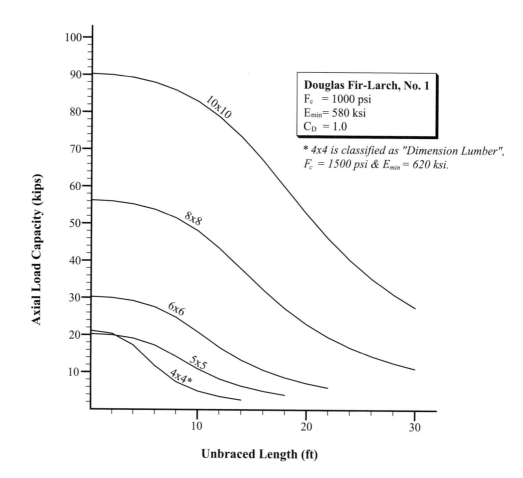

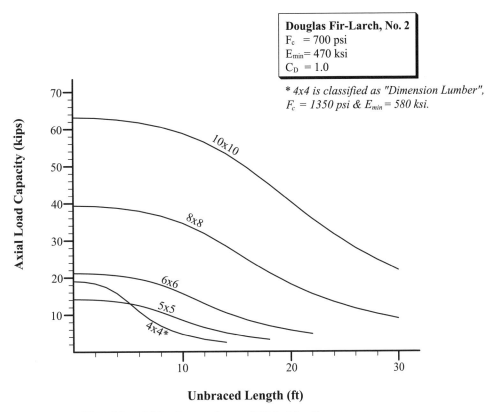

FIGURE B.11 Allowable axial loads on columns (DF-L, No. 2).

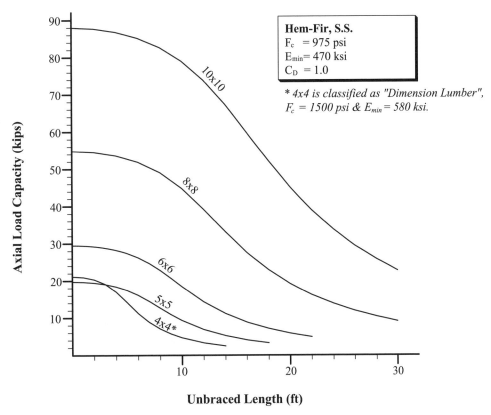

FIGURE B.12 Allowable axial loads on columns (hem-fir, Sel. Str.).

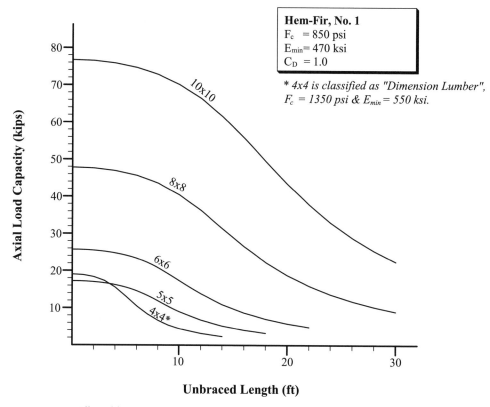

FIGURE B.13 Allowable axial loads on columns (hem-fir, No. 1).

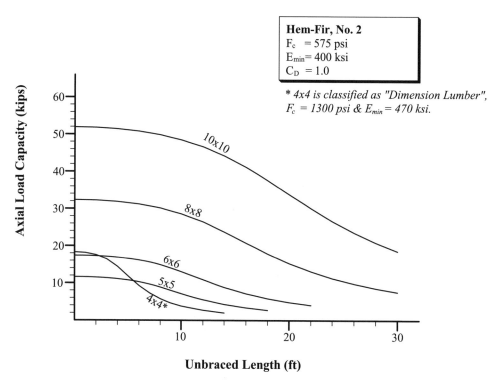

FIGURE B.14 Allowable axial loads on columns (hem-fir, No. 2).

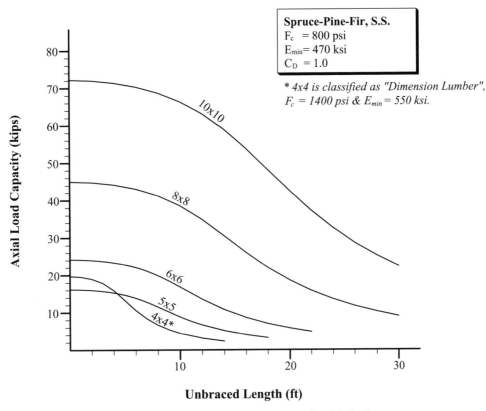

FIGURE B.15 Allowable axial loads on columns (spruce-pine-fir, Sel. Str.).

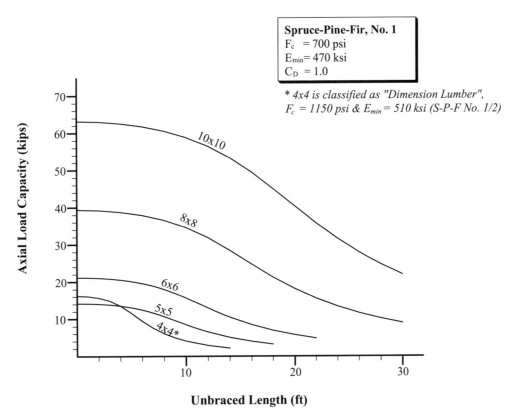

FIGURE B.16 Allowable axial loads on columns (spruce-pine-fir, No. 1).

FIGURE B.17
Allowable axial loads on columns (spruce-pine-fir, No. 2).

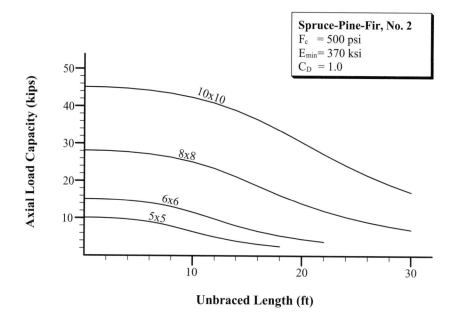

FIGURE B.18
Allowable axial loads on columns (southern pine, Sel. Str.).

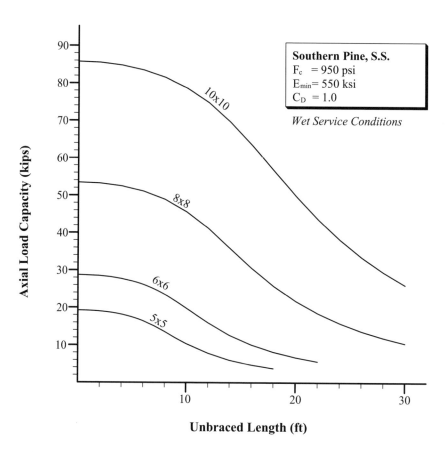

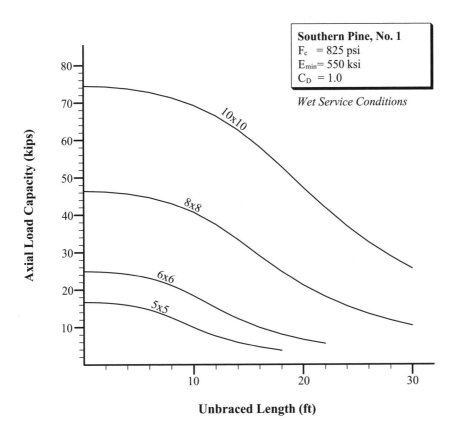

FIGURE B.19
Allowable axial loads on columns (southern pine, No. 1).

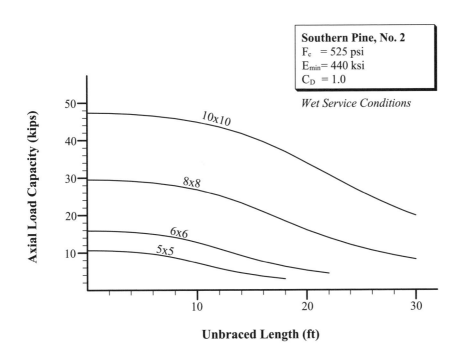

FIGURE B.20
Allowable axial loads on columns (southern pine, No. 2).

FIGURE B.21
Combined axial and bending loads on 2 × 6 wall studs (DF-L, Sel. Str., $C_D = 1.6$).

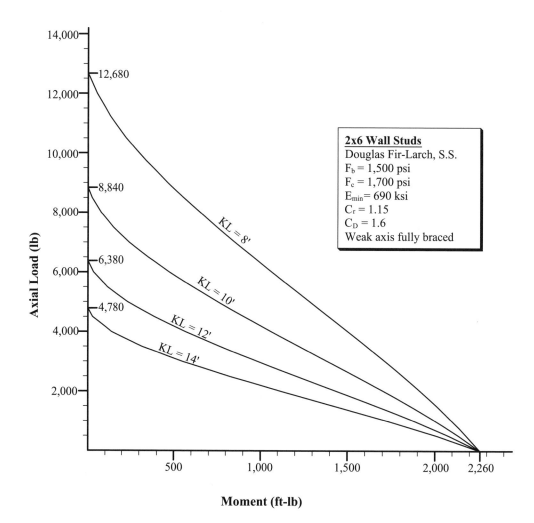

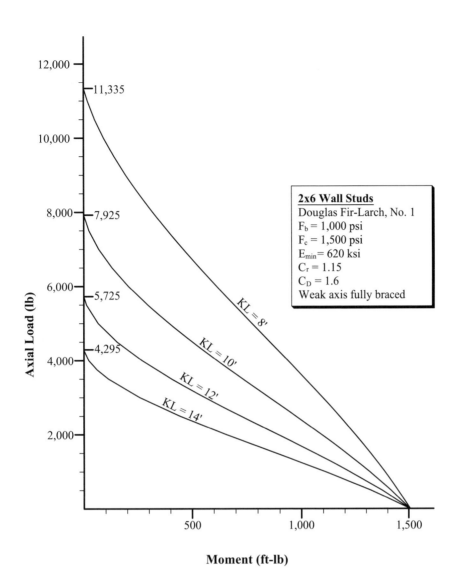

FIGURE B.22
Combined axial and bending loads on 2 × 6 wall studs (DF-L, No. 1, $C_D = 1.6$).

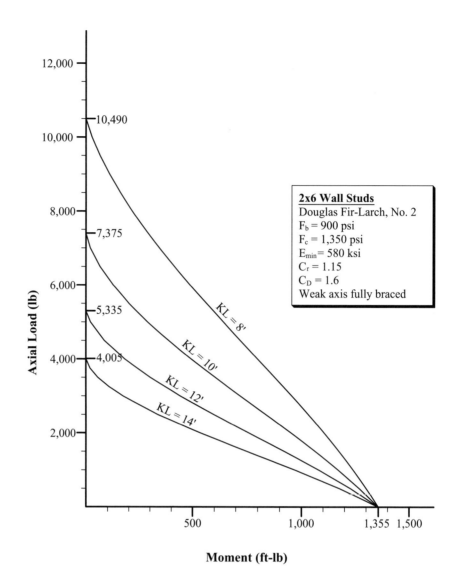

FIGURE B.23 Combined axial and bending loads on 2 × 6 wall studs (DF-L, No. 2, $C_D = 1.6$).

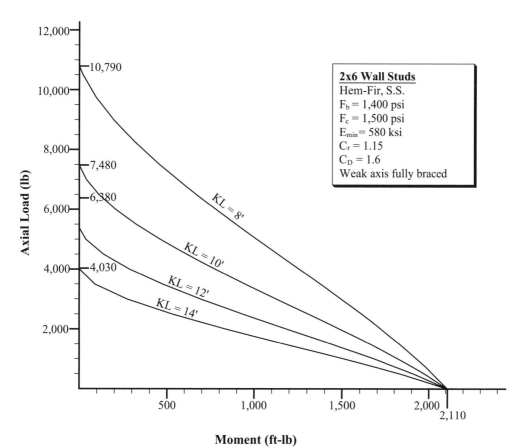

FIGURE B.24
Combined axial and bending loads on 2 × 6 wall studs (hem-fir, Sel. Str., $C_D = 1.6$).

FIGURE B.25
Combined axial and bending loads on 2 × 6 wall studs (hem-fir, No. 1, $C_D = 1.6$).

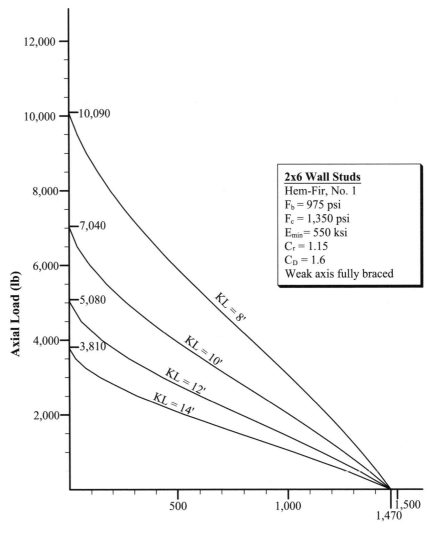

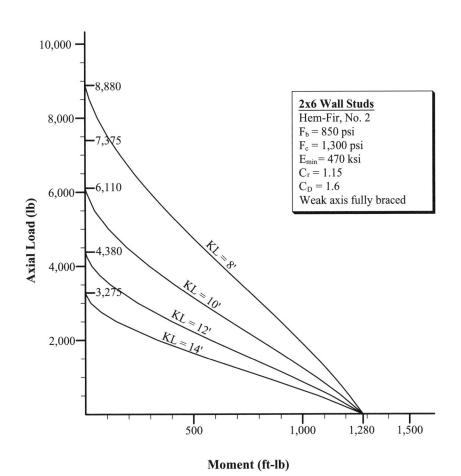

FIGURE B.26
Combined axial and bending loads on 2 × 6 wall studs (hem-fir, No. 2, $C_D = 1.6$).

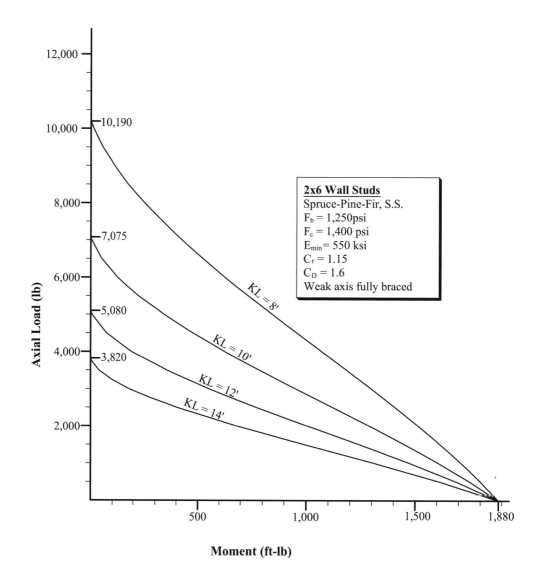

FIGURE B.27
Combined axial and bending loads on 2 × 6 wall studs (spruce-pine-fir, Sel. Str., $C_D = 1.6$).

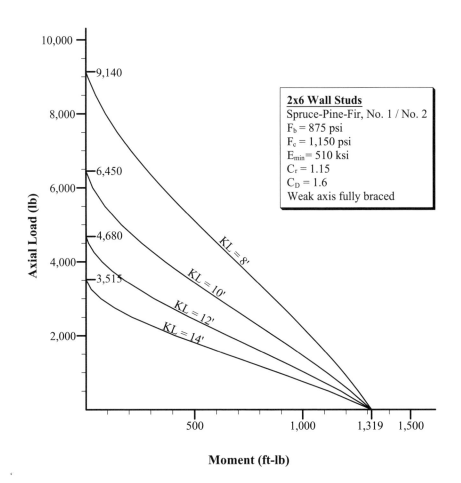

FIGURE B.28
Combined axial and bending loads on 2 × 6 wall studs (spruce-pine-fir, No. 1/No. 2, $C_D = 1.6$).

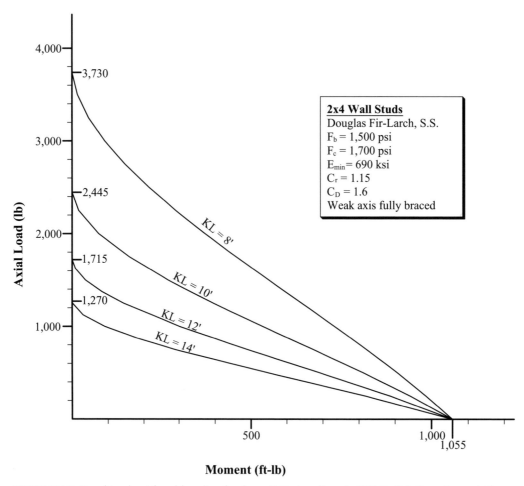

FIGURE B.29 Combined axial and bending loads on 2 × 4 wall studs (DF-L, Sel. Str., $C_D = 1.6$).

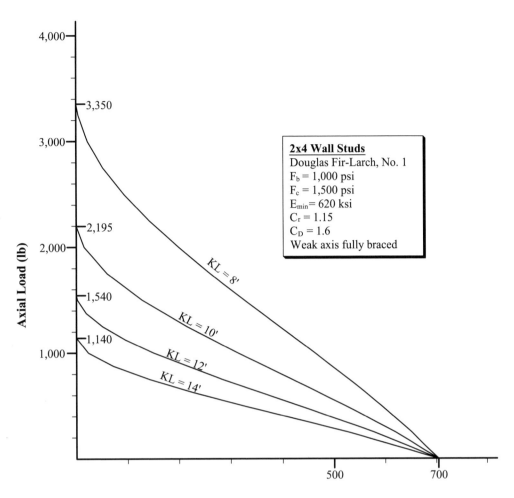

FIGURE B.30
Combined axial and bending loads on 2 × 4 wall studs (DF-L, No. 1, $C_D = 1.6$).

FIGURE B.31
Combined axial and bending loads on 2 × 4 wall studs (DF-L, No. 2, $C_D = 1.6$).

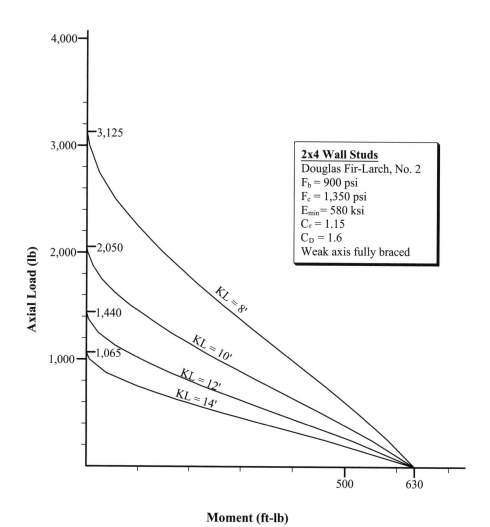

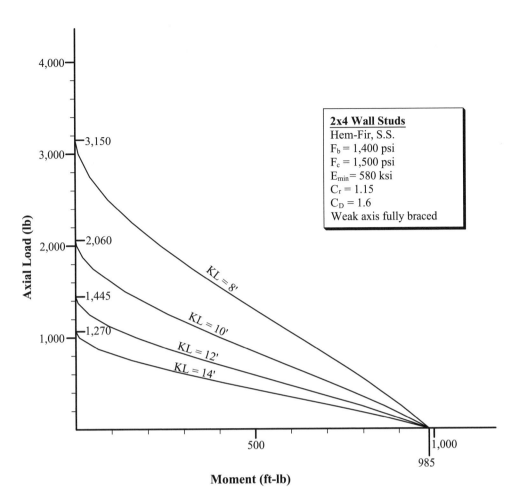

FIGURE B.32
Combined axial and bending loads on 2 × 4 wall studs (hem-fir, Sel. Str., $C_D = 1.6$).

FIGURE B.33
Combined axial and bending loads on 2 × 4 wall studs (hem-fir, No. 1, $C_D = 1.6$).

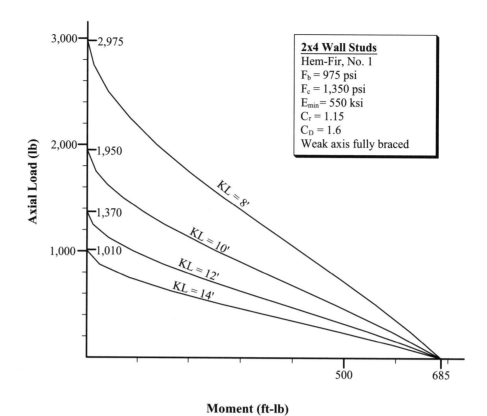

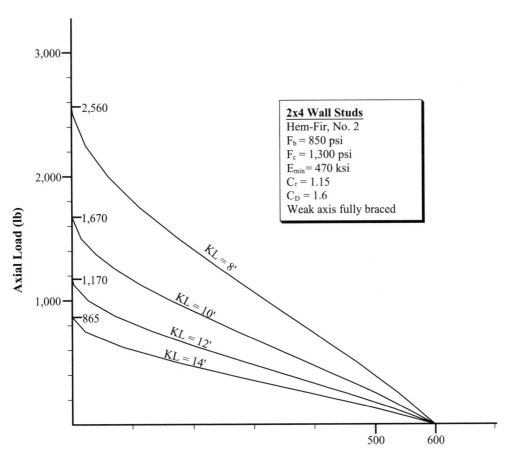

FIGURE B.34 Combined axial and bending loads on 2 × 4 wall studs (hem-fir, No. 2, $C_D = 1.6$).

FIGURE B.35
Combined axial and bending loads on 2 × 4 wall studs (spruce-pine-fir, Sel. Str., $C_D = 1.6$).

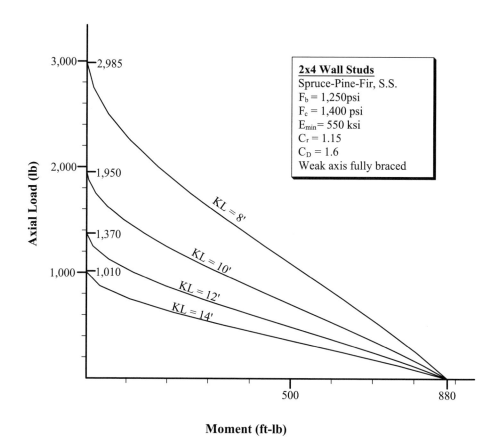

FIGURE B.36
Combined axial and bending loads on 2 × 4 wall studs (spruce-pine-fir, No. 1/No. 2, $C_D = 1.6$).

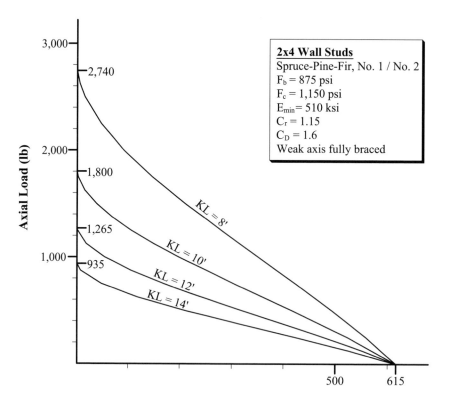

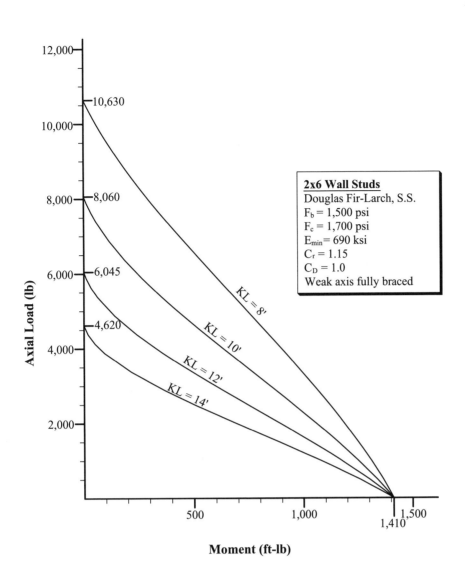

FIGURE B.37
Combined axial and bending loads on 2 × 6 wall studs (DF-L, Sel. Str., $C_D = 1.0$).

FIGURE B.38
Combined axial and bending loads on 2 × 6 wall studs (DF-L, No. 1, $C_D = 1.0$).

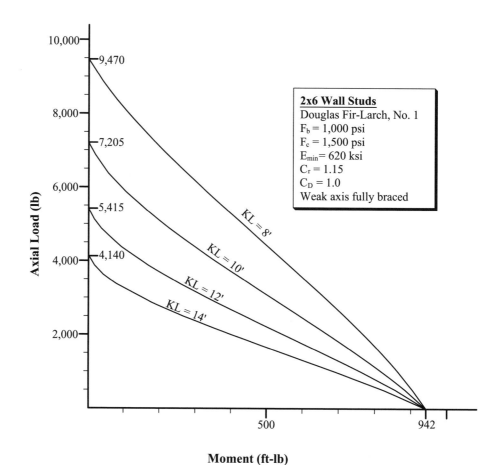

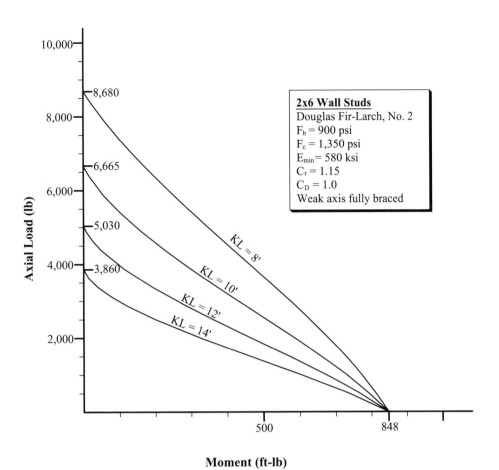

FIGURE B.39
Combined axial and bending loads on 2 × 6 wall studs (DF-L, No. 2, $C_D = 1.0$).

FIGURE B.40
Combined axial and bending loads on 2 × 6 wall studs (hem-fir, Sel. Str., $C_D = 1.0$).

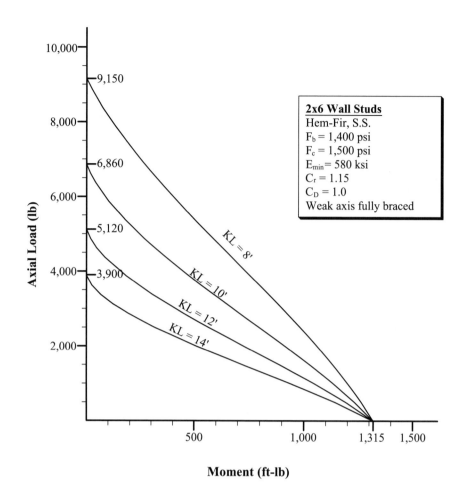

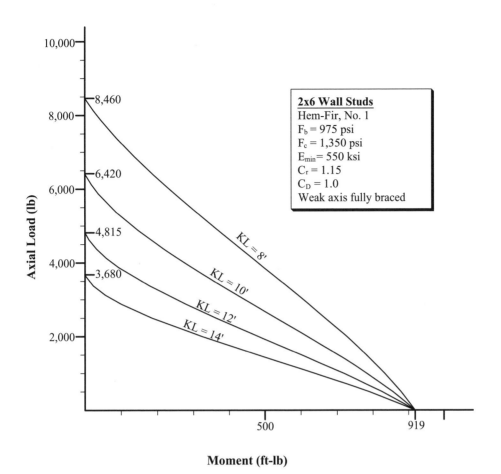

FIGURE B.41 Combined axial and bending loads on 2 × 6 wall studs (hem-fir, No. 1, $C_D = 1.0$).

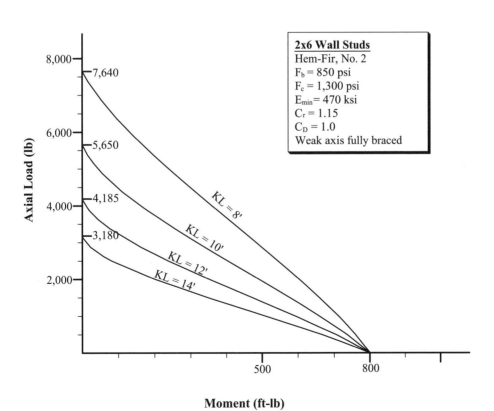

FIGURE B.42 Combined axial and bending loads on 2 × 6 wall studs (hem-fir, No. 2, $C_D = 1.0$).

FIGURE B.43
Combined axial and bending loads on 2 × 6 wall studs (spruce-pine-fir, Sel. Str., $C_D = 1.0$).

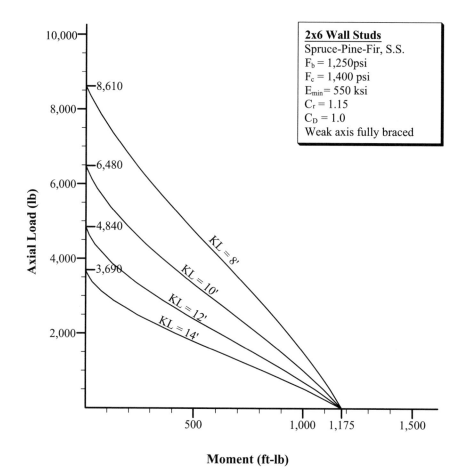

FIGURE B.44
Combined axial and bending loads on 2 × 6 wall studs (spruce-pine-fir, No. 1/No. 2, $C_D = 1.0$).

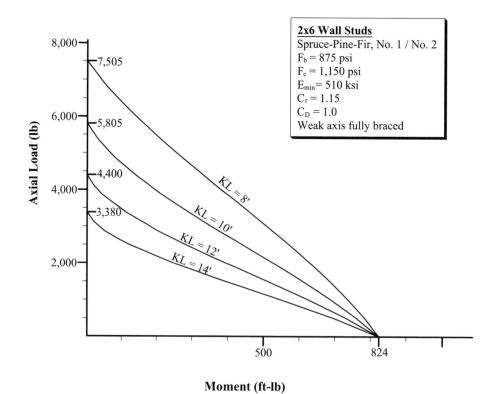

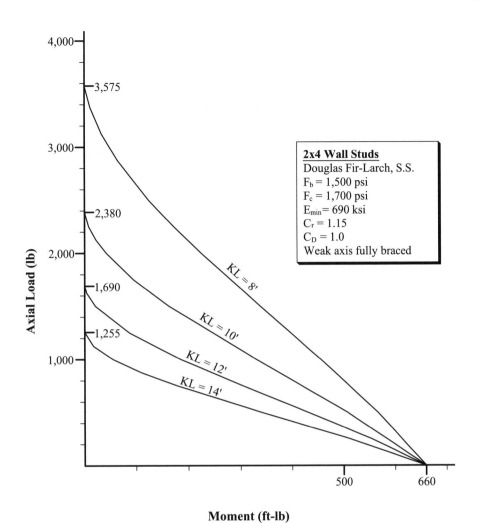

FIGURE B.45
Combined axial and bending loads on 2 × 4 wall studs (DF-L, Sel. Str., $C_D = 1.0$).

FIGURE B.46
Combined axial and bending loads on 2 × 4 wall studs (DF-L, No. 1, $C_D = 1.0$).

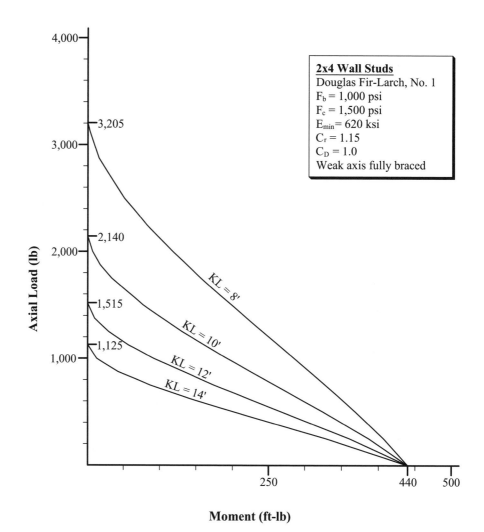

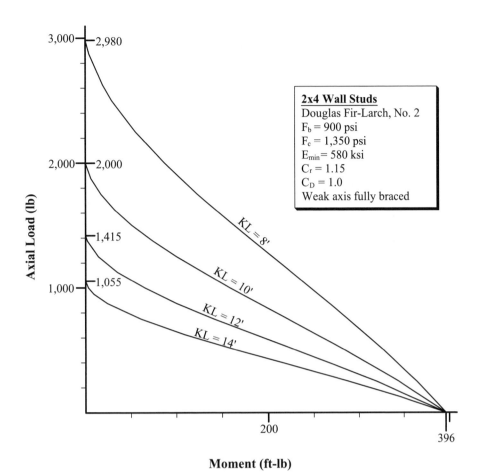

FIGURE B.47 Combined axial and bending loads on 2 × 4 wall studs (DF-L, No. 2, $C_D = 1.0$).

FIGURE B.48
Combined axial and bending loads on 2 × 4 wall studs (hem-fir, Sel. Str., $C_D = 1.0$).

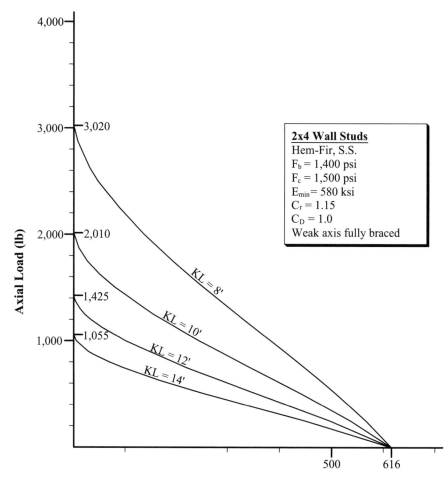

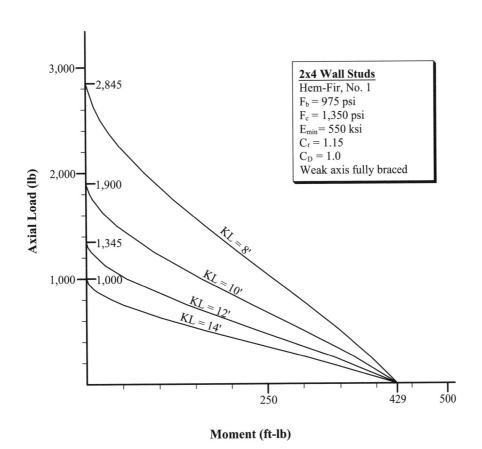

FIGURE B.49
Combined axial and bending loads on 2 × 4 wall studs (hem-fir, No. 1, $C_D = 1.0$).

FIGURE B.50
Combined axial and bending loads on 2 × 4 wall studs (hem-fir, No. 2, $C_D = 1.0$).

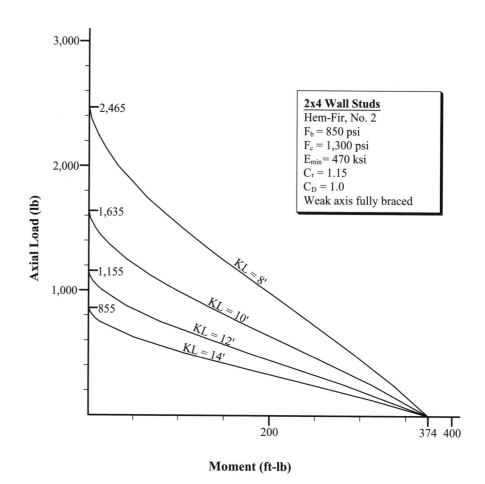

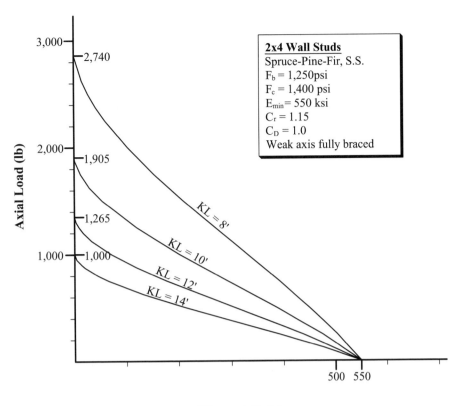

FIGURE B.51
Combined axial and bending loads on 2 × 4 wall studs (spruce-pine-fir, Sel. Str., $C_D = 1.0$).

FIGURE B.52 bined axial and bending loads on 2 × 4 wall studs (spruce-pine-fir, No. 1/No. 2, $C_D = 1.0$).

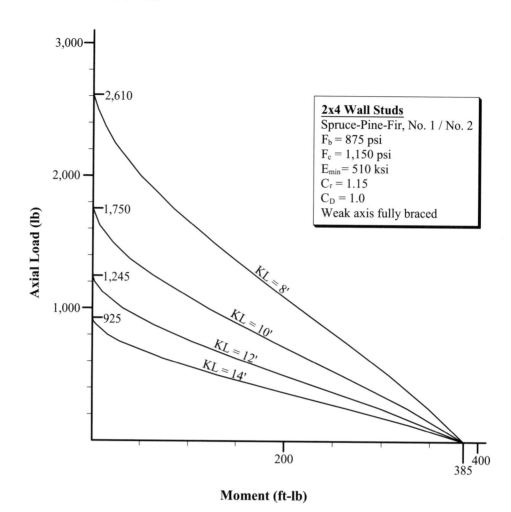

Index

A

Adjustment factors
    connections, 249–251
    glulam, 58–66
    sawn lumber, 58–66
Allowable stress design (ASD), 57
Allowable stresses, design example, 60, 70–77
Anchor bolts
    design example, 240–241, 338–342
    uplift, 338
Annual rings, 12
Axes, in wood, 15–17
Axial loads
    compression, *see* Columns
    tension, *see* Tension members

B

Balloon-type construction, 2
Base shear
    winds loads, 49–50, 287
    seismic loads, 52–53, 283–284
Beam stability factor, C_L, 62–63
    design examples, 102, 109, 304, 309
Beam-columns, *see* Combined stresses
Beams and stringers (B&S), 14
Beams
    bearing stress, *see* Bearing stress
    bending and compression, *see* Combined stresses
    bending and tension, *see* Combined stresses
    bending stress
        adjustment factors, 58–66
        beam stability factor, C_L, 62–63
        curvature factor, C_c, 65
        flat use factor, C_{fu}, 62, 118
        incising factor, C_i, 62
        lateral stability, 8, 62–63, 117–118
        load duration factor, C_D, 59–60
        notches, 116, 118–119, 122
        repetitive member factor, C_r, 61
        size factor, C_F, 61
        temperature factor, C_t, 62
        volume factor, C_V, 64
        weak axis, 62
        wet service factor, C_M, 61
    bridging, 4, 8
    camber, 42, 84, 111
    cantilever, 7, 64, 84
    creep, 14, 39, 83, 84
    deflection, 39–42
    effective length, 63–64
    flatwise loading, 62, 114
    flitch beam, 128–131
    header, 126–127, 307
    lateral bracing, 117–118
    modulus of elasticity
        E, 66
        E_{min}, 66
    notching, 116, 121, 122
    shear stress, 82–83
        adjustment factors, 58
        notches, 118–119, 122
    size factor, C_F, 61
    slenderness ratio, R_B, 63
    stringer, 28, 31, 113–116
    unbraced length, 63–66
    vibrations, 84, 128–142
Bearing area factor, C_b, 65
Bearing length, 65, 85
Bearing stress, 65–69
    angle to the grain, 122–125
    parallel to grain (end grain), 122
    perpendicular to grain, 65–69, 85, 99
Bending stress, *see* Beams, bending stress
Blocking
    beams, 4, 8, 253
    diaphragms, 185–187, 192, 209–210
    shearwalls, 217–218

Board measure (board foot), 19–20
Bolted connections
    adjustment factors, 249–252
    angle to the grain, 260
    bending yield strength, 258–259
    bolt slenderness, 252
    design example
        anchor bolts, 240–241, 338–342
        double shear, 263
        single shear, 253, 255, 257
        wood-to-steel, 254
        wood-to-wood, 253, 256
    double shear, 259, 263
    dowel bearing strength, 253, 258–261
    geometry factor, C_Δ, 250–251, 256, 257
    group action factor, C_g, 250, 254, 255
    load duration factor, C_D, 249
    penetration, 251–252
    shrinkage, 249, 273
    single shear, 254, 255, 257
    spacing requirements, 250–252
    temperature factor, C_t, 250
    typical connections, 273–275
    wet service factor, C_M, 249
    yield limit equations, 252–253, 260
Bottom plates, 3, 8
Bound water, 11–13
Buckling stiffness factor, C_T, 65
Building codes, 1, 20–22
Built-up members
    beams, 6
    columns, 153, 157, 159–165

C

Camber, see Beams, camber
Cantilever beams, see Beams, cantilever
Cell structure, 11–13
Cellulose, 11, 19
Checks, 15
Chords
    diaphragms, see Diaphragms, chords
    shearwalls, see Shearwalls, chords
Clearwood tests, 58
Collar tie, 3, 8–11
Collector, see Drag strut
Column stability factor, C_P, 64, 157–159
Columns
    adjustment factors, 58
    bending, see Combined stresses
    bracing, 158–159
    buckling, 158–159
    buckling stiffness factor, C_T, 65
    built-up, 157, 159, 161–165
    column stability factor, C_P, 64, 158, 159
    design examples, 160–164, 311–314
        shearwall chord, 339–344
    eccentricity, 178
    effective length factor, 64, 153–158
    euler formula, 153
    slenderness ratio, 157–158
Combined stresses
    bending and compression, 162–166, 167–170, 171–178
    bending and tension, 151–153, 154–157
    biaxial bending, 166
    interaction equation, 152, 166
Common nails, 249
Compression parallel to grain, see Columns
Compression perpendicular to grain, 11, 16–18, 69
    adjustment factors
        bearing area factor, C_b, 65
        temperature factor, C_t, 62
        wet service factor, C_M, 61
Connections
    bolts, see Bolted connections
    hold-down, 218–219, 240, 337–341
    hurricane, 299–300
    joist hangers, 273
    lag screws, see Lag screw connections
    nails, see Nail connections
    non-preferred, 274–275
    penetration, 251–252
    pre-engineered, 270
    preferred, 274–275
    spacing requirements, 250–252
    strap, 240–241, 338
    wood screws, see Wood screw connections
    withdrawal, see Withdrawal connections
    yield limit, see Yield limit equations
Continuous load path, 189–191, 219–220
Creep, see Beams, creep
Cripple stud, 6–7
Curvature factor, C_c, 65

D

Dead loads, 25–28, 347–349
Decay, 12, 15
Decking, 14, 61–63, 118–120
    design example, 119–121
    layup patterns, 118
Defects, 12–13, 15–18
Deflection, 39–42
Density of wood, 19, 347
Design aids, 350
Detailing of connections, 273–275
Diaphragm factor, C_{di}, 249–251
Diaphragms
    anchorage, 189–190, 192
    aspect ratio, 196–197
    beam action, 183, 190–191, 193–194, 213

blocked, 187, 192, 195–196, 209
chords, 189–193, 200
    lap splice connection, 208–209, 254–257, 331, 335–337
drag strut, 189–195
flexible, 196–198
load cases, 192–196
nail, 187, 192
open front structure, 197
openings, 197–199
proportions, 196–197
rigid, 196–200
shear diagram, 191, 194, 198, 210, 211
shear strength, 192, 204, 205
unblocked, 192, 195–196, 210
unit shear, 190–195, 198
Dimension lumber, 14
Dowel bearing strength, 253, 258–261
Drag strut, 189–195, 200, 213
    force calculation, 192, 194, 208, 212, 331
    force diagram, 195, 206, 207, 213, 332–334
Dressed lumber, 14
Dry (moisture content), 11, 13–15, 19
Duration of load, see Load duration factor

E

Eccentric loading, 165, 167
Edge distance, 250–252
Effective length
    beams, 63–64
    columns, 64, 153–158
Embedment
    anchor rods, 240, 260
    connections, see Connections, penetration
End distance, 250–252
End grain factor, C_{eg}, 249, 251, 253
End grain loading, 249, 251, 252, 268
End joints–glulam, 67
Equilibrium moisture content, 13, 18
Euler critical buckling stress
    beams, 63
    columns, 153
Exclusion value (5 percent), 58, 158

F

Fabricated wood components
    i-joists, 3, 84, 117
    laminated veneer lumber (LVL), 5, 124, 243, 311
    parallel strand lumber (PSL), 5, 124
    open web trusses, 6
Factor of safety, 57, 66, 158, 224
Fiber saturation point, 13, 18
Finger joints, 67
Flat use factor, C_{fu}, 62, 118

Flexible diaphragms, see Diaphragms
Flitch beam, 128–131
Floor live load, 7, 25, 26, 38
Floor sheathing, 185–190, 317–319
Free water, 11, 13
Fungi, 15

G

Geometry factor, C_Δ, 249–250, 256
Glued laminated timber (Glulam)
    adjustment factors, 58–66
        volume factor, C_v, 64
    appearance grades, 67
    axial combinations, 67–69
    bending combinations, 67–69
    camber, 42, 84, 111
    combination symbols, 67–69
    compression zone in tension, 69
    design examples, 73–75, 92–96, 106–112, 301–307
    fabrication
        edge joints, 67
        end joints, 67
        glue, 61, 66–68
        mixed species, 66–68
        moisture content, 13
    grades, 67
    laminations
        compression, 69
        E-rated, 67
        tension, 69
        thickness, 66
        visually graded, 67
Grade stamp
    sawn lumber, 15, 18
    sheathing, 184–185, 216–217
Grading, 15–18
Grading agencies, 17–18
Grain orientation, 11–17, 59
Green lumber, 13, 20
Group action factor, C_g, 249–250
    design examples, 254, 255
Growth rings, 11–12

H

Hankinson formula, 123, 258, 259
Hardwood, 11, 12
Header, 4, 6
    design examples, 126–127, 307
Heartwood, 11, 12
Hip rafter, 5
Hold-down anchors, 219, 240–241, 299–300, 337–338, 340–343
Hold-down straps, 240, 337–338
Holes, in beams, see Beams, notches
Horizontal diaphragms, see Diaphragms

Horizontal plane, 27–29, 36
Horizontal shear, 82
Horizontal thrust, 3, 9, 28–30

I

I-joist, 84, 117
    design example (vibrations), 138
Incising factor, C_i, 62
In-grade testing, 58
Interaction equation, *see* Combined stresses

J

Jack rafters, 3, 5
Jack stud, 6, 7
Joist
    double or triple joists, 3, 4
    header and trimmer joists, 4
    in-line joists, 3, 4
    lapped joists, 3, 4
    rim joist, 4

K

Kiln-dried, 13, 66
King stud, 6, 7
Knee brace, 267, 268
Knots, 15, 58

L

Lag screws
    adjustment factors, 249–252
    bending yield strength, 258–259
    combined shear and withdrawal, 258, 270
    dimensions, 261
    design examples, 265, 269, 271
    Hankinson formula, 258
    penetration, 251–252
    spacing requirements, 250–252
    withdrawal, 249, 268, 269
    wood-to-metal, 265, 271
    wood-to-wood, 269
    yield modes, 252–253, 260
Laminated veneer lumber (LVL), 5, 126, 243, 311
Laminations
    glulam, 66–69
    plywood, 183–184
Lap splice (top plates), 208, 256, 257, 335–337
Lap splice (plywood), 242–245
Lateral loads, 42–53
    seismic, 45, 52–53, 283–284
    wind, 43–45, 46–51, 284–293
Lateral stability (bending), 8, 62–63, 117–118
Let-in bracing, 10
Lignin, 11, 16, 82

Load and resistance factor design (LRFD), 57
Load combinations, 25–26, 42, 60, 69, 75–77
Load duration factor, C_D, 14, 58–60, 249
    connections, 249
    load combinations, 60, 69, 75–77
    values, 60
Load/slip modulus, γ, 134–135
Loads
    combinations, 25–26, 42, 60, 69, 75–77
    dead loads, 25–28, 347–349
    duration, 14, 58–60, 249
    floor live load, 7, 25, 26, 38
    floor live load reduction, 35–39
    load path, 189–191, 219–220
    partitions, 26, 27
    roof, 25, 27–28, 30–32, 34
    seismic, 45, 283–284
    snow, 25–30, 32–38
    tributary areas, 28–33
    tributary width, 28–33
    wind, 43–45, 284–293
Lumber
    adjustment factors, 58
    allowable stresses, 57
    beams & stringers (B&S), 14
    classification, 13–16
    decking, 14, 61–63, 118–120
    defects, 12–13, 15–18
    density, 19, 347
    dimension lumber, 14
    dry, 11, 13–15, 19
    grade, 15–18
    green, 13, 20
    growth characteristics, 11–12
    machine evaluated (MEL), 18
    machine stress rated (MSR), 18
    moisture content, 11, 13–15, 19
    posts & timbers (P&T), 5, 14
    seasoned, 13, 18, 39
    shrinkage, 13–16, 18–22
    size classification, 14
    stress grades, 17, 18, 22, 118
    visually graded, 15, 17, 18

M

Machine evaluated lumber (MEL), 18
Machine stress rated (MSR), 18
Modulus of elasticity, E, 66
    adjustment factors, 58
    E_{min}, 66
    exclusion value, 58, 158
Moisture content
    effects
        connections, 249–250

creep (deflection), see Beams, creep
	lumber size, 11, 13–15, 19
	equilibrium moisture content, 13, 18
	fiber saturation, 13, 18
	green, 13, 20
	seasoned, 13, 18
	wet service factor, C_M, 61, 249
Moment magnification, see P-delta effects

N

Nail connections
	adjustment factors, 249–252
	bending yield strength, 258–259
	design example
		hold-down strap, 337–338
		knee brace, 267
		lap splice, 335–336
		plywood lap splice, 243
		toenail, 212
		wood-to-wood, 255, 262, 267
	dowel bearing strength, 253, 258–261
	penetration, 251–252
	prebored holes, 251
	sheathing
		diaphragms, 187, 192
		minimum nailing, 187, 217–218
		shearwalls, 217–218
	splitting, 250–251
	toenail, 252–253, 272
	withdrawal, 249, 252, 268, 270
	yield limit equations, 252–253, 260
	yield modes, 259
Natural frequency, 128, 132
NDS Code, 1
NDS Supplement (NDS-S), 1
Net area, 146–148
Normalized load method, 69, 75–77
Notching, see Beams, notches

O

Orientation of wood grain, 11–17, 59
Oriented strand board (OSB), 9, 11, 183
Overturning, 10, 11, 220, 223–227, 322–331

P

P-delta effects, 162
Parallel strand lumber (PSL), 5, 124
Parallel to grain, see Orientation of wood grain
Perpendicular to grain, see Orientation of wood grain
Penetration, 251–252
Platform construction, 2
Plywood
	axis orientation, 183–184, 188
	blocking, 185–187, 192, 209–210
	cross-laminations, 183–184
	diaphragms, see Diaphragms
	edge support, 183, 185–187
	exposure categories, 184
	exterior, 184–185
	floor sheathing, 183, 185–190
	glue, 184–185, 187
	grade, 183–185
	interior, 184–185
	product Standard, 183
	roof sheathing, 183–190
	shearwalls, see Shearwalls
	span rating, 185–186
	species groups, 183–185
	tongue-and-groove, 186
	veneer, 183–185
	wall sheathing, 183–184, 216–217
Posts & timbers (P&T), 5, 14
Purlins, 9, 106

R

Racking, 10
Rafter framing, 3–5
Repetitive member factor, C_r
Resisting moment, 224–227
Ridge beam, 3–5
Ridge board, 3–4, 9
Rigid diaphragm, 197–200
Roof sheathing, see Plywood
Rough sawn lumber, 14
Row spacing, 250–252
R, seismic factor, 52, 217

S

Sapwood, 11, 12
Sawn lumber, see Lumber
Scarf joint, 67
Seasoned lumber, 13, 18, 39
	air drying, 13
	kiln-drying, 13, 66
Section properties, 14
Seismic loads, see Loads
Shakes, 15
Shear
	beams, see Beams
	connections, see Connections
Shearwalls
	anchorage, 218–219, 240–241, 338–342
	aspect ratio, 219–221
	blocking, 217–218
	chords, 217, 218, 223–227, 235–241, 322, 325–331, 339–344
	drywall (GWB), 216–217, 242
	hold-downs, 219, 240–241, 299–300, 337–338, 340–343

Shearwalls (*continued*)
   nailing requirements, 217–218
   overturning, 10, 11, 220, 223–227, 322–331
   perforated, 219
   segmented, 219
   unit shear, 219, 234, 237, 242, 320–322, 332–334
   uplift (combined with shear), 242–244
Sheathing, *see* Plywood
Shrinkage, 13–16, 18–22
   calculation, 13, 19–21
   checking, 15
Sill plates, 2, 3, 7
Size classification, *see* Lumber, size classification
Size factor, C_F, 61
Slenderness ratio
   beam, R_B, 63
   column, 157–158
Slip modulus (vibrations), 134–135
Snow load, 25–30, 32–38
Spacing requirements (fasteners), 250–252
Species, 12–13, 19, 183–185
Split, 15, 19, 248–252
Stress adjustment factors, *see* Adjustment factors
Stress grades, *see* Lumber, stress grades
Stringer, *see* Beams, stringer
Strut, *see* Drag strut
Studs, *see* Wall studs

T

Temperature factor, C_t, 62, 250
Tension connections, 275
Tension members
   adjustment factors, 58, 145–146, 147–151
   bending plus tension, *see* Combined stresses
   net area, 146–148
Tension perpendicular to grain, 16, 17
Toenail, *see* Nail connections, toenail
Toenail factor, C_{tn}, 252–253
Tongue-and-grooved, 14, 118, 135, 186
Top plates, 2, 7, 208, 256, 257, 335–337
Torsion of horizontal diaphragms, 197–200
Tributary area, 28–34
Tributary width, 28–34
Trusses
   bridging, 180
   design examples, 149–150, 154–157, 167–170, 291–300
   profiles, 6
Two-layer floor system, 186, 189, 190

U

Unbalanced live loads, 7, 298, 32, 35
Unbraced length
   beams, 63–66
   columns, 153, 157–159
Underlayment, 186, 189
Unit shear
   diaphragms, *see* Diaphragms, unit shear
   shearwalls, *see* Shearwall, unit shear
Uplift connector
   roof, 299, 300
   wall, 240–241, 338

V

Valley rafter, 3–5
Veneer, 183–185
Vibrations, 131–142
   design examples, 137–142
Visually graded lumber, *see* Lumber
Volume factor, C_V, 64

W

Wall sheathing, *see* Plywood, wall sheathing
Wall studs, 2, 3, 6, 7, 171–178, 312–316
   cripple, 6–7
   jack, 6–7
   king, 6 7
Wane, 15
Warping, 13–15
Wet service factor, C_M, 61, 249
Wind loads, 43–45, 46–51, 284–293
   components and cladding, 43–45, 287–291
Withdrawal connections
   lag screws, 249, 268, 269
   nails, 249, 252, 268, 270
   wood screws, 249, 268, 269, 270
Wood axes, *see* Axes, wood
Wood screws
   bending yield strength, 258–259
   design examples, 265–266, 269, 271
   dowel bearing strength, 253, 258–261
   penetration, 251–252
   prebored holes, 251
   withdrawal, 249, 268, 269, 270
   yield limit equations, 252–253, 260
   yield modes, 259
Wood structures, types, 1

Y

Yield limit equations, 257–260